The constituents of the plasma state contribute to and are influenced by electric and magnetic fields, leading to a rich array of physical properties. *Plasma Physics* presents a wide ranging exposition of this 'fourth state of matter' that is scholarly, practical and eminently readable.

A basic knowledge of mathematics and physics is required to fully appreciate this text that weaves together general theory, special cases, and narrative explanations. Students will appreciate the frequent examples and the problem sets at the end of each chapter. Much of this material is drawn from astrophysics.

Plasma Physics provides astrophysicists with a unique resource, and laboratory physicists with an exciting alternative approach to this complex and fascinating field.

T0180441

PLASMA PHYSICS

STANFORD–CAMBRIDGE PROGRAM

The Stanford–Cambridge Program is an innovative publishing venture resulting from the collaboration between Cambridge University Press and Stanford University and its Press.

The Program provides a new international imprint for the teaching and communication of pure and applied sciences. Drawing on Stanford's eminent faculty and associated institutions, books within the Program reflect the high quality of teaching and research at Stanford University.

The Program includes textbooks at undergraduate and graduate level, and research monographs, across a broad range of the sciences.

Cambridge University Press publishes and distributes books in the Stanford–Cambridge Program throughout the world.

PLASMA PHYSICS

An introduction to the theory of astrophysical,
geophysical, and laboratory plasmas

PETER A. STURROCK

Center for Space Science and Astrophysics, Stanford University

CAMBRIDGE
UNIVERSITY PRESS

Published by the Press Syndicate of the University of Cambridge
The Pitt Building, Trumpington Street, Cambridge CB2 1RP
40 West 20th Street, New York, NY 10011-4211, USA
10 Stamford Road, Oakleigh, Melbourne 3166, Australia

First published 1994
Reprinted 1996

A catalogue record for this book is available from the British Library

Library of Congress cataloguing in publication data
Sturrock, Peter A. (Peter Andrew)
Plasma physics : an introduction to the theory of astrophysical,
geophysical, and laboratory plasmas / Peter A. Sturrock.
p.   cm.
Includes bibliographical references and index.
ISBN 0 521 44350 4. – ISBN 0 521 44810 7 (pbk)
1. Plasma (Ionized gases) 2. Magnetohydrodynamics. I. Title.
QC718.S76 1994
530.4'4–dc20   93–3418   CIP

ISBN 0 521 44350 4   hardback
ISBN 0 521 44810 7   paperback

Transferred to digital printing 2003

Contents

Preface

This book is based on a series of lectures that has been given at Stanford University, for longer than I care to remember, to graduate students from several departments: Aeronautics and Astronautics, Applied Physics, Electrical Engineering, Mechanical Engineering, and Physics. The course has also formed part of the Astronomy Course Program and of the Space Science Program.

The course has changed over the years, beginning as a three-quarter sequence emphasizing laboratory and geophysical plasmas, and evolving into a two-quarter sequence emphasizing solar and other astrophysical applications. Selected material has also been offered as a one-quarter course. The course has been much improved by input from many students (in fact, the first set of lecture notes was produced by students in the class) and from a sequence of dedicated teaching assistants, notably, in recent years, Dr Anton Bergmann, Ms Lisa Porter and Dr Yuri Taranenko.

For invaluable assistance in the preparation of this text, I am indebted to Mrs Louise Meyers-Norney, who entered the text, to Dr James and Mrs Maria Klimchuk, who entered the equations, and to Dr Taeil Bai and Mr David Faust, who helped prepare the figures. Thanks are due also to Dr George Field, Dr Robert Helliwell, Dr Eric Priest and Dr Gerard Van Hoven, who kindly reviewed some of the chapters, and to Dr Simon Mitton and Ms Fiona Thomson of Cambridge University Press for their generous support.

Some of the problems were developed as part of this course, but some were taken from other books. I regret that I did not keep records of the sources of all problems, and it is therefore inevitable that some problems are reproduced in the following pages without proper acknowledgement. For this delinquency I offer my apologies to the original authors. I would greatly appreciate being informed of these errors of omission – and also, of course, of errors of commission.

Finally, it is a pleasure to acknowledge my debt to my good friend Dr William Thompson, who first introduced me to plasma physics, and to thank Dr Marshall Rosenbluth, who has kindly permitted me to quote not only from his physics but also from his wit.

1

Introduction

The plasma state is sometimes referred to as the 'fourth' state of matter. As a solid is heated, it first goes through a transition in which bonds between adjacent molecules are loosened but not entirely broken, and the matter moves into the liquid state. As the matter is heated further, bonds holding adjacent particles close together are completely broken so that molecules can move more or less independently and the liquid becomes a gas. Further heating will lead to the dissociation of molecules into their constituent atoms. However, further heating may also lead to the ionization of the molecules or atoms of the gas, so that the gas then comprises neutral particles, ions and electrons. Although there is no sharp phase transition between the state of a simple neutral gas and the plasma state, the plasma state may nevertheless be regarded as part of the sequence solid–liquid–gas–plasma.

Since the plasma state includes free positive and negative charges, and since movements of these charges produce electrical currents, it is clear that the constituents of the plasma state will be influenced by electric and magnetic fields, and that the plasma can also produce electric and magnetic fields. Hence, in discussing the properties of a plasma, it is essential to regard the electromagnetic field as an integral part of the plasma system. This fact leads to a rich – indeed bewildering – array of properties of the plasma state.

The interaction of just two charged particles involves the electric and magnetic fields that they produce, but nevertheless the interaction is a very simple one, that of a simple orbit or a binary collision. A gas must be sufficiently highly ionized (and even then there are limits on the density and temperature) in order for the plasma to function as a system that is tightly knit by its electromagnetic interaction. These requirements will be explored in Chapter 2.

When a plasma is very tenuous – i.e., has very low density – the way that particles interact with the macroscopic electromagnetic field becomes much more important than the way that they interact with the microscopic electromagnetic fields associated with individual particles. In this situation, it becomes profitable to study the orbits of charged particles in large-scale electric and magnetic fields. This is known as 'orbit theory' that will be discussed in Chapters 3, 4 and 5.

Orbit theory is of crucial importance in the design of fusion reactors, in which it is necessary to confine a high-temperature plasma for sufficiently long periods of time that fusion reactions can occur and produce energy at a greater rate than is required to maintain the system. Orbit theory is also the key to understanding the existence of the Van Allen radiation belts in the Earth's magnetosphere (Hess, 1968). In the radiation-belt regions, charged particles would escape freely from the neighborhood of the Earth, if it were not for the presence of the Earth's magnetic field. As it is, the interaction of charged particles with the Earth's magnetic field is such that charged particles are 'trapped' in certain regions of space. The same holds true in the Sun's outer atmosphere, called the 'corona,' where charged particles can be trapped in localized regions such as the 'flux tubes' joining nearby sunspots of opposite polarity. No doubt the same considerations apply in the atmospheres of other stars, and also in the halo of our galaxy and the halos of other galaxies.

Since the charged particles of a plasma are tightly coupled to the electric and magnetic fields that permeate the plasma, the properties of electromagnetic waves in a plasma will differ significantly from those of electromagnetic waves in free space. Although charged particles can move freely along a magnetic field, their motion transverse to the field is very much inhibited; as a result, the plasma behaves as an anisotropic wave-propagating medium. Furthermore, the tendency of electrons to move rapidly in response to an electric field in such a way as to cancel that field has the tendency of preventing the propagation of waves of frequencies below a critical frequency called the 'plasma frequency.' The properties of electromagnetic waves propagating in a magnetized plasma are discussed briefly in Chapters 6 and 7. We find, for instance, that waves of very low frequency can propagate along the magnetic field in the Earth's magnetosphere; they have the property that high-frequency waves travel faster than low-frequency waves so that, if the signal is made audible, one hears a whistle of decreasing frequency (Helliwell, 1965). Such signals were picked up by telecommunication systems during World War I, but it took the subsequent development of the magneto-ionic theory to provide the correct explanation of what are now known as 'whistlers.'

Several characteristic forms of radio noise are produced by the sun (Kundu,

1965). Two of these are known as 'type II radio bursts' and 'type III radio bursts.' In each case, the frequency of the signal decreases with time. It is now known that type III bursts are produced by the excitation of oscillations ('plasma oscillations') produced by an electron beam travelling through the solar corona. The mechanism of Type II bursts is very similar, except that the electrons are accelerated, and excite plasma oscillations, in the neighborhood of a moving shock front. The mechanism by which a beam produces oscillations is discussed in Chapter 8, and some of the properties of these oscillations are further discussed in Chapter 9.

Although a low-density plasma tends to be confined by a magnetic field, such confinement can never be perfect. Collisions between particles make it possible for particles to diffuse across a magnetic field, and they also lead to 'pitch-angle scattering' that also limits the effectiveness of trapping in fusion devices, planetary magnetospheres, coronal flux tubes, etc. Since collisions also tend to dissipate the highly ordered motions involved in wave propagation, collisions will lead to wave dissipation. Some of the properties of collisions in a fully ionized plasma are discussed in Chapter 10.

If a magneto-plasma system is subject to disturbances that are sufficiently slow and of sufficiently large scale, it would be possible for the electrons to prevent the build-up of any electric field in a frame that moves with the plasma. In this situation, the higher frequency modes of oscillation and wave propagation do not play a role in the behavior of the system. The behavior then simplifies considerably and is called the 'MHD' or 'magnetohydrodynamic' approximation to the behavior of the system.

A notable property of MHD behavior was discovered by Hannes Alfvén in 1942 (Alfvén, 1942). It was shown by Michael Faraday in the 19th century that the stresses in a magnetic field comprise a pressure transverse to the field and a tension along the field. Alfvén found that certain waves that can propagate along the magnetic field in a magnetoplasma are analogous to waves propagating along a stretched string: these are now known as 'Alfvén waves.' The speeds of waves that propagate in MHD systems are determined by the sound speed and the Alfvén speed. The solar corona must be heated by some flux of nonthermal energy originating at the photosphere (Athay, 1976). One possibility is that the energy is carried in the form of Alfvén waves. The MHD equations and some of the properties of these equations are developed in Chapters 11 and 12, and some of the properties of MHD waves are discussed in Chapter 14.

In some astrophysical situations, the MHD equations are applicable but can be further simplified. In the solar corona, for instance, the gravitational force and gas pressure have only slight effects on the magnetic-field configuration. On the other hand, the corona is highly conducting so that the field is normally not in a potential (current-free) state. In this case, the Lorentz force must vanish everywhere, so that the current must everywhere be parallel to the magnetic field. Such configurations are called 'force-free.' They play a very important role in solar physics, since a force-free magnetic field contains more energy than the corresponding potential field with the same boundary conditions. It is believed that the excess energy is released at the time of a solar flare (Tandberg-Hanssen and Emslie, 1988). Some of the properties of force-free magnetic-field configurations are discussed in Chapter 13.

One of the most important recurring problems of plasma physics is that of stability. A certain situation may theoretically be possible, but it might not occur – or, if it occurs, it might not survive – in practice or in nature because it is unstable. Stability considerations will therefore play an important role in determining permissible designs for plasma containment systems such as tokamaks (Furth, 1968). Parker (1955) pointed out sometime ago that a magnetic flux tube trapped below the photosphere may become unstable to what he calls the 'buoyancy instability.' It is likely that this instability plays a key role in the development of active regions on the sun. Magnetohydrodynamic stability is discussed briefly in two chapters. In Chapter 15, we show how the stability of a system may be analysed in terms of the field equations and the equation of motion. In Chapter 16, we show that the properties of MHD systems may be derived from a Lagrangian function embodied in a variation principle. This approach makes it possible to calculate the energy and momentum associated with a disturbance and so to derive an 'energy principle' determining the stability or instability of a system.

Instabilities such as the two-stream instability and MHD instabilities can occur in an 'ideal' plasma in which collisions play no role. By contrast, there are other instabilities that exist by virtue of the collisional behavior of the plasma. The finite resistivity of a collisional plasma makes possible the resistive decay of any currents in the system. This is normally a very slow process in astrophysical settings. However, Dungey (1953) realized that, in certain configurations, resistive diffusion can lead to changes in the system that lead to strong concentrations of the current that then speed up the diffusion process. This is now regarded as an example of a 'resistive instability.' One of the most important examples of a resistive instability is the 'tearing-mode instability,' analysed by Furth, Killeen & Rosenbluth (1963), that is believed to play a key role in the release of stored energy in some astrophysical phenomena such as solar flares (Tandberg-Hanssen and Emslie, 1988). This topic is discussed in Chapter 17.

The behavior of a complex system can often be regarded as the sum of two different types of behavior: coherent or collective behavior, and noise. Since collisions between particles are effectively unpredictable, collisions contribute to the noise in a system. When instabilities occur, small random fluctuations are rapidly amplified, so that the system exhibits large-amplitude noise, sometimes called 'turbulence.' The behavior of charged particles in a turbulent plasma differs markedly from the behavior in a static plasma. Fermi (1949) proposed that cosmic rays are accelerated by the interaction of charged particles with randomly moving magnetized plasma clouds. Fermi's specific model has now been superseded by a number of different models. Nevertheless, the basic idea of what is now known as 'Fermi acceleration' retains a prominent role in astrophysics. The basic concepts are first presented in a very simple form in Chapter 18, where we discuss some of the properties of particles moving in stochastic electromagnetic fields.

Stochastic acceleration can be regarded as the transfer of energy from waves to particles. The reverse process also can occur: single particles can emit radiation (such as synchrotron radiation if they are moving in the presence of a magnetic field) and a plasma can, acting collectively, generate waves (for instance, a two-stream instability develops electrostatic waves). Quantum mechanics yields simple mathematical relations (due to Einstein) between spontaneous emission, stimulated emission and absorption. By adopting a quantum-mechanical viewpoint, similar relations can be obtained between corresponding processes in a plasma. By considering progressively larger dimensions, Planck's constant becomes negligible, and the equations may then be expressed in classical form. One of the equations so derived is an equation that can be used to study stochastic acceleration. This development relates the acceleration process to a corresponding radiation process. These equations are derived in Chapter 19.

As the reader will have noticed, each chapter or group of chapters is merely an introduction to a large area of research. This book is very much an 'introduction' to plasma physics. There are many advanced textbooks and collections of articles that deal with more advanced plasma theory, and with applications of this theory to fusion reactors and to astrophysical phenomena. Some of these books and articles, especially those dealing with astrophysical applications, are listed throughout the book for possible further study by the reader.

2

Basic concepts

2.1 Collective effects

The term 'plasma' was introduced by Tonks and Langmuir (1929) 'to designate that portion of an arc-type discharge in which the densities of ions and electrons are high but substantially equal.' In discussing oscillations of this region, Tonks and Langmuir noted that 'when the electrons oscillate, the positive ions behave like a rigid jelly . . .' It is possible that Tonks and Langmuir saw an analogy between this behavior of an arc discharge and the behavior of the components of blood, in which the particle-like corpuscles move through the fluid plasma. The term 'plasma' is now used quite generally to refer to quasi-neutral assemblies of charged particles, and 'plasma physics' is the study of the behavior of these systems. In some situations, the ions may be regarded as a stationary fluid, but in general this is not the case. Just as quantum mechanics shows how it is possible for matter to exhibit both particle-like and wave-like properties, so plasma physics shows how it is possible for matter to behave both as a collection of particles and as a fluid.

We adopt the modified Gaussian system of units for electromagnetic quantities, that is to say, electric quantities such as charge, potential and field strength will be measured in electrostatic units, and magnetic quantities such as current, potential and field strength will be measured in electromagnetic units. With this choice of units, Maxwell's equations take the following pleasing form

$$\nabla \times \mathbf{E} = -\frac{1}{c}\frac{\partial \mathbf{B}}{\partial t}, \qquad (2.1.1)$$

$$\nabla \cdot \mathbf{B} = 0, \qquad (2.1.2)$$

$$\nabla \cdot \mathbf{E} = 4\pi \zeta \qquad (2.1.3)$$

and

$$\nabla \times \mathbf{B} = \frac{1}{c} \frac{\partial \mathbf{E}}{\partial t} + 4\pi \mathbf{j}. \tag{2.1.4}$$

We use ζ for the charge density (in order to reserve ρ for mass density), $\mathbf{j}$ is the current density in emu, $\mathbf{E}$ is the electric field strength in esu, and $\mathbf{B}$ is the magnetic field strength in emu. These equations are not all independent. Equations (2.1.1) and (2.1.2) are related. Furthermore, the conservation equation

$$\frac{1}{c} \frac{\partial \zeta}{\partial t} + \nabla \cdot \mathbf{j} = 0 \tag{2.1.5}$$

relates (2.1.3) and (2.1.4). For further discussion about units and information relating modified Gaussian units to other systems, please see Appendix A.

For the time being, let us consider a simple plasma composed of fully ionized hydrogen. We write the charge, mass and number density of electrons as $-e$, m_e, n_e, and of protons as e, m_i, n_i. Then the mass density, charge density and current density are given by

$$\rho = n_e m_e + n_i m_i, \tag{2.1.6}$$

$$\zeta = e(n_i - n_e) \tag{2.1.7}$$

and

$$\mathbf{j} = \frac{e}{c}(n_i \mathbf{v}_i - n_e \mathbf{v}_e). \tag{2.1.8}$$

In equations (2.1.8) $\mathbf{v}_e$ and $\mathbf{v}_i$ represent the suitably averaged velocities of electrons and protons.

2.2 Charge neutrality and the Debye length

We consider a plasma which, initially, comprises a uniform density n_0 of both protons and electrons. There is no net charge density, so that we may assume that there is initially no electric field. We now ask how the charge density behaves if there is an attempt to change it. As a particularly simple example, let us suppose that the proton density is changed from n_0 to $(1-\delta)n_0$ in the region $-L < x < L$. If L is sufficiently small, the electric field due to this change will be so small that the electrons will suffer negligible disturbance. On the other hand, if L is sufficiently large, the change will have a drastic effect on the electron distribution. We wish to estimate the range of L at which the transition takes place.

If there were no change in electron density, (2.1.3) would lead to the following equation for the electric potential:

Basic concepts

$$\frac{d^2\phi}{dx^2} = 4\pi\delta n_0 e. \tag{2.2.1}$$

Assuming that the plasma as a whole (outside the disturbance) is maintained at potential $\phi = 0$, the appropriate solution of (2.2.1) is

$$\left.\begin{array}{ll} \phi = 2\pi\delta n_0 e(x^2 - L^2), & |x| < L, \\ \phi = 0 & , |x| > L. \end{array}\right\} \tag{2.2.2}$$

Hence, in particular,

$$\phi(0) = -2\pi\delta n_0 e L^2. \tag{2.2.3}$$

Now suppose that the plasma has temperature T, so that each particle has mean kinetic energy $\frac{1}{2}kT$ in each degree of freedom. If $\phi(0)$ is so small that an electron of average thermal energy can easily reach $x = 0$, there will be only a small change in the state of the plasma. On the other hand, if $\phi(0)$ is so large that very few electrons can reach $x = 0$, then there will be a drastic change in the state of the plasma. Hence the requirement that the plasma remain 'quasi-neutral' is that

$$\tfrac{1}{2}kT > 2\pi\delta n_0 e^2 L^2. \tag{2.2.4}$$

This can be rewritten in the form

$$\delta < (\lambda_D/L)^2, \tag{2.2.5}$$

where we have introduced the quantity λ_D, the 'Debye length,' defined by

$$\lambda_D{}^2 = \frac{kT}{4\pi n e^2}. \tag{2.2.6}$$

Numerically, this may be expressed as

$$\lambda_D = 10^{0.84} n^{-1/2} T^{1/2}. \tag{2.2.7}$$

Suppose, for instance, that $\delta = -1$, so that protons are completely removed from the range $-L < x < L$. Then we see that the plasma remains quasi-neutral if $L \ll \lambda_D$. If $L \gg \lambda_D$, the region indicated behaves as a vacuum.

One may also use (2.2.5) to estimate the degree of neutrality of plasmas under certain situations. For instance, consider the solar corona. Since the proton mass is much larger than the electron mass, the gravitational field acts mainly on the protons. The effect of the gravitational field is, of course, to lead to a barometric structure in which the density decreases exponentially with height. As we will see later, this scale height is given by

$$H = \frac{kT}{m_{av}g} \qquad (2.2.8)$$

where m_{av} is the average particle mass, approximately $\frac{1}{2}m_p$, and g is the gravitational acceleration. For the Sun, $g = 10^{4.44}\,\text{cm s}^{-2}$ so that (2.2.8) becomes, numerically,

$$H = 10^{3.78}\,T. \qquad (2.2.9)$$

The solar corona has a temperature of approximately $T = 10^6\,\text{K}$, so that $H = 10^{9.8}\,\text{cm}$. A typical density of the corona is $n = 10^8\,\text{cm}^{-3}$ so that, from (2.2.7), $\lambda_D = 10^{-0.2}\,\text{cm}$. We now see, from (2.2.5), that we must expect the *average* departure from quasi-neutrality to satisfy the condition $\delta < 10^{-20}$.

2.3 Debye shielding

In the above calculations, we have considered the response of electrons to a possible charge inequality, but ignored the response of ions. We must expect both species to respond to an electric field. The role of both electrons and ions will be seen in the next problem to be considered. Suppose that a point charge q is introduced into a thermal uniform plasma. We allow the plasma to settle down to a steady state after the charge is introduced, and we then examine this steady state. We expect that an electric field, described by potential ϕ, will develop. This field will necessarily be spherically symmetric with respect to the charge. Hence $\phi = \phi(r)$, if r is the radius measured from the position of the test charge. Now the electron and ion densities will be determined by the Maxwell–Boltzmann distributions, so that the changes in density are given by the Boltzmann factors:

$$\left.\begin{aligned} n_e &= n_{e,o}\exp\left(\frac{e\phi}{kT_e}\right) \\[2mm] n_s &= n_{s,o}\exp\left(-\frac{Z_s e\phi}{kT_s}\right). \end{aligned}\right\} \qquad (2.3.1)$$

We are now assuming that there are several species of ions, each with temperature T_s and charge $Z_s e$.

Since the unperturbed state has zero charge density, so that

$$n_{e,o} = \sum Z_s n_{s,o}, \qquad (2.3.2)$$

we see that Poisson's equation becomes

$$\nabla^2\phi = -4\pi\left[-(n_e - n_{e,o})e + \sum Z_s(n_s - n_{s,o})e\right]. \qquad (2.3.3)$$

On expanding the exponentials in equations (2.3.1) and keeping only terms linear in ϕ, (2.3.3) becomes

$$\nabla^2\phi = \lambda_D^{-2}\phi, \tag{2.3.4}$$

where

$$\lambda_D^{-2} = \lambda_{D,e}^{-2} + \sum \lambda_{D,s}^{-2}. \tag{2.3.5}$$

In this equation, $\lambda_{D,e}$ is the Debye length for electrons, given by (2.2.6), and $\lambda_{D,s}$ is the Debye length for each species of ions, given by

$$\lambda_{D,s}^2 = \frac{kT_s}{4\pi Z_s^2 n_s e^2}. \tag{2.3.6}$$

It is interesting to note that, if there is only one species of ion and if $Z > 1$ and if the electrons and ions have the same temperature, then the ion contribution to Debye shielding is greater than that of electrons, since $\lambda_{D,s}^{-2}$ is larger than $\lambda_{D,e}^{-2}$.

On using spherical coordinates, (2.3.4) becomes

$$\frac{1}{r}\frac{d^2}{dr^2}(r\phi) = \lambda_D^{-2}\phi, \tag{2.3.7}$$

away from the singularity at $r=0$. We could include the effect of the test charge by adding the term $4\pi q\delta(x)\delta(y)\delta(z)$ to the right-hand side of (2.3.7). The appropriate solution of this equation is

$$\phi = qr^{-1}e^{-r/\lambda_D}, \tag{2.3.8}$$

since this expression for ϕ satisfies (2.3.7) for $r>0$, and behaves in the correct way as $r \to 0$.

We see that, for $r \ll \lambda_D$, the electric field of the test charge is unaffected by the presence of a plasma. However, for $r > \lambda_D$, the electric field is attenuated by the exponential factor. The test charge is said to be 'shielded' by the plasma, the shielding distance being measured by λ_D.

We should bear in mind that this calculation has been made for a steady state. If the test charge is moving, we must expect the resulting disturbance of the plasma to differ from the state that we have just found. If the test charge is moving with a speed slow compared with the electron thermal velocity, it is probable that electrons will set up a shielding pattern similar to that just calculated. However, if the speed were, say, larger than the ion thermal velocity, the ions would not be able to set up a spherically-symmetrical state. The calculation of the electric field of a moving charge is the Cerenkov problem (Melrose, 1980).

2.4 The plasma parameter

In order for a system to be regarded as a plasma, it is usually required that the medium should be almost neutral, almost everywhere, almost all the time. We have found that, for this to be the case, the scale of the system must be large compared with the Debye length. There is another requirement that we now consider.

In deriving the expression for the screening of a test charge by the plasma, we made an assumption that was not brought out explicitly. We described the state of the electron gas by a density n_e and calculated the effect of electrons on the electric field near the test charge in terms of this density. The implicit assumption was that we may, to good approximation, replace the actual distribution of point electrons by a fictitious fluid with the same average charge density. Hence, in order for our analysis of Debye shielding to be valid, it is certainly necessary that there should be many electrons within a sphere of radius λ_D. Since each species has its own Debye length, we therefore introduce the parameters

$$\Lambda_e = n_e \lambda_{D,e}{}^3, \quad \Lambda_s = n_s \lambda_{D,s}{}^3, \qquad (2.4.1)$$

i.e. $$\Lambda_e = \left(\frac{k}{4\pi e^2}\right)^{3/2} n_e{}^{-1/2} T_e{}^{3/2}, \quad \Lambda_s = \left(\frac{k}{4\pi e^2}\right)^{3/2} Z_s{}^{-3} n_s{}^{-1/2} T_s{}^{3/2}. \qquad (2.4.2)$$

Numerically, these formulas become

$$\Lambda_e = 10^{2.52} n_e{}^{-1/2} T_e{}^{3/2}, \quad \Lambda_s = 10^{2.52} Z_s{}^{-3} n_s{}^{-1/2} T_s{}^{3/2}. \qquad (2.4.3)$$

Λ_e and Λ_s are referred to as the 'plasma parameters' of the electrons and of the ion species, respectively. It is interesting to note that, although Λ is the number of particles per Debye cube, Λ in fact decreases as the density increases. We also see that, if Z is large compared to unity, the condition that Λ_s be large is a more severe requirement than the condition that Λ_e be large.

We see that, in the calculation of the previous section, we implicitly assumed that the plasma parameters are large numbers. This assumption runs through most calculations of plasma physics. Hence we can say that, *in order for an assembly of charge particles to be considered a plasma, the plasma parameters should be large compared with unity.* We now explore some of the further consequences of this requirement.

Consider two electrons, each of energy $\frac{3}{2}kT$, moving in head-on collision. They will stop when the separation has been reduced to the value b, where

$$2 \cdot \frac{3}{2} kT = \frac{e^2}{b} \qquad (2.4.4)$$

so that

$$b = \frac{e^2}{3kT}. \qquad (2.4.5)$$

We may now compare this distance of closest approach b with the mean interparticle distance $n_e^{-1/3}$. We find that the ratio $\mathfrak{R}_1$ defined by

$$\mathfrak{R}_1 = \frac{b}{n_e^{-1/3}}, \qquad (2.4.6)$$

is expressible as

$$\mathfrak{R}_1 = \frac{1}{12\pi} \Lambda_e^{-2/3}. \qquad (2.4.7)$$

Hence in a plasma, for which $\Lambda_e \gg 1$, $\mathfrak{R}_1 \ll 1$. That is, the distance of closest approach is much smaller than the mean interparticle distance.

This result has some interesting consequences. At any instant, an electron will be experiencing a large-angle collision only if another electron (or some other charged particle) is within a distance of order b. Hence the probability P_2 that, at any instant, an electron is undergoing a large-angle collision with another electron is given, approximately, by

$$P_2 = n_e b^3. \qquad (2.4.8)$$

We find, from (2.4.5), that this may be expressed as

$$P_2 = (12\pi)^{-3}\Lambda_e^{-2}. \qquad (2.4.9)$$

Since $\Lambda_e \gg 1$, we see that, in a plasma, most electrons, for most of the time, are not undergoing large-angle collisions.

The probability P_3 that, at any instant, an electron is taking part in a large-angle encounter simultaneously with two other electrons is given by

$$P_3 = (n_e b^3)^2 = P_2^2. \qquad (2.4.10)$$

Hence, in a plasma, the effects of three-body collisions are negligible compared with those of two-body collisions.

Let us now consider the effect of the nearest neighbor of an electron. The mean separation of electrons is $n_e^{-1/3}$ so that the expected electric potential is of order $en_e^{1/3}$ and the mean potential energy due to this interaction is $e^2 n_e^{1/3}$. Hence the ratio $\mathfrak{R}_2$ of this mean potential energy to the mean kinetic energy $\frac{3}{2}kT_e$ is given by

$$\mathfrak{R}_2 = \frac{e^2 n_e^{1/3}}{\frac{3}{2}kT_e}. \qquad (2.4.11)$$

We find that this may be expressed as

$$\mathcal{R}_2 = \frac{1}{6\pi} \Lambda_e^{-2/3}. \tag{2.4.12}$$

Since, in a plasma, $\mathcal{R}_2 \ll 1$, we see that the effect of the nearest neighbor on an electron is small. In particular, at any instant, an electron is undergoing small-angle deflections by neighboring electrons.

Finally, we consider the thermal excitation of the 'collective modes' of a plasma and compare the energy density of these modes with the kinetic energy density. In a cube of size L, collective behavior may be analyzed into normal modes with wave numbers $k_r (r = 1, 2, 3)$ where

$$k_r = n_r \frac{2\pi}{L}. \tag{2.4.13}$$

However, as we may infer from the previous section, and as we shall see subsequently, collective modes of an electron gas have wave numbers less than λ_D^{-1}. Hence the total number of modes is, approximately, $[L/(2\pi\lambda_{D,e})]^3$. In thermal equilibrium each of these modes has, on the average, energy $\frac{1}{2}kT$ so that the energy density u_c in 'collective' modes will be, approximately,

$$u_c = \frac{1}{2} kT_e (2\pi\lambda_{D,e})^{-3}. \tag{2.4.14}$$

We wish to compare this with the kinetic energy density of electrons u_T, given by

$$u_T = \frac{3}{2} n_e kT_e. \tag{2.4.15}$$

If we define

$$\mathcal{R}_3 = \frac{u_c}{u_T}, \tag{2.4.16}$$

we find that

$$\mathcal{R}_3 = \frac{1}{24\pi^3} \Lambda_e^{-1}. \tag{2.4.17}$$

Hence, since $\Lambda_e \gg 1$, only a small fraction of the thermal energy of a plasma is in collective modes.

One half of the energy in collective modes will be stored in electric-field fluctuations of which the typical wavelength is of order $2\pi\lambda_{D,e}$. If the rms (root-mean-square) fluctuation in electron density is written as Δn_e, and if we write

$$\mathfrak{R}_4 = \frac{\Delta n_e}{n_e},$$

(2.4.18)

we find that $\mathfrak{R}_4$ may be expressed as

$$\mathfrak{R}_4 = \frac{1}{4\pi^{3/2}} \Lambda_e^{-1/2}.$$

(2.4.19)

Hence we see that, in a plasma for which $\Lambda_e \gg 1$, thermal fluctuations do not appreciably disturb charge neutrality.

2.5 Plasma oscillations

Probably the simplest example of collective behavior in a plasma is that of plasma oscillations, which we now consider very briefly. Imagine a gas that is fully ionized, infinite in extent, uniform, and of such low temperature that we may – for present purposes – ignore the thermal motion of ions and electrons. We also assume that there is no magnetic field and that the ion mass is sufficiently large, compared with the electron mass, so that we may ignore the motion of ions in what follows. In later chapters, we shall see what modifications occur if some of these restrictions are relaxed.

We now consider one-dimensional motion in the direction of the x-axis. If the sheet of electrons which was originally at position x is displaced to position $x + \xi$, if the sheets near $x = -\infty$ and $x = +\infty$ are undisturbed, and if the ordering of the electron sheets is unaffected by this displacement, we may evaluate the electric field seen by the displaced electron sheet by noting that there is an excess positive charge $ne\xi$ per unit area to the left of each sheet (and a deficit of the same magnitude to the right of the sheet), where the initial density is n. Since the total electric charge to the left of the sheet near $x = \infty$ is unchanged, there will be no electric field at $x = \infty$ and, similarly, no electric field at $x = -\infty$. Hence the electric field at a displaced charge sheet is given by

$$E = 4\pi ne\xi.$$

(2.5.1)

The equation of motion of the electron sheet is the equation of motion of any one electron in the sheet:

$$m_e \frac{d^2\xi}{dt^2} = -eE.$$

(2.5.2)

Hence we obtain the following differential equation for $\xi(t)$:

$$\frac{d^2\xi}{dt^2} + \omega_{p,e}^2 \xi = 0$$

(2.5.3)

where

$$\omega_{p,e}^2 = \frac{4\pi ne^2}{m_e}. \tag{2.5.4}$$

In this formula, ω_p is the 'radian' electron plasma frequency. Since the oscillation frequency is independent of the spatial form of the wave, it is independent of the wave number k of a sinusoidal wave. Hence the group velocity, defined by

$$u = \frac{d\omega}{dk}, \tag{2.5.5}$$

will be zero. (For information concerning group velocity, please see Appendix B.) If ν_p is the plasma frequency in Hz, so that $\omega_p = 2\pi\nu_p$, we find that, numerically,

$$\nu_{p,e} = 10^{3.95} n_e^{1/2}, \tag{2.5.6}$$

where we now write the electron density as n_e, anticipating that there will be similar plasma frequencies for each ion species.

Although these oscillations, which were termed 'plasma oscillations' by Tonks and Langmuir (1929) in their seminal article, are (according to our calculation) undamped, almost any complication of the model will lead to damping. For example, if the amplitude is sufficiently large that electron sheets do cross each other, the motion is no longer simple harmonic, and the oscillations tend to wash out in the course of time. Collisions between particles lead to damping, and so also does the effect of finite temperature, as we shall see in Chapter 9.

Even if the plasma is taken to be uniform and of zero temperature, we find that the above result (that there is no damping) depends crucially on the assumption that the motion is planar. To see this, we may carry out a similar analysis for cylindrical oscillations. If the electrons initially on a cylinder of radius r are displaced to a cylinder of radius $r + \xi$, evaluation of the electric field at the displaced position by means of (2.1.3) leads to the following equation of motion,

$$m_e \frac{d^2\xi}{dt^2} = -\frac{2\pi n_e e^2}{r+\xi} [(r+\xi)^2 - r^2], \tag{2.5.7}$$

provided there is no cross-over. If we write $\alpha = \xi/r$, (2.5.7) may be rewritten as

$$\frac{d^2\alpha}{dt^2} + \omega_p^2 \left(\frac{\alpha + \frac{1}{2}\alpha^2}{1+\alpha} \right) = 0. \tag{2.5.8}$$

Since this equation is now nonlinear in the amplitude α, it is clear that the oscillation is no longer simple harmonic. It is also clear that the period of the oscillation depends upon its amplitude. Hence, unless all particles have the same value of α initially, particles with different equilibrium radii will have different periods. Hence, since neighboring charge cylinders have different oscillation frequencies, we must expect that cross-over will eventually occur. Moreover, since the electric field of any radius is due to the cumulative effect of the displacements of cylindrical charged sheets of all internal radii, and since these charged sheets all have different frequencies, the electric field at any radius is the integral of a continuum of contributions of varying frequency. In such a situation, the resulting phase mixing eventually leads to a decay in the amplitude of the electric field at any point. This concept is closely related to that of 'Landau damping' that we shall study in Chapter 9.

As we see from (2.5.6), the plasma frequency is likely to be quite high. In a fusion device, for which n_e may be of order 10^{16} cm^{-3}, $\nu_{p,e} \sim 10^{12}$ Hz. If these oscillations were to produce radiation in the form of electromagnetic waves, these waves would have a wavelength of only 0.3 mm. In the solar corona $n_e \sim 10^8$ cm^{-3}, so that $\nu_{p,e} \sim 10^8$ Hz: that is, the radiation would have a frequency of order 100 MHz, with wavelengths in the meter range. Such radiation is in fact produced in the solar corona and is recorded on sweep-frequency radio receivers. A typical record is shown in Fig. 2.1. We see that the frequency–time trace has two components, with a frequency ratio of approximately 2 : 1. For each component, the frequency decreases as a function of time, with a time scale of seconds. Such a radio burst is known as a 'type III' radio burst. It is believed that such a burst is caused by a stream

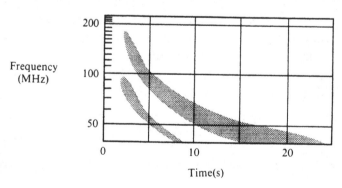

Fig. 2.1. Dynamic spectrum of a typical harmonic type III burst in the 40 to 240 MHz range. (Figure reproduced with kind permission from Wild, Murray and Rowe 1954.)

of high-energy electrons suddenly injected into the corona and traveling out into interplanetary space. As we shall see in Section 8.4, such an electron stream is unstable to the development of plasma oscillations. As a result of nonlinear processes, these oscillations give rise to radiation at the plasma frequency (ν_p) and also at its harmonic ($2\nu_p$).

There is another type of burst, known as a 'type II' radio burst, that is basically similar to a type III burst except that its time scale is longer, indicating that the disturbance generating this radiation is slower. It is believed that this disturbance is a shock wave – either a blast wave that is initiated by a sudden explosion in the corona and propagates outward freely, or a bow shock, the shock wave that runs ahead of an ejected mass of gas.

Problems

Problem 2.1. Find the Debye lengths $\lambda_{D,e}$, $\lambda_{D,p}$, the plasma parameters Λ_e, Λ_p, the plasma frequencies $\nu_{p,e}$, $\nu_{p,p}$, and the thermal velocities $v_{th} = (kT/m)^{1/2}$ for electrons and protons under the following conditions.

(a) A fusion machine: $n_e = n_p = 10^{16}\,\text{cm}^{-3}$, $T_e = T_p = 10^7\,\text{K}$.
(b) The Earth's magnetosphere: $n_e = n_p = 10^4\,\text{cm}^{-3}$, $T_e = T_p = 10^3\,\text{K}$.
(c) The center of the Sun: $n_e = n_p = 10^{26}\,\text{cm}^{-3}$, $T_e = T_p = 10^{7.2}\,\text{K}$.
(d) The solar corona: $n_e = n_p = 10^8\,\text{cm}^{-3}$, $T_e = T_p = 10^6\,\text{K}$.
(e) The solar wind: $n_e = n_p = 10\,\text{cm}^{-3}$, $T_e = T_p = 10^5\,\text{K}$.
(f) The atmosphere of a neutron star: $n_e = n_p = 10^{12}\,\text{cm}^{-3}$, $T_e = T_p = 10^7\,\text{K}$.

Can all of these cases qualify to be described as 'plasmas'?

Problem 2.2. Prove that the net charge in the Debye shielding cloud exactly cancels the charge Q of the test charge.

Problem 2.3. Consider a sphere of radius R containing equal densities n_0 of electrons of temperature T_0 and of ions that are taken to have infinite mass. Now suppose that a Maxwell demon moves the electrons to the surface of the sphere, distributing them uniformly over the surface. The ions are left in their original positions.

(a) What is the potential energy? (How much energy must be expended by the demon?)
(b) Calculate the electric potential and electric field as functions of radius.
(c) Suppose that the potential energy is found to be the same as the total kinetic energy of the electrons in their original configuration. Can you relate the radius R to the electron Debye length? (This is known as *Birdsall's problem*.)

Problem 2.4. Modify the discussion (Section 2.5) of the plasma frequency using a slab geometry. Assume that the ions are singly charged and have finite mass m_i rather than infinite mass, and show that the slabs oscillate at a frequency given by

$$\omega_p{}^2 = \frac{4\pi ne^2}{m_e} + \frac{4\pi ne^2}{m_i} .$$

Problem 2.5. Show that, in a nonrelativistic plasma ($kT \ll m_e c^2$), the mutual Coulomb (electrostatic) force between two typical particles is much more important than the mutual Lorentz (magnetic) force.

3
Orbit theory – uniform fields

In Chapter 1, we pointed out that a plasma exhibits both collective behavior and individual-particle behavior. In Chapter 2, we examined some of the elementary forms of collective behavior and determined a criterion for such behavior to be manifested in a plasma. In this chapter, we examine some aspects of the behavior of individual particles in several different configurations of electric, magnetic and gravitational fields.

3.1 Particle motion in a static, uniform magnetic field

The rate of change of the energy U of a particle in an electromagnetic field is given by

$$\frac{\mathrm{d}U}{\mathrm{d}t} = \mathbf{v} \cdot \left(q\mathbf{E} + \frac{q}{c}\mathbf{v} \times \mathbf{B} \right) = q\mathbf{v} \cdot \mathbf{E}. \tag{3.1.1}$$

Only an electric field can change the energy of a particle. Hence the energy of a particle moving in a purely magnetic field will be a constant of the motion. The relativistic equation of motion may be written as

$$\frac{\mathrm{d}}{\mathrm{d}t}(\gamma m\mathbf{v}) = q\mathbf{E} + \frac{q}{c}\mathbf{v} \times \mathbf{B}, \tag{3.1.2}$$

where

$$\gamma = (1 - \beta^2)^{-1/2} \tag{3.1.3}$$

and

$$\beta = \frac{v}{c}. \tag{3.1.4}$$

Since in a static magnetic field $\gamma = $ constant, (3.1.2) becomes

$$\frac{d\mathbf{v}}{dt} = \frac{q}{\gamma mc} \mathbf{v} \times \mathbf{B}. \tag{3.1.5}$$

One convenient way to examine the consequences of equations such as the above is to adopt a special coordinate system. If we adopt coordinates x_1, x_2, x_3 such that $B = (0, 0, B)$, (3.1.5) becomes

$$\frac{dv_1}{dt} = \frac{qB}{\gamma mc} v_2, \tag{3.1.6}$$

$$\frac{dv_2}{dt} = -\frac{qB}{\gamma mc} v_1, \tag{3.1.7}$$

and

$$\frac{dv_3}{dt} = 0. \tag{3.1.8}$$

We see that $v_\parallel$, the component of $\mathbf{v}$ parallel to $\mathbf{B}$, is a constant.

On introducing the symbol ω_g for the 'gyrofrequency' or 'cyclotron frequency',

$$\omega_g = \frac{|q|B}{\gamma mc}, \tag{3.1.9}$$

and ε for the sign of the charge,

$$\varepsilon = \frac{q}{|q|}, \tag{3.1.10}$$

we find that (3.1.6) and (3.1.7) are satisfied by

$$v_1 = v_\perp \cos(\omega_g t), \tag{3.1.11}$$

$$v_2 = -\varepsilon v_\perp \sin(\omega_g t), \tag{3.1.12}$$

where $v_\perp$ is the magnitude of the velocity component perpendicular to B:

$$v_\perp = (v_1^2 + v_2^2)^{1/2}. \tag{3.1.13}$$

Some authors incorrectly use the term 'Larmor frequency' for the frequency just defined. The Larmor frequency is one half the gyrofrequency (Goldstein, 1950).

For an appropriate choice of the coordinate system, the particle displacement may be expressed as

$$x_1 = r_\perp \sin(\omega_g t), \tag{3.1.14}$$

$$x_2 = \varepsilon r_\perp \cos(\omega_g t), \tag{3.1.15}$$

and

$$x_3 = v_\parallel t, \tag{3.1.16}$$

where the gyroradius $r_\perp$ is defined by

$$r_\perp = \frac{v_\perp}{\omega_g}. \tag{3.1.17}$$

These equations show that the projection of particle motion on the x_1-x_2 plane, looking along the direction of the magnetic field, is left-hand circular motion for positive ions and right-hand circular motion for electrons. Hence, if we consider electromagnetic waves of frequency comparable with the electron gyrofrequency, we can see that right-hand polarized waves will interact more strongly with electrons than will left-hand polarized waves. This will have important implications for the properties of electromagnetic waves propagating in a magnetized plasma.

Electromagnetic radiation is produced when a charged particle is accelerated. In the case of synchrotron radiation, this acceleration is due to the centripetal acceleration of a charged particle moving in a curved orbit. On noting that the acceleration of a particle may be written as

$$\frac{d\mathbf{v}}{dt} = \frac{ds}{dt}\frac{d\mathbf{v}}{ds} = v_t \frac{d\mathbf{v}}{ds}, \tag{3.1.18}$$

where s denotes arc length and v_t is the total speed of the particle, we see that

$$\left|\frac{d\mathbf{v}}{dt}\right| = \frac{v_t^2}{r_c}, \tag{3.1.19}$$

where the instantaneous radius of curvature r_c is given by

$$r_c = \frac{v_t}{\left|\dfrac{d\mathbf{v}}{ds}\right|}. \tag{3.1.20}$$

If we now write

$$v_\perp = v_t \sin\theta, \quad v_\parallel = v_t \cos\theta, \tag{3.1.21}$$

so that θ is the angle between the velocity vector and the magnetic field, we see that $r_\perp$ may be expressed as

$$r_\perp = r_g \sin\theta, \tag{3.1.22}$$

where

$$r_g = \frac{v_t}{\omega_g},$$ (3.1.23)

and

$$r_c = r_g \operatorname{cosec} \theta.$$ (3.1.24)

As θ becomes smaller, the radius of the orbit (as measured by $r_\perp$) becomes smaller, but the radius of curvature r_c becomes larger. Hence if we consider radiation from electrons with an isotropic distribution of velocity vectors, most of the radiation will be due to particles moving almost normal to the magnetic field.

3.2 Particle motion in electric and magnetic fields

We now consider the nonrelativistic motion of a charged particle moving through static uniform electric and magnetic fields. We see from the nonrelativistic form of (3.1.2) that the velocity components satisfy the equations

$$\frac{dv_\parallel}{dt} = \frac{q}{m} E_\parallel$$ (3.2.1)

and

$$\frac{d\mathbf{v}_\perp}{dt} = \frac{q}{m} \left(\mathbf{E}_\perp + \frac{1}{c} \mathbf{v}_\perp \times \mathbf{B} \right),$$ (3.2.2)

where $\mathbf{v}_\perp$ is the part of the velocity vector normal to the magnetic field. Clearly, the particle accelerates freely in response to the electric field component along the magnetic field.

One method of determining the motion resulting from (3.2.2) is to consider the possibility that the transverse velocity can be separated into a constant component and a time-varying component, writing

$$\mathbf{v}_\perp(t) = \mathbf{v}_d + \mathbf{v}_g(t).$$ (3.2.3)

We will see that $\mathbf{v}_d$ corresponds to 'drift' motion, and $\mathbf{v}_g$ corresponds to 'gyro' motion.

The equation of motion (3.2.2) now becomes

$$\frac{d\mathbf{v}_g}{dt} = \frac{q}{m} \left(\mathbf{E}_\perp + \frac{1}{c} \mathbf{v}_d \times \mathbf{B} + \frac{1}{c} \mathbf{v}_g \times \mathbf{B} \right).$$ (3.2.4)

On separating out the time-independent and the time-dependent terms in this equation, we obtain

$$\mathbf{E}_{\perp} + \frac{1}{c}\mathbf{v}_d \times \mathbf{B} = 0 \qquad (3.2.5)$$

and

$$\frac{d\mathbf{v}_g}{dt} = \frac{q}{mc}(\mathbf{v}_g \times \mathbf{B}). \qquad (3.2.6)$$

On operating on (3.2.5) with a cross product with respect to **B**, we find that

$$\mathbf{v}_d = c\frac{\mathbf{E} \times \mathbf{B}}{B^2}. \qquad (3.2.7)$$

This shows that our nonrelativistic theory is valid only if $|\mathbf{E}| \ll |\mathbf{B}|$.

If we assume that $E_{\parallel} = 0$, we may obtain the above result more generally for relativistic motion. This may be seen by considering a transformation from the original coordinate frame to a frame moving with a drift velocity given by (3.2.7). (See, for instance Jackson (1962) p. 380.) In this frame, the electric field vanishes if (3.2.5) is satisfied, so that the motion is determined by (3.2.6), with m replaced by γm. Hence we see that, even in the relativistic case, the motion may be broken down into drift motion given by (3.2.7) and gyromotion provided that $|\mathbf{E}| < |\mathbf{B}|$.

We should note that the drift velocity, given by (3.2.7), depends on neither the charge nor the mass of the particle. (On the other hand, the derivation is valid only for particles with nonzero charge.) We see, in Fig. 3.1, the form of the orbital motion of ions and electrons in the crossed electric and magnetic field geometry. As a consequence, we may note that if a plasma comprises an equal number density of protons and electrons, the drift motion will not lead to the development of a nonzero current density. (We may also understand this result by referring to (11.5.11) that gives the equation of motion

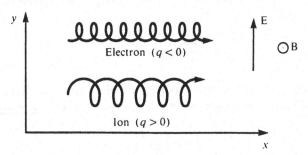

Fig. 3.1. Schematic representation of the orbits of ions and electrons in crossed electric and magnetic fields, showing that both species drift in the same direction, consistent with (3.2.7).

of a charged fluid. Since there is no net acceleration, the term $\rho d\mathbf{U}/dt$ is zero. Since protons and electrons have equal number densities, $\zeta = 0$. Hence it follows from (11.5.11) that $\mathbf{j}_\perp = 0$. Note that this result can also be verified for a relativistic plasma.)

3.3 Particle motion in magnetic and gravitational fields

We now consider the nonrelativistic motion of particles moving under the combined influence of uniform magnetic and gravitational fields. The equation of motion may now be expressed as

$$\frac{d\mathbf{v}}{dt} = \mathbf{g} + \frac{q}{mc}\,\mathbf{v} \times \mathbf{B}, \tag{3.3.1}$$

where $m\mathbf{g}$ is the gravitational force on a particle of mass m. This equation is identical to that for motion in an 'effective' electric field and the same magnetic field if we make the identification

$$\mathbf{E}_{\text{eff}} = \frac{m}{q}\,\mathbf{g}. \tag{3.3.2}$$

Hence we see from (3.2.7) that the drift-component of the motion is now given by

$$\mathbf{v}_d = \frac{mc}{q}\frac{\mathbf{g} \times \mathbf{B}}{B^2}. \tag{3.3.3}$$

There is also freely accelerated motion in the direction of $\mathbf{B}$, if $\mathbf{g} \cdot \mathbf{B} \neq 0$.

We note that the drift velocity now depends on both charge and mass. In particular, electrons and positively charged ions will drift in *opposite* directions, as shown in Fig. 3.2. Hence the combined effect of gravitational and magnetic fields is such as to produce a current in the system. For the simple case of a uniform plasma consisting only of protons and electrons, the current density is given by

$$\mathbf{j} = -\frac{ne}{c}\mathbf{v}_{e,d} + \frac{ne}{c}\mathbf{v}_{p,d}, \tag{3.3.4}$$

where $\mathbf{v}_{e,d}$ and $\mathbf{v}_{p,d}$ are the drift velocities of electrons and protons respectively. On substituting for these velocities from (3.3.3), we find that

$$\mathbf{j} = \rho\frac{\mathbf{g} \times \mathbf{B}}{B^2}. \tag{3.3.5}$$

Unlike most plasma systems for which the larger electron velocity, due to the smaller electron mass, has the result that most of the current is carried

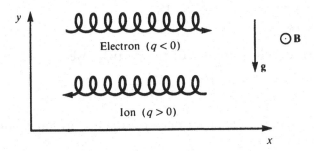

Fig. 3.2. Schematic representation of the orbits of ions and electrons in crossed gravitational and magnetic fields. In this case, ions and electrons drift in opposite directions, resulting in a net current density.

by electrons, we see that in the present system most of the current is carried by the ions.

We also see that the net force density on the plasma, given by

$$\mathbf{F} = \mathbf{j} \times \mathbf{B} + \rho \mathbf{g}, \tag{3.3.6}$$

vanishes since $\mathbf{j}$ is given by (3.3.5). This is as we would expect, since we know that there is no time-independent component of acceleration perpendicular to the magnetic field.

3.4 Particle motion in a time-varying uniform magnetic field

Since a magnetic field does no work on the particle, particle energy in a static magnetic field is conserved. However, if the magnetic field is time-dependent, there must be an accompanying electric field that satisfies (2.1.1), i.e.

$$\nabla \times \mathbf{E} = -\frac{1}{c} \frac{\partial \mathbf{B}}{\partial t}. \tag{3.4.1}$$

Clearly the electric field cannot be uniform in space. Hence we must expect that the electric field will be such as to change the particle energy.

We here consider nonrelativistic motion, and focus only on the motion normal to the magnetic field. If

$$U_\perp = \tfrac{1}{2} m v_\perp{}^2, \tag{3.4.2}$$

we see that

$$\frac{\mathrm{d}U_\perp}{\mathrm{d}t} = q\mathbf{v}_\perp \cdot \mathbf{E}. \tag{3.4.3}$$

On writing

$$\mathbf{v} = \frac{d\mathbf{X}}{dt},$$

(3.4.4)

where $\mathbf{X}(t)$ denotes the trajectory of the particle, we see that

$$\frac{dU_\perp}{dt} = q \frac{d\mathbf{X}}{dt} \cdot \mathbf{E},$$

(3.4.5)

so that the total change in $U_\perp$ over one cycle of the orbital motion is given by

$$\Delta U_\perp = \int_0^P dt \, q \frac{d\mathbf{X}}{dt} \cdot \mathbf{E},$$

(3.4.6)

where P is the period of the motion.

The variation in the magnetic field implies a time variation in both gyroradius and gyroperiod, so that the orbit will not be closed. However, we now make the assumption that the change in $\mathbf{B}$ during one period is small compared with the magnitude of $\mathbf{B}$, i.e.,

$$P \left| \frac{dB}{dt} \right| = 2\pi \omega_g^{-1} \left| \frac{dB}{dt} \right| \ll |\mathbf{B}|.$$

(3.4.7)

With this assumption, the time integral may be replaced by a line integral taken over a fictitious circular orbit of the particle:

$$\Delta U_\perp = \oint q \, d\mathbf{X} \cdot \mathbf{E}.$$

(3.4.8)

By using Stokes' theorem, this may be expressed as

$$\Delta U_\perp = -q \int d\mathbf{S} \cdot (\nabla \times \mathbf{E}).$$

(3.4.9)

The minus sign appears in (3.4.9) since the contour integral in (3.4.8) follows the particle motion. For a positively charged particle moving in a magnetic field in the x_3 direction, this orbit is a *left*-hand circular motion with respect to the x_3 axis. We now see, from (3.4.1), that

$$\Delta U_\perp = \frac{|q|}{c} \int d\mathbf{S} \cdot \frac{\partial \mathbf{B}}{\partial t}.$$

(3.4.10)

We now write $|q|$ in this equation since the change in energy is in fact independent of the sign of the charge.

Since we are assuming that the magnetic field is uniform, (3.4.10) is expressible as

$$\Delta U_\perp = |q| \frac{\pi r_\perp^2}{c} \frac{dB}{dt}.$$

(3.4.11)

Hence the rate of change of energy is given by

$$\frac{dU_\perp}{dt} = \frac{\Delta U_\perp}{P} = \frac{1}{2}\frac{|q|}{c}\omega_g r_\perp^2 \frac{dB}{dt}, \tag{3.4.12}$$

from which we may verify that

$$\frac{1}{U_\perp}\frac{dU_\perp}{dt} = \frac{1}{B}\frac{dB}{dt}. \tag{3.4.13}$$

We may define the 'magnetic moment' of a charged particle by

$$\mu = \frac{U_\perp}{B}. \tag{3.4.14}$$

Equation (3.4.13) now shows that

$$\frac{d\mu}{dt} = 0, \tag{3.4.15}$$

so that

$$\frac{dU_\perp}{dt} = \mu\frac{dB}{dt}. \tag{3.4.16}$$

Although we write (3.4.15) as an equality, it must be remembered that the entire calculation is approximate, being based on the inequality of (3.4.7). In fact, μ is only *approximately* constant. Equation (3.4.15) is a good representation of the behavior of the particle only if the rate of change of magnetic field is slow compared with the gyromotion, and if the variation of **B** does not contain a significant harmonic component of frequency $2\omega_g$.

On comparing (3.4.12) and (3.4.16), we see that

$$\mu = \frac{1}{2}\frac{|q|}{c}\omega_g r_\perp^2. \tag{3.4.17}$$

This may be expressed alternatively as

$$\mu = \frac{1}{2\pi}\frac{q^2}{mc^2}\Phi, \tag{3.4.18}$$

where

$$\Phi = \pi r_\perp^2 B. \tag{3.4.19}$$

Hence we see that (3.4.15) implies also that

$$\frac{d\Phi}{dt} = 0, \tag{3.4.20}$$

where Φ indicates the magnetic flux embraced by the particle orbit.

The quantity μ (or, equivalently, Φ) is an example of an 'adiabatic invariant.' The quantity is not strictly invariant, but it is approximately invariant if the rates of change of the parameters of the system are sufficiently slow and if the variation is aperiodic. In the next chapter, we will meet other forms of adiabatic invariant. The quantity μ is sometimes referred to as the 'first adiabatic invariant.'

Problems

Problem 3.1. Calculate the electron and proton gyrofrequencies for the following systems, and compare them with the plasma frequencies found in Problem 2.1. Find the gyroradii of particles moving at their thermal velocities.

(a) A fusion machine: $n_e = n_p = 10^{16}$ cm^{-3}, $T_e = T_p = 10^7$ K, $B = 10^4$ G.
(b) The Earth's magnetosphere: $n_e = n_p = 10^4$ cm^{-3}, $T_e = T_p = 10^3$ K, $B = 10^{-2}$ G.
(c) The center of the Sun: $n_e = n_p = 10^{26}$ cm^{-3}, $T_e = T_p = 10^{7.2}$ K, $B = 10^6$ G.
(d) The solar corona: $n_e = n_p = 10^8$ cm^{-3}, $T_e = T_p = 10^6$ K, $B = 1$ G.
(e) The solar wind: $n_e = n_p = 10$ cm^{-3}, $T_e = T_p = 10^5$ K, $B = 10^{-5}$ G.
(f) The atmosphere of a neutron star: $n_e = n_p = 10^{12}$ cm^{-3}, $T_e = T_p = 10^7$ K, $B = 10^{12}$ G.

Problem 3.2. Consider the nonrelativistic motion of particles of charge q and mass m in crossed static electric and magnetic fields. Take the electric field to be $\mathbf{E} = (0, E, 0)$, and the magnetic field to be $\mathbf{B} = (0, 0, B)$.

(a) Assuming that q, E and B are all positive, sketch the orbit of a typical particle.
(b) For the particular case that a particle is initially at rest at the origin, determine the form of the orbit. Is this an example of the orbit of a 'typical' particle that you drew in (a)?

Problem 3.3. Consider orthogonal uniform electric and magnetic fields, $\mathbf{E} = (0, E, 0)$ and $\mathbf{B} = (0, 0, B)$.

(a) Assuming that $E < B$, find the velocity of the moving frame in which the electric field is zero. What is the (relativistically correct) motion of the particle in this frame? What would be the corresponding motion in the rest frame? Relate this finding to the drift velocity found in Chapter 3.
(b) Assuming that $E > B$, find the velocity of the moving frame in which the magnetic field is zero. What is the motion of the particle in this frame? What would be the corresponding motion in the rest frame?

Problem 3.4. Consider a uniform plasma consisting of a density n each of electrons and protons in orthogonal uniform gravitational and magnetic fields $\mathbf{g} = (0, -g, 0)$ and $\mathbf{B} = (0, 0, B)$.

(a) Calculate the drift velocities of each species for the following situations
  (i) In the Earth's magnetosphere at $r = 2R_\oplus$, assuming that the surface field strength is 1G.
  (ii) Near the surface of the Sun, in an active region where $B = 10^2$ G.
  (iii) In the magnetosphere of a neutron star, assumed to be of the same mass as the Sun, at $r = 2R_\odot$, assuming that the surface field strength is 10^{12} G and $R_* = 10^6$ cm.
(b) Verify that in each case the gravitational force and Lorentz force acting on the plasma cancel out.

Problem 3.5. Consider a cylindrically symmetric coil that produces a field that is virtually uniform within a certain volume. At time $t = t_0$, the field has a small initial value B_0, and a particle is gyrating in a small circular orbit of radius r_0 at a distance R_0 from the axis of symmetry. The magnetic field increases slowly and linearly in time to reach a value B_1 at time t_1.

(a) Calculate E_ϕ as a function of radius and time.
(b) Calculate the drift velocity due to the combined action of the electric and magnetic fields.
(c) Hence calculate R_1, the final value of the distance of the particle from the axis of symmetry.
(d) What quantity (in addition to the magnetic moment) is invariant as a result of this drift motion?
(e) Is your analysis relativistically correct?

Problem 3.6. Consider the (nonrelativistic) motion of an electron in a uniform time-varying magnetic field produced by a coil of cylindrical symmetry that gives rise to an electric field of cylindrical symmetry. Assuming that there is no motion in the z-directions, show that the equations of motion are, in terms of cylindrical polar coordinates,

$$\ddot{r} - r\dot{\phi}^2 = \frac{-e}{m_e c} r\dot{\phi} B \tag{1}$$

and

$$r\ddot{\phi} + 2\dot{r}\dot{\phi} = \frac{-e}{m_e c}\left(\frac{-\dot{B}r}{2} - \dot{r}B\right). \tag{2}$$

If the magnetic field grows from zero and if the electron is initially at rest, show that the equations become

$$\dot{\phi} = \frac{1}{2}\frac{e}{m_e c} B \tag{3}$$

and

$$\ddot{r} + \frac{1}{4}\left(\frac{e}{m_e c}\right)^2 B^2 r = 0. \tag{4}$$

For the particular case that

$$B(t) = bt \tag{5}$$

(b=constant), show that (4) becomes

$$\frac{d^2 r}{du^2} + \frac{1}{2u}\frac{dr}{du} + r = 0, \tag{6}$$

where

$$u = \frac{e\dot{B}_0}{4m_e c}t^2.$$

This may be solved in terms of Bessel functions of order 1/4. Show that, in the asymptotic approximation, the magnetic moment of the electron is constant. Also study the variation of the magnetic moment for small values of t. After what time is the variation of the magnetic moment less than one per cent?

Problem 3.7. Consider the motion of a charged particle in a time-varying uniform magnetic field $\mathbf{B} = (0, 0, B(t))$. Assume that the field is nearly constant, with a superposed small sinusoidal fluctuation,

$$\mathbf{B}(t) = B_0 \hat{\mathbf{z}}(1 + \varepsilon \sin 2\Omega t)$$

where Ω is the gyrofrequency corresponding to the field strength B_0. If the system has plane symmetry so that $E_y = 0$ and E_x is an odd function of y, and if the particle is initially moving in a circular orbit centered in the plane $y = 0$, find an expression for the increase of energy in one orbit. Hence find an approximate expression for the rate of change of energy of the electron due to the periodic variation of the magnetic field.
What do you infer about the magnetic moment?

Problem 3.8. Consider, in nonrelativistic theory, a particle that is gyrating in a circular orbit in a substantially uniform magnetic field.

(a) Evaluate the volume of magnetic field that has the same energy as the kinetic energy of the particle.
(b) Consider the cylinder that has this volume and has the same radius as the orbit of the particle. What is the height of this cylinder?
(c) Do you recognize this expression?

Problem 3.9. Consider a distribution of particles that all have the same initial kinetic energy U_0 but have velocity vectors that are distributed isotropically. Consider the following sequence of events:

(i)   The magnetic field strength is increased slowly by the factor λ.
(ii) The particle distribution is isotropized (for instance, by mutual collisions or by scattering by weak magnetic inhomogeneities).
(iii) The magnetic field strength is slowly returned to its original value.
(iv) The distribution is again isotropized.

(a) If the final mean energy is U_1, express U_1 in terms of U_0 and λ.
(b) Show that $U_1 > U_0$.
(c) The particles in fact have a distribution of energy. What is the maximum energy and what is the minimum energy?

4

Adiabatic invariants

4.1 General adiabatic invariants

We have already encountered the first adiabatic invariant μ, the magnetic moment, that arose in a nonrelativistic treatment of particle motion in a magnetic field. It is useful to consider a more general development of the concept of adiabatic invariance for application to other situations, including relativistic particle motion.

Consider a system described by dynamical variables q_i and a Lagrangian function $L(q_i, \dot{q}_i, t)$. The canonical momentum variables are defined by

$$p_i = \frac{\partial L}{\partial \dot{q}_i},\qquad (4.1.1)$$

and the dynamical equations are given by

$$\frac{\mathrm{d}p_i}{\mathrm{d}t} = \frac{\partial L}{\partial q_i}.\qquad (4.1.2)$$

We consider a closed family of solutions enumerated by the variable κ that we assume runs from 0 to 2π:

$$\left.\begin{aligned} p_i &= P_i(\kappa, t), \\ q_i &= Q_i(\kappa, t). \end{aligned}\right\}\qquad (4.1.3)$$

If we now consider the quantity defined by

$$J = \oint \mathrm{d}\kappa\, P_i \frac{\partial Q_i}{\partial \kappa},\qquad (4.1.4)$$

we find that the total time derivative of J is given by

$$\frac{\mathrm{d}J}{\mathrm{d}t} = \oint \mathrm{d}\kappa \left[\frac{\mathrm{d}P_i}{\mathrm{d}t} \frac{\partial Q_i}{\partial \kappa} + P_i \frac{\partial}{\partial \kappa}\left(\frac{\mathrm{d}Q_i}{\mathrm{d}t}\right) \right],\qquad (4.1.5)$$

where we have changed the order of differentiation in the second term. In view of (4.1.1) and (4.1.2), this may be expressed as

$$\frac{dJ}{dt} = \oint d\kappa \left[\frac{\partial L}{\partial q_i} \frac{\partial q_i}{\partial \kappa} + \frac{\partial L}{\partial \dot{q}_i} \frac{\partial \dot{q}_i}{\partial \kappa} \right], \qquad (4.1.6)$$

where it is to be understood that the actual values of q_i and $\dot{q}_i$ are Q_i and $\dot{Q}_i$.

Since L itself does not depend upon κ, we see that

$$\frac{dJ}{dt} = \oint d\kappa \frac{dL}{d\kappa} = 0, \qquad (4.1.7)$$

so that J is a strict invariant of the system. It is known as the Poincaré invariant.

Note that we made no assumption about the significance of the parameter κ. One possibility is that the system exhibits periodic motion with frequency ω. Then κ could be a phase factor that enters a description of the form

$$q_i = Q_i(\omega t + \kappa). \qquad (4.1.8)$$

Normally we expect to have strictly periodic motion only if L is time independent. Then the constancy of J (which, as we shall see, is related to energy) would not be particularly significant.

However, now suppose that L depends on time but in such a way that it varies slowly and aperiodically with respect to the oscillatory behavior under consideration. If we consider that the motion begins at time $t = 0$, then, for $t \approx 0$,

$$q_i \approx Q_{i,0}(\omega_0 t + \kappa_0), \qquad (4.1.9)$$

where κ_0 determines the initial phase at which the system begins its motion. However, for given functional form $Q_{i,0}$ and given phase κ_0, the initial conditions of the system are completely determined so that κ_0 enumerates a closed set of dynamical evolutions of the system for all time.

Hence J_0, defined by

$$J_0 = \oint d\kappa \, P_i \frac{\partial q_i}{\partial \kappa_0}, \qquad (4.1.10)$$

is a Poincaré invariant, so that it is a strict invariant of the system.

Now suppose that, as a result of the slow and aperiodic variation of L, the system slowly evolves through a sequence of oscillatory states, so that the evolution of q_i may be written as

$$q_i = Q_i(\omega(t)t + \kappa(t), t). \qquad (4.1.11)$$

We suppose that the frequency $\omega(t)$ and the phase factor $\kappa(t)$ are slowly varying functions of time, and that the dependence of q_i upon time, as indicated
by the last term in parentheses, is also a slow variation – representing, for
instance, the slow variation in the amplitude of the oscillation. Then, clearly,
we could at any time form the quantity

$$\tilde{J}(t) = \oint \mathrm{d}\kappa \, p_i \frac{\partial q_i}{\partial \kappa}. \qquad (4.1.12)$$

This is not a Poincaré invariant, so there is no obvious reason why this
should be an invariant, but it is normally found that $\kappa(t)$ is simply related
to the initial phase κ_0:

$$\kappa(t) = \kappa_0 + f(t), \qquad (4.1.13)$$

where $f(t)$ is a slowly varying function of time. In particular, if the variation
of L is slow and aperiodic, we will normally find that the instantaneous phase
at time t is related to the initial phase at time $t=0$ by (4.1.13) where $f(t)$ is
a slowly varying function of t. In this case, we see from (4.1.7) that

$$\frac{\mathrm{d}\tilde{J}(t)}{\mathrm{d}t} = 0. \qquad (4.1.14)$$

$\tilde{J}(t)$ is an invariant, not a strict invariant, but an 'adiabatic invariant.' For
further discussion of this point, see Sturrock (1955).

   In order to see the difference between the Poincaré invariant and the
adiabatic invariant, we might consider the simple case of a system with only
one degree of freedom, that is then described by the variables q and p.
Suppose that, at time $t=0$, a closed family of solutions is that shown in
Fig. 4.1(a). Some time later, for a complex system, this set of states of the
system may have evolved to that shown in Fig. 4.1(b). The shape is quite different, but the area enclosed by the contour has necessarily remained
constant.

   Now suppose that, at $t \approx 0$, the system is going through simple periodic
motion, as indicated by the dotted lines in Fig. 4.2(a). Suppose also that the
initial closed family of solutions is one that corresponds to a particular
amplitude of oscillation (and therefore to a particular energy) as indicated
by the solid line in Fig. 4.2(a). If the system evolves in time and then settles
down to a steady state, the system may once again exhibit periodic behavior
as indicated by the broken lines in Fig. 4.2(b). In general, there is no reason
to expect that the initial family of solutions corresponds to a new family lying
on a constant-amplitude contour. However, if the system evolves slowly and
aperiodically, then it is a good approximation that a constant-amplitude

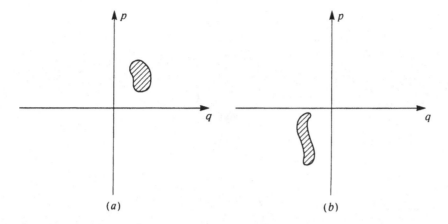

Fig. 4.1. We consider a closed family of trajectories in p, q, t space: (a) shows the intersections of those trajectories with the plane $t = 0$, and (b) shows the intersections of the trajectories with a plane representing a later time.

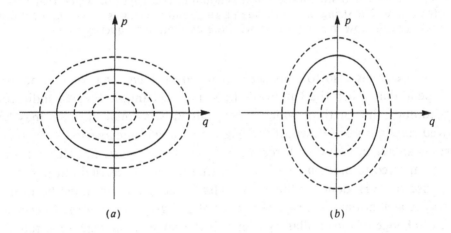

Fig. 4.2. We now consider a single trajectory. In (a), the system is in an almost steady state so that the representative point in phase space maps out a closed contour. In (b), the system has evolved to another almost steady state; the representative point still maps out a closed contour, and the area of the closed contour in (b) is the same as that of the closed contour in (a).

family of solutions will remain a constant-amplitude family of solutions. Then the new family will be represented by the heavy contour in Fig. 4.2(b), namely, the constant-amplitude contour that embraces the same area as the original contour in Fig. 4.2(a).

A comparison of Fig. 4.1 and Fig. 4.2 is a comparison of the concepts of the Poincaré invariant and the adiabatic invariant.

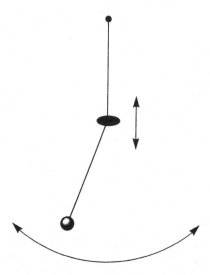

Fig. 4.3. Schematic representation of a pendulum, the length of which that varies slowly in time. The string passes through a ring, and the ring is moved up or down at a rate that is slow in comparison with the oscillation frequency.

It is interesting also to consider a classical problem, that of the motion of a pendulum when the length of the string is changed. Fig. 4.3 indicates a pendulum, the length of which is determined by a small ring that may be moved vertically up or down. If the ring moves slowly and aperiodically, then there is an average upward force on the ring due to the string. Hence if the ring is allowed to move upward slowly, that motion is carried out partly in response to work done by the string. Hence the pendulum must be losing energy. Calculation of the rate of change of energy shows that the fractional rate of change of the oscillatory energy is the same as the fractional rate of change of the frequency. Hence the 'action,' defined by

$$J_A = \frac{U}{\omega}, \qquad (4.1.15)$$

where U is the oscillatory energy and ω is the frequency, is an adiabatic invariant.

It is easy to see that this quantity would not be an invariant if the motion of the ring were arbitrarily rapid. If the ring were to move up rapidly at the instant the string is vertical, there will be no change in the energy of the pendulum. Hence, if the motion of the ring were composed of many very small sudden movements, each movement occurring when the string is vertical, there would be no change in the energy of the pendulum. Hence we see why, in this problem, it is necessary that the motion be slow and aperiodic in order for the action to remain approximately constant.

4.2 The first adiabatic invariant: magnetic moment

The nonrelativistic motion of a particle in a magnetic field may be described by the Lagrangian function

$$L = \frac{1}{2} m \dot{r}^2 + \frac{1}{2} m r^2 \dot{\phi}^2 + \frac{q}{c} r \dot{\phi} A_\phi, \qquad (4.2.1)$$

if we adopt polar coordinates z, r, ϕ, ignore motion in the z direction, and assume that the magnetic field is symmetric about the origin. If the field is uniform, then

$$A_\phi = \frac{1}{2} B(t) r, \qquad (4.2.2)$$

where we allow for a slow variation of the strength of the field with time.

The concept of cylindrical symmetry of a uniform magnetic field may seem odd. If the field is static, the concept makes no sense. However, if the field is time-dependent, the concept is significant. We know that a time-dependent magnetic field necessarily leads to the development of an electric field. Hence we are in fact assuming that the *electric* field is of cylindrical symmetry.

Combining (4.2.1) and (4.2.2) into

$$L = \frac{1}{2} m \dot{r}^2 + \frac{1}{2} m r^2 \dot{\phi}^2 + \frac{1}{2} \frac{q}{c} B r^2 \dot{\phi}, \qquad (4.2.3)$$

we see that

$$p_r = m \dot{r}, \quad p_\phi = m r^2 \dot{\phi} + \frac{1}{2} \frac{q}{c} B r^2. \qquad (4.2.4)$$

The equation for p_r becomes

$$m \frac{\mathrm{d}^2 r}{\mathrm{d}t^2} = m r \dot{\phi}^2 + \frac{q}{c} B r \dot{\phi}, \qquad (4.2.5)$$

and the equation for p_ϕ becomes

$$\frac{\mathrm{d}p_\phi}{\mathrm{d}t} = 0. \qquad (4.2.6)$$

If B varies only slowly in time then, at any instant, (4.2.5) is satisfied approximately if $r \approx$ constant and

$$\dot{\phi} = -\varepsilon \Omega, \qquad (4.2.7)$$

where we now use Ω, in place of ω_g, for the gyrofrequency defined by

$$\Omega = \frac{|q|B}{mc}.$$ (4.2.8)

Since the system varies only slowly with time, we expect that initially circular motion about the origin will remain circular, so that

$$\phi(t) = -\varepsilon(\Omega(t)t + \kappa(t)),$$ (4.2.9)

where the phase function will vary only slowly with time. Hence we can form an adiabatic invariant from

$$J = \oint d\kappa \left(p_r \frac{\partial r}{\partial \kappa} + p_\phi \frac{\partial \phi}{\partial \kappa} \right).$$ (4.2.10)

Since the motion is almost circular, we may neglect the term involving p_r. Hence we obtain

$$J = -2\pi\varepsilon p_\phi.$$ (4.2.11)

On using (4.2.7) in (4.2.4), we obtain

$$J = \pi m r^2 \Omega.$$ (4.2.12)

Since the transverse energy is given by

$$U_\perp = \tfrac{1}{2} m r^2 \Omega^2,$$ (4.2.13)

we see that

$$J = 2\pi \frac{U_\perp}{\Omega}.$$ (4.2.14)

This expression for the adiabatic invariant is related to the magnetic moment obtained in Chapter 3 by

$$J = 2\pi \frac{mc}{|q|} \mu.$$ (4.2.15)

4.3 Relativistic form of the first adiabatic invariant

In relativistic theory, we must replace (4.2.1) by

$$L = mc^2 \left\{ 1 - \left[1 - \frac{\dot{z}^2 + \dot{r}^2 + r^2\dot{\phi}^2}{c^2} \right]^{1/2} \right\} + \frac{q}{c} (\dot{z} A_z + \dot{r} A_r + r\dot{\phi} A_\phi), \quad (4.3.1)$$

where we now allow for the possibility of motion in the z direction. Hence we find that

$$p_z = \gamma m \dot{z} + \frac{q}{c} A_z,$$

$$p_r = \gamma m \dot{r} + \frac{q}{c} A_r,$$

$$p_\phi = \gamma m r^2 \dot{\phi} + \frac{q}{c} r A_\phi. \tag{4.3.2}$$

If we now repeat the argument of the previous section for the present relativistic case, we find that the equation of motion for the radial coordinate leads to

$$\dot{\phi} = -\varepsilon \Omega \equiv -\varepsilon \frac{|q|B}{\gamma mc} \tag{4.3.3}$$

for the case that the motion is almost circular. If we introduce the phase factor by writing

$$\phi(t) = -\varepsilon (\Omega(t)t + \kappa(t)), \tag{4.3.4}$$

we find that (4.2.12) is now replaced by

$$J = \pi \gamma m r^2 \Omega. \tag{4.3.5}$$

The two terms of p_ϕ in (4.3.2) in fact give rise to two contributions to the adiabatic invariant. We find that we may write

$$J = 2\pi H - \frac{|q|}{c} \Phi, \tag{4.3.6}$$

where

$$H = \gamma m r^2 \Omega, \tag{4.3.7}$$

so that H is the kinetic angular momentum, and

$$\Phi = \pi r^2 B. \tag{4.3.8}$$

However, these two terms are related by

$$\pi H = \frac{|q|}{c} \Phi. \tag{4.3.9}$$

Hence the fact that J is an adiabatic invariant also guarantees the fact that H and Φ also are adiabatic invariants. We see that

$$J = \pi H \quad \text{and} \quad J = \frac{|q|}{c} \Phi. \tag{4.3.10}$$

In the nonrelativistic case, we saw that the transverse kinetic energy varies in proportion to the gyrofrequency, and therefore in proportion to the magnetic field strength. This is not true in the relativistic case. If we write

$$\beta_\perp = \frac{r\dot{\phi}}{c},\tag{4.3.11}$$

we find from (4.3.5) that

$$\gamma\beta_\perp \propto B^{1/2}.\tag{4.3.12}$$

In the nonrelativistic case, this gives us once more the result that $\beta_\perp^2$ is proportional to B. However, in the ultra-relativistic case, if there is no z motion so that $\beta_\perp \approx 1$, we find that the energy is proportional to $B^{1/2}$, not to B.

4.4 The second adiabatic invariant: the bounce invariant

Consider, to start with, the motion of a charged particle in a static magnetic field, allowing for a slow spatial variation in the strength of the field. We adopt relativistic theory, since it is no more difficult than nonrelativistic theory. Since, in a static field, $\gamma = $ constant, we see from (4.3.12) that

$$\beta_\perp{}^2 = \beta^2\frac{B}{B_R},\tag{4.4.1}$$

where B_R is a constant determined by the initial conditions of the orbit. Since

$$\beta^2 = \beta_\parallel{}^2 + \beta_\perp{}^2,\tag{4.4.2}$$

we see that

$$\beta_\parallel{}^2 = \beta^2\left(1 - \frac{B}{B_R}\right).\tag{4.4.3}$$

This indicates that $\beta_\parallel = 0$ wherever $B = B_R$. Furthermore, it is obviously not allowable that the particle should move into a region where $B > B_R$. This indicates that the charged particle will be *reflected* wherever B attains the value B_R. That is, the particle will behave as if it had been reflected by a 'mirror.' Such a field configuration is therefore termed a 'magnetic mirror.'

   It is clear that the motion along the magnetic field will cease when the particle arrives at the point where $B = B_R$, but it is perhaps not so obvious that the particle will be reflected. It may therefore be helpful to look at the problem in a slightly different way. The motion described by (4.4.3) is the same as that of a fictitious particle for which the total energy is expressible in the form

$$U_t = \tfrac{1}{2}\gamma m v_\parallel^2 + V, \qquad (4.4.4)$$

where V is the 'potential energy' defined by

$$V = \frac{1}{2}\gamma m v_t^2 \frac{B}{B_R}. \qquad (4.4.5)$$

We use U_t for the total energy, v_t for the total speed, and we are adopting a fictitious nonrelativistic model, even if the actual motion is relativistic. Equation (4.4.4) is clearly the energy equation of a particle moving in a potential well described by the potential energy $V(s)$. We know that the form (4.4.4) of the total energy leads to the equation of motion

$$\gamma m \frac{dv_\parallel}{dt} = -\frac{\partial V}{\partial s}. \qquad (4.4.6)$$

Hence, as long as dB/ds is nonzero at the reflection point, we expect that particles will be reflected from that point. That is, the field in that region will indeed behave as a 'magnetic mirror.'

Let us now consider the case that the magnetic field, as a function of distance s along a field line, varies in strength as shown in Fig. 4.4. We assume that B has its minimum value B_m at $s = s_0$, and that it has maxima $B_{M,1}$ and $B_{M,2}$ at positions $s = s_1$ and $s = s_2$, respectively. Then the trapping properties of this configuration are determined by the lesser value of $B_{M,1}$, $B_{M,2}$, that we refer to as B_M. If we write

$$\beta_\perp = \beta \sin\theta, \quad \beta_\parallel = \beta \cos\theta, \qquad (4.4.7)$$

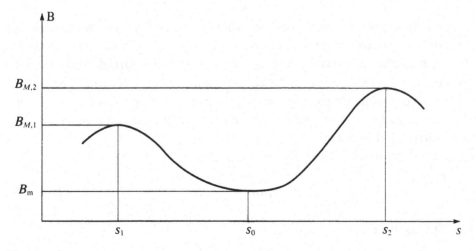

Fig. 4.4. Example of magnetic field strength, as a function of position, that leads to a 'magnetic bottle.'

and if θ_0 is the value of θ at $s=s_0$, we see that a particle will be trapped if and only if $\theta_0 > \theta_L$, where

$$\sin \theta_L = \left(\frac{B_m}{B_M}\right)^{1/2}, \tag{4.4.8}$$

so that $\theta_0 = \theta_L$ may be said to define the the 'loss cone' of the system.

If we consider that an isotropic particle distribution is suddenly introduced at $s=s_0$, the fraction of particles that will be lost, F_L, is given by

$$F_L = \frac{1}{2\pi} \int_0^{\theta_L} 2\pi \sin \theta \, d\theta = 1 - \cos \theta_L, \tag{4.4.9}$$

that is, by

$$F_L = 1 - \left(1 - \frac{B_m}{B_M}\right)^{1/2} \tag{4.4.10}$$

It is clear that any particles that are trapped in a magnetic mirror will undergo a second type of oscillatory motion, in addition to their gyromotion, namely their 'bouncing' motion between the reflection points. Hence we can assign a second adiabatic invariant to the motion of these particles by using (4.1.12).

We can, for convenience, re-write this expression as an integral over time, as in (4.4.11), where the integral is taken by following the particle through one oscillation:

$$J_2 = \oint p_i \, dq_i. \tag{4.4.11}$$

However, if any confusion arises, one should remember that the integral is really defined as an integral over phase. For instance, where a system is multiply periodic (as is true in the present case, since the particle also exhibits gyromotion about the magnetic field lines), an integral over phase provides a simple expression for each adiabatic invariant, whereas an integral over time would lead to a confusing mixture of contributions related to all the periodicities of the system.

Since, in rectangular Cartesian coordinates,

$$p_r = \gamma m v_r + \frac{q}{c} A_r, \tag{4.4.12}$$

(4.4.11) may be expressed as

$$J_2 = \oint \left[\gamma m v_\parallel + \frac{q}{c} A_\parallel\right] ds, \tag{4.4.13}$$

where $v_\parallel$ and $A_\parallel$ are the components of $\mathbf{v}$ and $\mathbf{A}$ in the direction of the element of arc length ds. It is clear that the second term involving $A_\parallel$ will vanish, since the sign of $A_\parallel$ changes with the direction of motion along the field line. We can also see this result by noting that the second term really represents the quantity given by

$$\frac{q}{c} \oint \mathbf{A} \cdot d\mathbf{x} = \frac{q}{c} \Phi, \tag{4.4.14}$$

where Φ is the flux included in the contour mapped out as the particle moves through a complete bounce motion. However, since the particle is moving to and fro along the same field line, the area embraced by the contour is zero so that $\Phi = 0$.

Furthermore, we see that the first term in (4.4.13) includes two equal contributions, one from motion from s_1 to s_2, and the other from the reverse motion. Hence we may write (4.4.13) as

$$J_2 = 2\gamma m \int_{s_1}^{s_2} v_\parallel \, ds. \tag{4.4.15}$$

On using (4.4.3), we see that this is expressible as

$$J_2 = 2mc\gamma\beta I \tag{4.4.16}$$

where

$$I = \int_{s_1}^{s_2} \left(1 - \frac{B(s)}{B_R} \right)^{1/2} ds. \tag{4.4.17}$$

If the magnetic field is static, so that γ and β are constants, our calculation shows that the geometrical quantity I is an adiabatic invariant. However, if the magnetic field varies slowly in time, so that the energy of particles may change, then the quantity given by (4.4.16) is an adiabatic invariant, but the quantity given by (4.4.17) is not.

The existence of the first and second adiabatic invariants have important consequences concerning the trapping of particles in complex magnetic fields, as we shall now see.

4.5 Magnetic traps

Consider the magnetic field configuration shown in Fig. 4.5, a configuration that is not assumed to have any particular type of symmetry. Consider the motion of a particle that starts out with a given velocity vector so that we know the initial value of the first adiabatic invariant, that is approximately

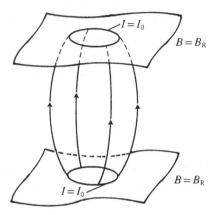

Fig. 4.5. A magnetic field configuration that does not have cylindrical symmetry but leads to the trapping of charged particles.

constant. If the field is static, so that the particle energy is a constant, we know that the particle is reflected at points where the magnetic field strength takes the value B_R.

As we shall see in the next chapter, a particle tends to drift in an inhomogeneous magnetic field, so that it will migrate to other magnetic-field lines. If we now construct two surfaces on which $B = B_R$, and between which $B < B_R$, we know that the particle will be reflected at these two surfaces and trapped between the surfaces.

However, since we are assuming that the magnetic field is static, we can go further. We know, from the properties of the second adiabatic invariant, that the particle is constrained to move in such a way that $I = $ constant, where I is defined by (4.4.16).

Hence, for each field line bounded by the two surfaces $B = B_R$, we may calculate the value of I. We may then identify the shell of magnetic-field lines for which $I = I_0$, where I_0 is the initial value of I. Hence we know that the particle will move in such a way that it is constrained to remain on the shell and that its motion is bounded by the two reflection surfaces. If the shell forms a closed surface, as indicated in Fig. 4.5, then the particle is constrained to remain on that surface, no matter what type of drift the magnetic-field gradients may imply.

Another example of a magnetic-field configuration that leads to trapping is that shown in Fig. 4.6, that of the magnetosphere of a body with a dipole magnetic field, such as the Earth. In this case, particles bounce from one reflection point to another in helical trajectories. In addition to this motion, there is (as we shall see in the next chapter) a drift around the earth due to

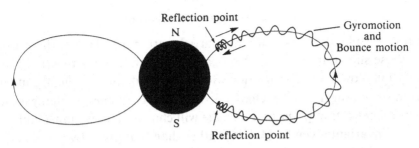

Fig. 4.6. A dipole-type magnetic field, such as that of the Earth, leads to particle trapping. Particles exhibit gyromotion around magnetic field lines, bounce motion along field lines between reflection points, and drift motion around the Earth.

the curvature of the magnetic field lines (curvature drift) and to the spatial variation of B with radius (gradient drift). Hence, for the particles trapped in the Earth's magnetosphere, there are three types of oscillations due to (a) the gyromotion, (b) the 'bounce' motion, and (c) the drift motion.

Another example is the trapping of particles in the Sun's magnetic field, that is much more complex than the approximately dipole field of the earth. In an active region, that inevitably contains surface magnetic fields of opposite polarities, and typically contains at least one pair of sunspots of opposite polarities, part of the magnetic field will be as shown in Fig. 4.7. The Sun produces some radio bursts (stationary Type IV microwave radio bursts) that are initiated by flares and are believed to be due to gyrosynchrotron radiation. Hence, such flares indicate that mildly relativistic electrons are somehow trapped at coronal heights in an active region. We see from Fig. 4.7 that, here again, if the field is static it is reasonable that particles should be trapped in the region above the two reflection surfaces.

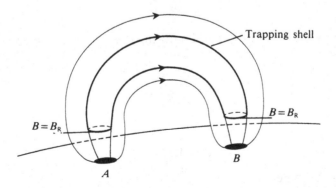

Fig. 4.7. A magnetic flux tube in a solar active region also provides for particle trapping. A: positive-polarity sunspot; B: negative-polarity sunspot.

4.6 The third adiabatic invariant

Let us now consider the possibility that particles are trapped in configurations such as those shown in Figs. 4.5 through 4.7, and that the magnetic field is not static but slowly varying in time. Some of the details in the arguments concerning trapping must be changed, since particle energy changes and therefore the reflection point of a particle will change, and we must consider the bounce invariant given by (4.4.16) rather than that given by (4.4.17). We note that, in such situations, there is a third periodicity associated with the motion, namely the periodic motion around the flux tube. In the case of the magnetosphere of the Earth, this corresponds to the periodic drifting of particles around the Earth. Once again, we may use (4.1.12) to associate an adiabatic invariant with this periodic motion. The equation for this invariant may be written as

$$J_3 = \oint \left(\mathbf{p} + \frac{q}{c} \mathbf{A} \right) \cdot d\mathbf{s}, \tag{4.6.1}$$

where the contour is now a phase contour related to this drifting motion, that is, a closed contour that lies on the surface $I = I_0$ and is directed in the same way as the drift motion.

The ratio of the first and second terms in (4.6.1) may be written approximately as

$$\mathcal{R} = \frac{2\pi m v_D R}{\frac{q}{c} \pi R^2 B}, \tag{4.6.2}$$

where v_D is the magnitude of the drift velocity around the tube. As we shall see in the next chapter, the magnitude of v_D is given approximately by

$$v_D \approx \frac{1}{2} \frac{mc}{qB} \frac{v_\perp{}^2}{R}, \tag{4.6.3}$$

considering only the gyromotion and ignoring the motion along the magnetic field, so that

$$\mathcal{R} \approx \left(\frac{v_\perp}{\Omega R} \right)^2. \tag{4.6.4}$$

However,

$$v_\perp = \Omega r_\perp, \tag{4.6.5}$$

so that (4.6.4) becomes

$$\mathcal{R} \approx \left(\frac{r_\perp}{R} \right)^2. \tag{4.6.6}$$

Since it is usually the case that $r_\perp \ll R$, we see that the first term in the integral (4.6.1) may be neglected in comparison with the second term. Hence, to good approximation, the third adiabatic invariant may be written as

$$J_3 = \frac{q}{c} \Phi,   (4.6.7)$$

where Φ is the magnetic flux embraced by the closed contour associated with the periodic drift motion around the trapping shell.

It is worth reminding ourselves of the conditions for the existence of these three invariants. The first adiabatic invariant is valid if the magnetic field varies on a time-scale long compared with the gyroperiod, and if the spatial gradients are characterized by lengths that are large compared with the gyroradius. (Throughout these discussions, we also include the condition that the variation should be aperiodic.) The second adiabatic invariant is valid if the time scale of variation is long compared with the bounce period, and the third adiabatic invariant is valid if the time scale for variation is long compared with the period of drift motion around the trapping shell.

For simplicity of discussion, we have been considering the motion and trapping properties of particles of given energy and given magnetic moment. In discussing a real plasma machine, or a real astrophysical situation, one would of course be concerned with a range of particle energy and a range of magnetic moment, and also perhaps with particles of more than one species. In these cases, all particles will not be confined. As shown in Section 4.4, particles with direction vectors inside the 'loss cone' will not be trapped. For a specific magnetic-field configuration, there may also be a similar restriction on the range of values of the second adiabatic invariant I for which particles are trapped. Such cases need to be investigated on a case-by-case basis.

Problems

Problem 4.1. Consider a mirror machine of length $2L$ with a mirror ratio of 10, so that $B(L) = B(-L) = 10 B(0)$. A group of $N (N \gg 1)$ electrons with an isotropic velocity distribution is released at the center of the machine. Ignoring collisions and the effect of space charge, how many electrons escape?

Problem 4.2. An electron with speed v_0 moves along the axis of a tube of length $2L$, in which the axial magnetic field has the form

$$B(z) = B_0 e^{K|z|}.$$

(a) Find an equation for $d\alpha/dt$, where α is the pitch angle of the trajectory.

(b) Find the minimum value of α_0, where α_0 is the value of α at $z = 0$, that is needed to ensure that the electron is trapped.

(c) Sketch $\alpha(t)$ and $v_z(t)$, where v_z is the component of velocity parallel to the axis.

(d) Find the bounce period as a function of α_0.

Problem 4.3. Suppose that electrons are injected at the center of the configuration described in Problem 4.2 with a pitch angle distribution $f(\alpha)$ of the form

$$f(\alpha)d\alpha \propto \alpha e^{-\alpha^2/\alpha_0^2}d\alpha.$$

Ignoring collisions, what fraction of the electrons escape?

5

Orbit theory

5.1 Particle motion in a static inhomogeneous magnetic field

The motion of a relativistic particle in a static magnetic field is given by (3.1.5). On introducing the vector

$$\mathbf{\Omega} = \frac{\Omega}{B} \mathbf{B}, \tag{5.1.1}$$

where the gyrofrequency is defined by (3.1.9), we see that the equation of motion may be written as

$$\frac{d\mathbf{v}}{dt} = \varepsilon \mathbf{v} \times \mathbf{\Omega}, \tag{5.1.2}$$

where $\varepsilon (= \pm 1)$ indicates the sign of the particle charge.

We suppose that the gyroradius is sufficiently small that the magnetic-field vector experienced by a particle changes by only a small amount in the time span of one gyroperiod. Then we may consider the motion of the particle in a volume of such a scale that the particle performs several gyrations while in its volume, but over which the magnetic field varies only slightly. Our aim is to calculate under these conditions the departure of the particle motion from the simple helical motion given by equations (3.1.11) to (3.1.16).

We choose to make our calculations in terms of a preferred coordinate system. We assume that the magnetic field is tangential to the x_3 axis at the origin. If the gyroradius were infinitesimally small, the particle would move along magnetic-field lines. Hence its motion could then be written in the form

$$
\begin{aligned}
v_1 &= v_\parallel^2 \Omega^{-1} \Omega_{1,3} t, \\
v_2 &= v_\parallel^2 \Omega^{-1} \Omega_{2,3} t, \\
v_3 &= v_\parallel,
\end{aligned}
\right\} \tag{5.1.3}
$$

to first order in t, where we introduce the notation

$$\Omega_{1,2} = \frac{\partial \Omega_1}{\partial x_2}, \text{ etc.} \tag{5.1.4}$$

We now assume that the gyroradius is small, so that $|x_{1,0}|$, $|x_{2,0}|$ are small, but not infinitesimally small. We must therefore add correction terms on the right-hand side of equations (5.1.3). We shall find that these correction terms may be expressed in the form

$$\left.\begin{array}{l} v_1 = v_\parallel^2 \Omega^{-1} \Omega_{1,3} t + v_\perp \cos(\Omega t) + \Delta v_1, \\[2mm] v_2 = v_\parallel^2 \Omega^{-1} \Omega_{2,3} t - \varepsilon v_\perp \sin(\Omega t) + \Delta v_2, \\[2mm] v_3 = v_\parallel + \Delta v_3. \end{array}\right\} \tag{5.1.5}$$

Similarly, the spatial position mapped out by a particle in its trajectory may be expressed as

$$\left.\begin{array}{l} x_1 = \tfrac{1}{2} v_\parallel^2 \Omega^{-1} \Omega_{1,3} t^2 + r_\perp \sin(\Omega t) + \Delta x_1, \\[2mm] x_2 = \tfrac{1}{2} v_\parallel^2 \Omega^{-1} \Omega_{2,3} t^2 + \varepsilon r_\perp \cos(\Omega t) + \Delta x_2, \\[2mm] x_3 = v_\parallel t + \Delta x_3. \end{array}\right\} \tag{5.1.6}$$

Clearly, the terms involving Ωt are the same as those that we found in Section 3.1 to describe the gyromotion of a particle in a uniform static magnetic field.

Certain terms in these expressions, such as those representing displacement along a curved magnetic-field line, are proportional to t or to some power of t. When we examine the equation of motion, we shall restrict our attention to a small volume in the neighborhood of the origin so that we shall consider that $t \to 0$, except of course in terms such as $\sin \Omega t$ and $\cos \Omega t$ that are rapidly varying.

As a further simplification, we shall restrict our attention to effects that are *linear* in the derivatives of the magnetic field. This is a valid approximation if the gyroradius is small compared with the length scales governing the spatial variation of the magnetic field. For this reason, terms such as $\Delta x_1 \Omega_{2,1}$ will be neglected, since Δx_1 itself is due to inhomogeneities so that this combination would be of second order in the ratio of the gyroradius to the gradient length scale.

Finally, in examining the motion as described by (5.1.5) and (5.1.6), we shall restrict our attention to the components of Δv_1, Δv_2, Δx_1, and Δx_2 that are non-oscillatory.

There are correction terms to the motion that involve harmonics of the gyrofrequency, but these are of no particular interest to us. We shall write

$$\langle \Delta v_1 \rangle = v_{D,1}, \quad \langle \Delta v_2 \rangle = v_{D,2}, \tag{5.1.7}$$

where the angular brackets imply averages over a gyroperiod. Clearly,

$$\langle \Delta x_1 \rangle = v_{D,1} t, \quad \langle \Delta x_2 \rangle = v_{D,2} t, \tag{5.1.8}$$

so that these terms may be neglected in our analysis that is restricted to the neighborhood of the origin. The quantities $v_{D,1}$, $v_{D,2}$ are the components of the particle drift motion transverse to the magnetic field that are due to the magnetic-field inhomogeneities.

We now consider the components of the equation of motion (5.1.2), i.e.,

$$\left.\begin{aligned}
\frac{dv_1}{dt} &= \varepsilon (v_2 \Omega_3 - v_3 \Omega_2), \\[2mm]
\frac{dv_2}{dt} &= \varepsilon (v_3 \Omega_1 - v_1 \Omega_3), \\[2mm]
\frac{dv_3}{dt} &= \varepsilon (v_1 \Omega_2 - v_2 \Omega_1).
\end{aligned}\right\} \tag{5.1.9}$$

We substitute into these equations the velocity as given by (5.1.5), where Ω_1, etc., are to be evaluated along the path of the moving particle. Hence, remembering that we are retaining only terms linear in t, except in the oscillatory terms $\sin \Omega t$, $\cos \Omega t$, we see that we will be introducing expressions such as

$$\Omega_1(\mathbf{x}) = \Omega_1(0) + r_\perp \sin(\Omega t) \Omega_{1,1}(0) + \varepsilon r_\perp \cos(\Omega t) \Omega_{1,2}(0). \tag{5.1.10}$$

With these guidelines, we find that equations (5.1.9) take the form

$$v_\parallel^2 \Omega^{-1} \Omega_{1,3} - \Omega v_\perp \sin(\Omega t) + \dot{v}_\perp \cos(\Omega t)$$

$$= \varepsilon (-\varepsilon v_\perp \sin(\Omega t) + \Delta v_2)(\Omega + \Omega_{3,1} r_\perp \sin(\Omega t) + \Omega_{3,2} \varepsilon r_\perp \cos(\Omega t))$$

$$- \varepsilon v_\parallel (\Omega_{2,1} r_\perp \sin(\Omega t) + \Omega_{2,2} \varepsilon r_\perp \cos(\Omega t)), \tag{5.1.11}$$

$$v_\parallel^2 \Omega^{-1} \Omega_{2,3} - \varepsilon \Omega v_\perp \cos(\Omega t) - \varepsilon \dot{v}_\perp \sin(\Omega t)$$

$$= \varepsilon v_\parallel (\Omega_{1,1} r_\perp \sin(\Omega t) + \Omega_{1,2} \varepsilon r_\perp \cos(\Omega t))$$

$$- \varepsilon (v_\perp \cos(\Omega t) + \Delta v_1)(\Omega + \Omega_{3,1} r_\perp \sin(\Omega t) + \Omega_{3,2} \varepsilon r_\perp \cos(\Omega t)), \tag{5.1.12}$$

$$\dot{v}_\parallel = \varepsilon\,(v_\perp \cos(\Omega t) + \Delta v_1)\,(\Omega_{2,1} r_\perp \sin(\Omega t) + \Omega_{2,2}\varepsilon r_\perp \cos(\Omega t))$$

$$-\varepsilon\,(-\varepsilon v_\perp \sin(\Omega t) + \Delta v_2)\,(\Omega_{1,1} r_\perp \sin(\Omega t) + \Omega_{1,2}\varepsilon r_\perp \cos(\Omega t)), \quad (5.1.13)$$

where $\dot{v}_\perp = dv_\perp/dt$, etc., and $\dot{v}_\parallel$ now represents the effect of Δv_3.

It is readily verified that the above equations are satisfied to zero order in the spatial derivatives, with the correction terms all set equal to zero. We now wish to extract from the above equations terms of the first order in the gradients, considering only terms that are non-oscillatory. In this way we obtain the following equations:

$$v_\parallel^2 \Omega^{-1}\Omega_{1,3} = \varepsilon\Omega v_{D,2} - \tfrac{1}{2} v_\perp r_\perp \Omega_{3,1} \qquad (5.1.14)$$

$$v_\parallel^2 \Omega^{-1}\Omega_{2,3} = -\varepsilon\Omega v_{D,1} - \tfrac{1}{2} v_\perp r_\perp \Omega_{3,2} \qquad (5.1.15)$$

$$\dot{v}_\parallel = \tfrac{1}{2} v_\perp r_\perp (\Omega_{1,1} + \Omega_{2,2}). \qquad (5.1.16)$$

Equations (5.1.14) and (5.1.15) give the following expressions for the drift-velocity components:

$$\left.\begin{aligned}
v_{D,1} &= -\varepsilon\Omega^{-2} v_\parallel^2 \Omega_{2,3} - \tfrac{1}{2}\varepsilon\Omega^{-2} v_\perp^2 \Omega_{3,2}, \\
v_{D,2} &= \varepsilon\Omega^{-2} v_\parallel^2 \Omega_{1,3} + \tfrac{1}{2}\varepsilon\Omega^{-2} v_\perp^2 \Omega_{3,1}.
\end{aligned}\right\} \qquad (5.1.17)$$

On referring back to (5.1.1), we see that these expressions may be rewritten as

$$\left.\begin{aligned}
v_{D,1} &= \frac{\gamma mc}{qB}\left(-v_\parallel^2 \kappa_2 - \frac{1}{2} v_\perp^2 B^{-1}\frac{\partial B_3}{\partial x_2}\right), \\
v_{D,2} &= \frac{\gamma mc}{qB}\left(v_\parallel^2 \kappa_1 + \frac{1}{2} v_\perp^2 B^{-1}\frac{\partial B_3}{\partial x_1}\right),
\end{aligned}\right\} \qquad (5.1.18)$$

where κ_1, κ_2, are the components of the curvature of the magnetic field in the x_1, x_2 directions, given by

$$\kappa_1 = B^{-1}\frac{\partial B_1}{\partial x_3}, \quad \kappa_2 = B^{-1}\frac{\partial B_2}{\partial x_3}. \qquad (5.1.19)$$

The first terms in the brackets of (5.1.18) are usually referred to as the 'curvature-drift' terms, and the second terms are usually known as the 'gradient-drift' terms, although clearly the curvature of the field lines is really a consequence of magnetic-field gradient.

We next consider the information to be obtained from (5.1.16). Since the magnetic field **B** is divergence-free, we see that

$$\Omega_{1,1} + \Omega_{2,2} + \Omega_{3,3} = 0, \qquad (5.1.20)$$

so that (5.1.16) may be rewritten as

$$\frac{dv_{\parallel}}{dt} = -\frac{1}{2}v_{\perp}r_{\perp}\Omega_{3,3}.$$ (5.1.21)

Since we are considering that the magnetic field is static,

$$\frac{1}{2}\frac{dv^2}{dt} \equiv v_{\perp}\frac{dv_{\perp}}{dt} + v_{\parallel}\frac{dv_{\parallel}}{dt} = 0,$$ (5.1.22)

so that (5.1.21) may be expressed as

$$v_{\perp}\frac{dv_{\perp}}{dt} = \frac{1}{2}v_{\perp}r_{\perp}v_{\parallel}\Omega_{3,3}.$$ (5.1.23)

Since $r_{\perp} = \Omega^{-1}v_{\perp}$, we now see that

$$\frac{1}{v_{\perp}}\frac{dv_{\perp}}{dt} = \frac{1}{2}\frac{1}{\Omega}\frac{d\Omega}{dt}.$$ (5.1.24)

This shows that

$$v_{\perp} \propto \Omega^{1/2} \quad \text{or} \quad v_{\perp} \propto B^{1/2}.$$ (5.1.25)

This is clearly consistent with (4.3.12), so that we have once more established the existence of the first adiabatic invariant. Whereas in Chapter 3 the invariant was derived for a magnetic field that varies slowly in time, we have here derived it for a magnetic field that varies slowly in space.

5.2 Discussion of orbit theory for a static inhomogeneous magnetic field

Equations (5.1.18) for the components of drift velocity of a charged particle in a non-uniform magnetic field were expressed in terms of a preferred coordinate system. Alternatively, we may express them in terms of vector quantities as follows,

$$\mathbf{v}_D = \frac{\gamma mc}{qB}\left(\frac{v_{\parallel}^2}{B}\mathbf{B}\times\boldsymbol{\kappa} + \frac{1}{2}\frac{v_{\perp}^2}{B^2}\mathbf{B}\times\nabla B\right),$$ (5.2.1)

where the curvature vector $\boldsymbol{\kappa}$ has components $(\kappa_1, \kappa_2, 0)$ that are defined, for the preferred frame, by (5.1.19). One may justify this equation by verifying that, with the particular coordinate system adopted in Section 5.1, (5.2.1) leads back to the equations (5.1.18).

The first term in this equation is known as the curvature drift and the second as the gradient or ∇B drift. Let us first consider the curvature term. The resulting drift velocity gives rise to a Lorentz force given by

$$\mathbf{F}_c = \frac{q}{c} \frac{\gamma m c}{q B} \frac{v_\parallel^2}{B} ((\mathbf{B} \times \boldsymbol{\kappa}) \times \mathbf{B}), \tag{5.2.2}$$

that is,

$$\mathbf{F}_c = \gamma m v_\parallel^2 \boldsymbol{\kappa}, \tag{5.2.3}$$

since $\boldsymbol{\kappa} \cdot \mathbf{B} = 0$. Thus we see that the curvature of the field lines gives rise to a drift velocity which in turn induces a Lorentz force acting to accelerate the particles so that they follow the curvature of the magnetic-field lines.

Let us now consider the ∇B term in (5.2.1). For convenience, we here adopt the nonrelativistic approximation. We see from (4.4.6) that the motion of a charged particle along the magnetic field is the same as if it were moving under the influence of a field with potential energy

$$V = \mu B. \tag{5.2.4}$$

We shall see that we can attribute the ∇B drift also to the influence of this potential.

Assuming this to be the case, the particle is moving under the effect of a force given by

$$\mathbf{F}_G = -\nabla V = -\mu \nabla B. \tag{5.2.5}$$

On referring to (3.3.3) for the drift of a charged particle in a gravitational field, we see that this will give rise to the following drift velocity:

$$\mathbf{V}_{D,G} = -\frac{c}{q} \frac{\mu}{B^2} \nabla B \times \mathbf{B}. \tag{5.2.6}$$

On recalling the definition of the magnetic moment, (3.4.14), we see that (5.2.6) becomes

$$\mathbf{v}_{D,G} = \frac{1}{2} \frac{m c v_\perp^2}{q B^3} \mathbf{B} \times \nabla B. \tag{5.2.7}$$

This is identical with the nonrelativistic form of the ∇B term in (5.2.1). Hence we can indeed attribute the ∇B drift to the influence of a fictitious force field for which the potential is given by (5.2.4).

We see from (5.2.1) that the signs of the curvature and ∇B drifts depend on the sign of the charge. Hence we must expect that these drifts will generate currents in a plasma in an inhomogeneous magnetic field. However, we must be careful in calculating current densities from results obtained from orbit theory, as we can see from the following thought experiment.

Consider a black-body cavity, the walls of which are at temperature T, and which is filled with an assembly of electrons in thermodynamic equilibrium with the cavity. The electron distribution will have uniform density and have a Maxwellian form in velocity space. Hence the current density will be zero everywhere.

Now suppose that the cavity is permeated by a non-uniform magnetic field. We may again assume that the electron gas is in thermodynamic equilibrium with the cavity. The velocity distribution is therefore unchanged, and the current density remains zero throughout the cavity. On the other hand, the electrons will experience a drift due to the processes we have been considering. Hence we arrive at the following paradox: electrons drift in the cavity, but the current density is zero. For the particular case that

$$B_1 = 0, \quad B_2 = 0, \quad B_3 = a + bx_1, \tag{5.2.8}$$

all electrons will drift in the $-x_2$ direction if a and b are positive and the cavity extends over the range $0 < x_1 < L$. We have the curious result that all electrons drift in the same direction, but there is zero current density.

On examining in detail the current density produced by the assembly of trajectories produced by the inhomogeneous magnetic field, we find that the current density is indeed zero even though all electrons drift in the same direction. Furthermore, as shown in Fig. 5.1, particles near the lower boundary of the cavity move in a direction opposite to that of the ∇B drift so that the total current in the x_2 direction can indeed be zero.

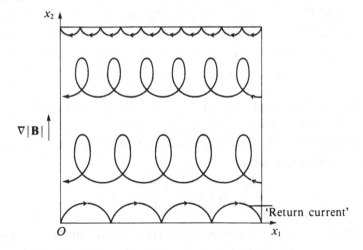

Fig. 5.1. Schematic representation of motion of electrons within a heat bath, in the presence of a non-uniform magnetic field. Electrons that do not intersect the upper or lower surfaces all drift from right to left. On the other hand, the current density is zero everywhere.

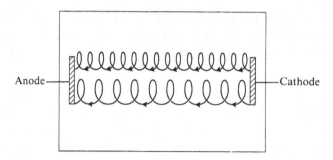

Fig. 5.2. Schematic representation of motion of electrons within a cavity, when they are drawn from a finite cathode and flow to a finite anode. In this case there is a net current from one electrode to the other. This system is not in thermodynamic equilibrium.

If, instead of considering a closed cavity, we had considered the motion of electrons generated by a finite cathode in a large cavity, as indicated in Fig. 5.2, then we would be correct in inferring that the ∇B drift gives rise to a current leaving the cathode.

5.3 Drifts in the Earth's magnetosphere

We consider the nonrelativistic motion of charged particles in the Earth's magnetosphere that was shown schematically in Fig. 4.6. We assume that a certain region of the magnetosphere contains a fully ionized plasma in which the electrons and the ions each have a bi-Maxwellian distribution such that motion along the magnetic field and transverse the magnetic field each have characteristic temperatures $T_\parallel$, $T_\perp$, with the consequence that

$$\tfrac{1}{2}m\langle v_\parallel{}^2\rangle = \tfrac{1}{2}kT_\parallel, \quad \tfrac{1}{2}m\langle v_\perp{}^2\rangle = kT_\perp. \tag{5.3.1}$$

The difference in these equations arises from the fact that motion transverse the magnetic field has two degrees of freedom.

We can now use (5.2.1) to determine the average drift velocity of particles with this velocity distribution:

$$\mathbf{v}_D = \frac{c}{q}\frac{kT_\parallel}{B^2}\mathbf{B}\times\boldsymbol{\kappa} + \frac{c}{q}\frac{kT_\perp}{B^3}\mathbf{B}\times\nabla B. \tag{5.3.2}$$

If we now consider that the plasma comprises equal numbers of electrons and protons, we may derive the following expression for the total ring current due to a distribution of finite cross-sectional area:

$$J = \int d^2 S\, n\left\{\frac{k(T_{p,\parallel}+T_{e,\parallel})}{B^2}\mathbf{B}\times\boldsymbol{\kappa} + \frac{k(T_{p,\perp}+T_{e,\perp})}{B^3}\mathbf{B}\times\nabla B\right\}. \tag{5.3.3}$$

We note, from our preceding discussion of the thought experiment, that we are entitled to calculate the total current of the trapped particles by using our expressions for drift velocity, but we would not be entitled to calculate the current density from those expressions.

5.4 Motion in a time-varying electric field

Let us now consider a system comprised of a static, uniform magnetic field in the x_3 direction and a time-varying, uniform electric field in the x_2 direction:

$$\mathbf{B} = (0, 0, B), \quad \mathbf{E} = (0, E_2(t), 0). \tag{5.4.1}$$

If the rate of change of the electric field is small compared with the gyro-frequency, we may study the orbits by considering a steady increase in the electric field. We therefore consider the form

$$E_2(t) = \dot{E}_{2,0} t. \tag{5.4.2}$$

From (3.2.7), we obtain the following expression for the drift velocity in the x_1 direction:

$$v_{D,1} = \frac{c\dot{E}_{2,0}}{B} t. \tag{5.4.3}$$

Since the drift motion now exhibits acceleration, we may speculate that this will give rise to a further drift motion. We may anticipate that the acceleration given by (5.4.3) will have an effect equivalent to the gravitational field with an x_1 component given by

$$g_1 = \frac{c\dot{E}_{2,0}}{B}. \tag{5.4.4}$$

If we now refer to (3.3.3), we see that we must expect that there will be an additional drift velocity in the x_2 direction (in the same direction as the electric field), given by

$$v_{P,2} = -\frac{mc^2}{qB^2} \dot{E}_{2,0}. \tag{5.4.5}$$

We shall now confirm that this drift does in fact occur; it is known as the 'polarization drift'.

The (nonrelativistic) equation of motion of this system is

$$\frac{d\mathbf{v}}{dt} = \frac{q}{m} \dot{\mathbf{E}}_0 t + \frac{q}{mc} \mathbf{v} \times \mathbf{B}. \tag{5.4.6}$$

We attempt to solve this equation by adopting a solution of the form

$$\mathbf{v} = \mathbf{v}_g(t) + \mathbf{a}_D t + \mathbf{v}_P, \tag{5.4.7}$$

in which $\mathbf{v}_g(t)$ represents the usual gyromotion and $\mathbf{a}_D t$ represents an acceleration term. On substituting (5.4.7) in (5.4.6), we obtain

$$\frac{d\mathbf{v}_g}{dt} + \mathbf{a}_D = \frac{q}{m}\dot{\mathbf{E}}_0 t + \frac{q}{mc}[\mathbf{v}_g \times \mathbf{B} + (\mathbf{a}_D t) \times \mathbf{B} + \mathbf{v}_p \times \mathbf{B}]. \tag{5.4.8}$$

We see that the terms involving $\mathbf{v}_g$ may be separated out and satisfy the same equation as (3.1.5). Terms linear in t lead to the equation

$$\dot{\mathbf{E}}_0 + \frac{1}{c}\mathbf{a}_D \times \mathbf{B} = 0, \tag{5.4.9}$$

and terms independent of t lead to

$$\mathbf{a}_D = \frac{q}{mc}\mathbf{v}_p \times \mathbf{B}. \tag{5.4.10}$$

Equation (5.4.10) leads to

$$\mathbf{v}_p = \frac{mc}{qB^2}\mathbf{B} \times \mathbf{a}_D, \tag{5.4.11}$$

and (5.4.9) leads to

$$\mathbf{a}_D = -c\frac{\mathbf{B} \times \dot{\mathbf{E}}_0}{B^2}. \tag{5.4.12}$$

Hence we see from (5.4.11) and (5.4.12) that

$$\mathbf{v}_p = \frac{mc^2}{qB^2}\dot{\mathbf{E}}_0. \tag{5.4.13}$$

Generalizing to an arbitrary time dependence for the electric field, we may replace (5.4.13) by

$$\mathbf{v}_p = \frac{mc^2}{qB^2}\frac{\partial \mathbf{E}}{\partial t}. \tag{5.4.14}$$

As indicated earlier, this is referred to as the 'polarization drift.'

In order to see why this term is used, we may refer back to the following Maxwell equation:

$$\frac{1}{c}\frac{\partial \mathbf{E}}{\partial t} + 4\pi\mathbf{j} = \nabla \times \mathbf{B}. \tag{5.4.15}$$

On using (5.4.14), and considering a plasma that may be composed of several different species, we find that (5.4.15) becomes

$$\frac{1}{c}\frac{\partial \mathbf{E}}{\partial t} + \frac{4\pi\rho c}{B^2}\frac{\partial \mathbf{E}}{\partial t} = \nabla \times \mathbf{B}. \tag{5.4.16}$$

This may be written equivalently as

$$\frac{1}{c}\frac{\partial \mathbf{D}}{\partial t} = \nabla \times \mathbf{B}, \tag{5.4.17}$$

where the familiar electric displacement vector $\mathbf{D}$ and electric polarization vector $\mathbf{P}$, related by

$$\mathbf{D} = \mathbf{E} + 4\pi\mathbf{P}, \tag{5.4.18}$$

may be expressed as

$$\mathbf{D} \equiv \varepsilon\mathbf{E}, \tag{5.4.19}$$

where the dielectric coefficient ε is given by

$$\varepsilon = \left(1 + \frac{4\pi\rho c^2}{B^2}\right), \tag{5.4.20}$$

and

$$\mathbf{P} = \frac{\rho c^2}{B^2}\mathbf{E}. \tag{5.4.21}$$

Hence it appears that the polarization current has the effect of modifying the dielectric constant from its vacuum value $\varepsilon = 1$.

Let us now consider a transverse electromagnetic wave propagating along the magnetic field. Then, including the effect of the dielectric constant ε, we find (following arguments to be developed in Chapter 6) that the dispersion relation will be

$$\omega^2 = \frac{c^2 k^2}{\varepsilon}. \tag{5.4.22}$$

On now substituting for the value for ε from (5.4.20), we see that this equation becomes

$$\omega^2 = \frac{c^2 k^2}{1 + \dfrac{c^2}{v_A{}^2}}, \tag{5.4.23}$$

where v_A, known as the 'Alfvén speed,' is defined by

$$v_A{}^2 = \frac{B^2}{4\pi\rho}. \tag{5.4.24}$$

If $v_A \ll c$, (5.4.23) becomes, approximately,

$$\omega^2 = v_A^2 k^2. \tag{5.4.25}$$

This is the MHD form of the dispersion equation for Alfvén waves, as we shall see in Chapter 14. Equation (5.4.23) is in fact a more accurate form of the dispersion equation for Alfvén waves, since it includes the effect of the displacement term in (2.14).

The energy density associated with an electric field in a material medium with dielectric coefficient ε is given by

$$u_E = \frac{1}{8\pi} \, \mathbf{D} \cdot \mathbf{E} = \frac{1}{8\pi} \, \varepsilon E^2, \tag{5.4.26}$$

(provided ε is frequency-independent). On using (5.4.20), we see that this becomes

$$u_E = \frac{1}{8\pi} E^2 + \frac{1}{2} \frac{\rho c^2}{B^2} E^2. \tag{5.4.27}$$

On now referring to (3.2.7), we see that this is identical to

$$u_E = \frac{1}{8\pi} E^2 + \frac{1}{2} \rho v_D^2. \tag{5.4.28}$$

The total energy density now comprises the energy density of the electric field and the kinetic energy density associated with the $\mathbf{E} \times \mathbf{B}$ drift. In a material medium, the additional term in the energy density arises from the potential energy associated with atomic and molecular forces required to pull the negative and positive charges apart. In a plasma, this is replaced by the kinetic energy density associated with the drift velocity.

5.5 Particle motion in a rapidly time-varying electromagnetic field

We consider the motion of a charged particle in an oscillatory electric field that has the form of a standing wave:

$$\mathbf{E}(\mathbf{x}, t) = \tilde{\omega} \mathbf{E}_0(\mathbf{x}) \sin(\omega t). \tag{5.5.1}$$

It is convenient to introduce the symbol $\tilde{\omega}$ as a small ordering parameter. The associated magnetic field is given by

$$\mathbf{B}(\mathbf{x}, t) = \tilde{\omega} \mathbf{B}_0(\mathbf{x}) \cos(\omega t) \tag{5.5.2}$$

where, as we see from (2.1.1)

$$\mathbf{B}_0 = \frac{c}{\omega} \, \nabla \times \mathbf{E}_0. \tag{5.5.3}$$

The nonrelativistic motion of a charged particle in these combined fields is determined by the equation of motion

$$m\frac{d^2x}{dt^2} = qE + \frac{q}{c}\frac{dx}{dt} \times B.$$ (5.5.4)

We seek an approximate solution of the equation of motion. We first express the position $x(t)$ of a particle in the form

$$x(t) = x_0 + \xi(x_0, t),$$ (5.5.5)

where x_0 may be taken to be the stationary position of the particle in the absence of the electromagnetic field, and ξ is the displacement of the particle from the original position due to the electromagnetic field.

We expand the displacement vector in powers of $\tilde{\omega}$:

$$\xi(x_0, t) = \tilde{\omega}\xi_1(x_0, t) + \tilde{\omega}^2\xi_2(x_0, t) + \ldots$$ (5.5.6)

On noting that

$$E(x, t) = E(x_0, t) + (\xi \cdot \nabla)E(x_0, t) + \ldots,$$ (5.5.7)

we see that

$$E(x, t) = \tilde{\omega}E_0 \sin(\omega t) + \tilde{\omega}^2\xi_1 \cdot \nabla E_0 \sin(\omega t) + \ldots$$ (5.5.8)

and

$$B(x, t) = \tilde{\omega}B_0 \cos(\omega t) + \tilde{\omega}^2\xi_1 \cdot \nabla B_0 \cos(\omega t) + \ldots,$$ (5.5.9)

in which all terms on the right-hand side of these two equations are to be evaluated at the position x_0.

We now substitute the expressions (5.5.5), (5.5.6), (5.5.8) and (5.5.9) in the equation of motion (5.5.4), and separate terms of equal power in the expansion parameter $\tilde{\omega}$. To first order in this parameter, we obtain

$$\frac{d^2\xi_1}{dt^2} = \frac{q}{m}E_0 \sin(\omega t).$$ (5.5.10)

Hence, to lowest order, the motion is described by

$$\xi_1 = -\frac{q}{m\omega^2}E_0 \sin(\omega t).$$ (5.5.11)

That is, it is simple harmonic motion about its original position.

Terms of second order lead to the equation

$$\frac{d^2\xi_2}{dt^2} = -\frac{q^2}{m^2\omega^2}E_0 \cdot \nabla E_0 \sin^2(\omega t) - \frac{q^2}{m^2c\omega}E_0 \times B_0 \cos^2(\omega t).$$ (5.5.12)

We are particularly interested in the secular contribution to this equation that may be written as

$$\left\langle \frac{d^2 \xi_2}{dt^2} \right\rangle = -\frac{q^2}{2m^2\omega^2} \mathbf{E}_0 \cdot \nabla \mathbf{E}_0 - \frac{q^2}{2m^2 c\omega} \mathbf{E}_0 \times \mathbf{B}_0. \tag{5.5.13}$$

On using (5.5.3) and noting that

$$\mathbf{E}_0 \cdot \nabla \mathbf{E}_0 + \mathbf{E}_0 \times (\nabla \times \mathbf{E}_0) = \tfrac{1}{2} \nabla E_0^2, \tag{5.5.14}$$

we find that (5.5.13) may be expressed as

$$\left\langle \frac{d^2 \xi_2}{dt^2} \right\rangle = -\frac{q^2}{4m^2\omega^2} \nabla (E_0^2) \tag{5.5.15}$$

or, equivalently, as

$$\left\langle \frac{d^2 \xi_2}{dt^2} \right\rangle = -\frac{q^2}{2m^2\omega^2} \nabla (E^2). \tag{5.5.16}$$

This mean second-order acceleration may be considered as arising from an equivalent mean second-order force density $\mathbf{F}$ given by

$$\mathbf{F} = -\frac{nq^2}{2m\omega^2} \nabla (E^2), \tag{5.5.17}$$

where n is the particle number density. This may be expressed in terms of the mean electric pressure

$$p_E = \frac{1}{8\pi} \langle E^2 \rangle \tag{5.5.18}$$

as follows

$$\mathbf{F} = -\left(\frac{\omega_p}{\omega}\right)^2 \nabla p_E, \tag{5.5.19}$$

where ω_p is the plasma frequency of the species.

We see that the effect of the oscillatory electric field is to move particles towards regions of lowest electric field strength. However, we see that the effect depends sensitively on the ratio ω_p/ω, so that the effect is greatest for low-frequency fields.

It is found, from spacecraft experiments, that there are regions in the solar wind that have high oscillatory electric fields. Let us consider such a situation, shown schematically in Fig. 5.3. The total force density on the gas will be given by the electric-field term (5.5.19) and by the gas pressure term, i.e., by

$$\mathbf{F} = -\left(\frac{\omega_p}{\omega}\right)^2 \nabla p_E - \nabla p_G \tag{5.5.20}$$

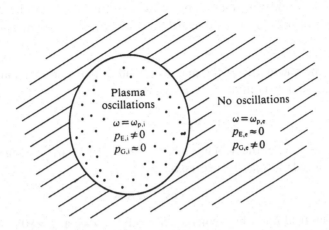

Fig. 5.3. Schematic representation of a 'cavity' in the solar wind. The 'internal' region contains only a low density of plasma but a high density of plasma oscillations. The 'external' region contans a higher density of plasma but a low density of plasma oscillations.

where p_G is the gas pressure. Hence, in place of the simple balance of gas pressure between the interior and the exterior, we will now have

$$-\left(\frac{\omega_{p,i}}{\omega}\right)^2 \nabla p_{E,i} - p_{G,i} = -\left(\frac{\omega_{p,e}}{\omega}\right)^2 \nabla p_{E,e} - p_{G,e} \qquad (5.5.21)$$

where i and e refer to the interior and exterior, respectively.

If, for simplicity, we take $p_{G,i} \approx 0$, and $p_{E,e} \approx 0$, and if we assume that the electric field is produced by plasma oscillations so that $\omega = \omega_{P,i}$, then (5.5.21) becomes

$$p_{E,i} = p_{G,e}. \qquad (5.5.22)$$

Typical conditions in the solar wind are that the electron and proton densities are each about $5\,cm^{-3}$, the electron temperature is about $10^5\,K$, and the proton temperature is substantially smaller. Hence the total gas pressure is about $10^{-10.2}\,dyne\,cm^{-2}$. If this gas pressure is balanced by electric pressure, then the electric field strength will be $10^{-4.4}\,esu$, i.e., $10^{-1.9}\,V\,cm^{-1}$.

Problems

Problem 5.1. The vector magnetic field at vector position **x** due to a magnetic dipole of vector dipole moment **m** at the origin is given by

$$\mathbf{B} = \frac{3(\mathbf{m}\cdot\mathbf{x})\mathbf{x} - \mathbf{x}^2}{|\mathbf{x}|^5}$$

To a good approximation, the magnetic field of the Earth can be considered to be due to a dipole of moment $10^{25.9}\,G\,cm^3$ directed towards the South pole.

(a) Show that the magnitude of the magnetic field at a distance r from the center of the Earth and at latitude λ is given by

$$B(r, \lambda) = \frac{m}{r^3}\sqrt{1 + 3\sin^2\lambda}$$

(b) Show that the form of the field line which intersects the Earth's surface of radius $R = R_0$ at latitude λ_0 may be expressed as

$$r = R_0 \cos^2\lambda / \cos^2\lambda_0$$

(c) Consider the field line which emerges from the surface at $\lambda = 80°$. What is the value of B at the surface? What are the maximum and minimum values of B, and at what values of r and λ do they occur?

(d) For the field line identified in (c), what is the minimum pitch angle necessary for electrons released at $\lambda = 0°$ to be trapped?

Problem 5.2. Consider the model of Problem 5.1, and the field line identified in part (c) of that problem.

(a) Calculate the curvature drift velocity and the grad-B drift velocity for an electron of energy 1 keV located where the field line intersects the equator.

(b) What energy would be necessary for the gyroradius of the electron to be equal to the Earth's radius?

(c) We define t_v, an estimate of the characteristic time of variation of the magnetic field, by

$$dB/dt = Bt_v^{-1}.$$

For an electron of energy 1 keV, find the range of values of t_v for which each of the three adiabatic invariants may be considered to be constant.

Problem 5.3. Consider a uniform plasma in a uniform magnetic field B_0. Consider the response of the electrons to a low-frequency electromagnetic field with electric and magnetic vectors given by the real parts of $E_1 e^{-i\omega t}$ and $B_1 e^{-i\omega}$, where $B_1 \ll B_0$. Calculate the polarization drift velocity and so derive the dispersion relation (5.4.23).

Problem 5.4. Consider a rectangular coordinate system with x_1 in the vertically upward direction. A slab of charge-neutral plasma is infinite in the x_1 and x_3 directions, and bounded in the x_2 direction. There is a gravitational field of strength g in the $-x_1$ direction, and a magnetic field of strength B in the x_3 direction. Suppose that B is sufficiently strong that all motion may be analyzed in terms of drift motions, and suppose that the plasma has attained a steady state.

(a) How do the electrons and ions drift in the x_2 direction due to the crossed gravitational and magnetic fields?
(b) Calculate the rate of accumulation of surface charge on the surfaces of the slab.
(c) Calculate as a function of time the electric field component E_2 that develops due to the surface charges.
(d) Calculate the time-dependent $\mathbf{E} \times \mathbf{B}$ drift motion in the x_1 direction.
(e) Calculate the polarization-drift motion in the x_2 direction.
(f) What is the resulting gravitational acceleration of the slab?

6

Electromagnetic waves in a cold electron plasma

A large part of plasma physics is concerned with the study of small fluctuations in the state of a plasma from a given equilibrium state. If the equilibrium system is uniform in space and stationary in time, then it is possible to study the fluctuations by Fourier analysis techniques. That is, we may understand the properties of the system by determining what types of waves propagate in that system. In some situations, we will find that the time variation is not purely periodic but may include either an exponential damping or an exponential growth. In the latter case, we say that the system is unstable.

As a first, simple approach to this complex area, we begin by studying wave propagation in a stationary electron–ion plasma, under the approximations that the ion motion may be neglected and the electron temperature may be neglected. In later chapters, these assumptions will be relaxed.

6.1 The wave equation

In discussing wave phenomena in a plasma, it is convenient to express the response of the plasma to an electromagnetic field by means of a dielectric tensor, that we introduced in the previous chapter. We begin with the relevant Maxwell equations:

$$\nabla \times \mathbf{B} = \frac{1}{c} \frac{\partial \mathbf{E}}{\partial t} + 4\pi \mathbf{j}, \tag{6.1.1}$$

$$\nabla \times \mathbf{E} = -\frac{1}{c} \frac{\partial \mathbf{B}}{\partial t}. \tag{6.1.2}$$

These equations determine the response of the electromagnetic field to currents in the plasma.

The fluctuations in the current in turn determine the fluctuations in electric charge density through (2.1.5).

In order to have a closed system of equations, we need to know how the electromagnetic field influences the current. In general, this is a nonlinear and a nonlocal relationship. For present purposes, we assume that the relationship may be adequately described in a form that is both local and linear:

$$\mathbf{j} = \sigma \cdot \mathbf{E} \quad \text{or} \quad j_r = \sigma_{rs} E_s. \tag{6.1.3}$$

This is usually referred to as a generalization of Ohm's law.

As we saw in Chapter 5, it is sometimes convenient to re-write (6.1.1) in the form

$$\nabla \times \mathbf{B} = \frac{1}{c} \frac{\partial \mathbf{D}}{\partial t}. \tag{6.1.4}$$

Then the current density $\mathbf{j}$ must be expressible as

$$\mathbf{j} = \frac{1}{4\pi c} \left(\frac{\partial \mathbf{D}}{\partial t} - \frac{\partial \mathbf{E}}{\partial t} \right). \tag{6.1.5}$$

Since we are assuming that the unperturbed state of the system is stationary, we could analyze the system by carrying out a Fourier transformation in time. For present purposes, we simply consider one component that has a time dependence of the form $\exp(-i\omega t)$. We will therefore introduce the following change in notation

$$\mathbf{E}(\mathbf{x}, t) \rightarrow \mathbf{E}(\mathbf{x}, \omega) e^{-i\omega t}, \text{ etc.} \tag{6.1.6}$$

Equation (6.1.5) now becomes

$$j_r = -\frac{i\omega}{4\pi c} (D_r - E_r). \tag{6.1.7}$$

On using (6.1.3), we now obtain the following relationship between the displacement vector $\mathbf{D}$ and the electric field $\mathbf{E}$:

$$D_r = E_r + i \frac{4\pi c}{\omega} \sigma_{rs} E_s. \tag{6.1.8}$$

It is conventional to write the relationship between these two variables in the form

$$D_r = K_{rs} E_s. \tag{6.1.9}$$

We therefore see from (6.1.8) that the 'dielectric tensor' K_{rs} is related to the conductivity tensor σ_{rs} as follows:

$$K_{rs} = \delta_{rs} + i \frac{4\pi c}{\omega} \sigma_{rs}. \tag{6.1.10}$$

On taking the curl of (6.1.2), using (6.1.4) and (6.1.9), we finally obtain the following wave equation for the electric field:

$$\nabla \times (\nabla \times \mathbf{E}) - \frac{\omega^2}{c^2} \mathsf{K} \cdot \mathbf{E} = 0. \tag{6.1.11}$$

This form of the wave equation is in fact applicable to a plasma containing many different species. Each species contributes to the current $\mathbf{j}$ so that, in (6.1.11), the dielectric tensor K will be the sum of several contributions, one from each species. Also note that the above equation is applicable to a system that is not uniform in space.

6.2 Waves in a cold electron plasma without a magnetic field

We now consider waves in a cold, uniform plasma free from magnetic field, under the assumption that collisions may be neglected. We also assume that the waves are of sufficiently high frequency that ion motion may be neglected. In this case, the relationship between current and electric field is to be derived from the electron equation of motion

$$m_e \frac{d\mathbf{v}}{dt} \equiv m_e \left(\frac{\partial \mathbf{v}}{\partial t} + \mathbf{v} \cdot \nabla \mathbf{v} \right) = -e\mathbf{E}. \tag{6.2.1}$$

Since we are treating fluctuations only in the linear approximation, the term $\mathbf{v} \cdot \nabla \mathbf{v}$ in (6.2.1) may be neglected. Hence, with the change of notation indicated by (6.1.6), the equation of motion becomes

$$\mathbf{v} = -\mathrm{i} \frac{e}{m_e \omega} \mathbf{E}. \tag{6.2.2}$$

Since the current density of electron gas is related to the velocity by

$$\mathbf{j} = -\frac{ne}{c} \mathbf{v}, \tag{6.2.3}$$

where $n(\mathbf{x})$ is the unperturbed electron density, we find that

$$\mathbf{j} = \mathrm{i} \frac{ne^2}{m_e c \omega} \mathbf{E}. \tag{6.2.4}$$

Hence, for a cold, stationary, collision-free electron gas, the conductivity tensor is expressible as

$$\sigma_{rs} = \mathrm{i} \frac{ne^2}{m_e c \omega} \delta_{rs}. \tag{6.2.5}$$

We may now use (6.1.10) to find the corresponding dielectric tensor. We find that this is expressible as

$$K_{rs} = \left(1 - \frac{\omega_p^2}{\omega^2}\right)\delta_{rs}, \tag{6.2.6}$$

where ω_p is the electron plasma frequency defined by (2.5.4).

With the form of the dielectric tensor given by (6.2.6), the wave equation (6.1.11) takes the form

$$\nabla \times (\nabla \times \mathbf{E}) - \left(\frac{\omega^2 - \omega_p^2}{c^2}\right)\mathbf{E} = 0. \tag{6.2.7}$$

Let us now consider a system that, in its unperturbed state, is not only stationary but also homogeneous. It is then possible to consider fluctuations by Fourier analyzing in space as well as in time. Hence we now replace (6.1.6) by

$$\mathbf{E}(\mathbf{x}, t) \rightarrow \mathbf{E}(\mathbf{k}, \omega)e^{i(\mathbf{k} \cdot \mathbf{x} - \omega t)}. \tag{6.2.8}$$

The wave equation (6.1.11) now becomes

$$\mathbf{k} \times (\mathbf{k} \times \mathbf{E}) + \frac{\omega^2}{c^2}\mathbf{K} \cdot \mathbf{E} = 0, \tag{6.2.9}$$

and the special case (6.2.7) becomes

$$\mathbf{k} \times (\mathbf{k} \times \mathbf{E}) + \left(\frac{\omega^2 - \omega_p^2}{c^2}\right)\mathbf{E} = 0. \tag{6.2.10}$$

The wave equation (6.2.10) leads to waves of nonzero amplitude only for certain relationships between the wave vector $\mathbf{k}$ and the frequency ω. This relationship is known as the 'dispersion relation.' One way to find this relationship is to express (6.2.9) in matrix form

$$M_{rs}E_s = 0, \tag{6.2.11}$$

and then note that the condition that the vector $\mathbf{E}$ be nonzero is that

$$\det(M_{rs}) = 0. \tag{6.2.12}$$

However, for the present problem we may proceed a little more simply by expressing the electric field in the form

$$\mathbf{E} = \mathbf{E}_\parallel + \mathbf{E}_\perp, \tag{6.2.13}$$

where $\mathbf{E}_\parallel$ and $\mathbf{E}_\perp$ are parallel and perpendicular, respectively, to the wave vector $\mathbf{k}$. On substituting this expression into (6.2.10) and using the relationships

$$\mathbf{k} \times \mathbf{E}_{\parallel} = 0, \quad \mathbf{k} \cdot \mathbf{E}_{\perp} = 0, \tag{6.2.14}$$

we find that the equation reduces to

$$-c^2 k^2 \mathbf{E}_{\perp} + (\omega^2 - \omega_p^2)(\mathbf{E}_{\parallel} + \mathbf{E}_{\perp}) = 0. \tag{6.2.15}$$

Clearly this equation separates into two distinct equations. One of these is the equation for $\mathbf{E}_{\parallel}$,

$$(\omega^2 - \omega_p^2)\mathbf{E}_{\parallel} = 0, \tag{6.2.16}$$

and the other the equation for $\mathbf{E}_{\perp}$:

$$(\omega^2 - \omega_p^2 - c^2 k^2)\mathbf{E}_{\perp} = 0. \tag{6.2.17}$$

That is, there are two different 'modes' of wave propagation. One is the 'longitudinal' mode, for which the electric field is parallel to the wave vector, governed by the dispersion relation

$$\omega^2 = \omega_p^2. \tag{6.2.18}$$

This mode corresponds to the plasma oscillations that were first introduced in Section 2.5.

The second mode is a transverse mode, for which the electric field is transverse to the wave vector, that is governed by the dispersion equation

$$\omega^2 = \omega_p^2 + c^2 k^2. \tag{6.2.19}$$

This dispersion relation is shown schematically in Fig. 6.1. It is clear that such waves can propagate only if the frequency is higher than the plasma frequency. We say that this mode has a 'cutoff' at $\omega = \omega_p$. It is clear from (6.1.2) that this mode involves both an electric and a magnetic field that are orthogonal to each other and to the wave vector $\mathbf{k}$. Hence the dispersion equation (6.2.19) represents an electromagnetic wave.

It is clear from (6.2.19) that

$$\text{if } \omega > \omega_p, \quad k = \pm c^{-1}(\omega^2 - \omega_p^2)^{1/2}. \tag{6.2.20}$$

Hence, for this range of frequency, the wave vector is real and the waves are propagating. The group velocity (see Appendix B), defined by

$$u_r = \frac{\partial \omega}{\partial k_r}, \tag{6.2.21}$$

has the value

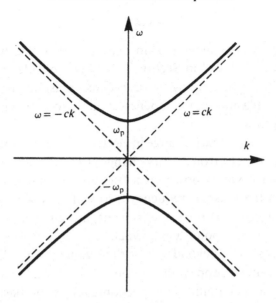

Fig. 6.1. Dispersion relation for electromagnetic waves in a cold electron plasma free from magnetic field.

$$u_r = c \left(1 - \frac{\omega_p^2}{\omega^2} \right)^{1/2} \frac{k_r}{k}, \qquad (6.2.22)$$

so that it tends to the speed of light as $\omega \to \infty$, but the group velocity tends to zero as $\omega \to \omega_p$.

We also see that

$$\text{if } \omega < \omega_p, \quad k = \pm ic^{-1}(\omega_p^2 - \omega^2)^{1/2}. \qquad (6.2.23)$$

In this case, the wave vector is imaginary so that the wave amplitude grows exponentially in one direction and decreases exponentially in the opposite direction. Growth in space may represent the effect of an instability, or it may simply denote that the wave is evanescent (see Appendix C). We see from (6.2.19) that the wave frequency is always real for real values of k, and this indicates that the system is stable. We could also have inferred that the system is stable from the fact that there is no source of energy in the system. Hence the waves that grow or decay exponentially in space must be evanescent waves.

We also see that

$$\text{if } \omega \ll \omega_p, \quad k = \pm ic^{-1}\omega_p. \qquad (6.2.24)$$

That is, the waves grow or decay exponentially in space with a characteristic length given by

$$\lambda_s = c\omega_p^{-1}. \qquad (6.2.25)$$

This is called the 'electromagnetic skin depth,' and is clearly analogous to the Debye length introduced in Section 2.2. If we consider waves traveling through empty space and impinging on a dense plasma, the wave will not propagate into the plasma, but will penetrate only to a distance comparable with the skin depth λ_s.

Rather than consider a sharp change in electron density, let us now consider the possibility that the electron density changes slowly in space. In particular, let us suppose that it varies slowly as a function of one spatial coordinate x, $n = n(x)$. Then the plasma frequency also will vary with x, as indicated schematically in Fig. 6.2. If the wave is propagating in the x direction, then we see that $k = 0$ at the point $x = x_c$ where $\omega_p = \omega$, ω being the wave frequency. For values of x beyond the critical value $x = x_c$, k is imaginary. Hence the wave cannot propagate beyond the location $x = x_c$. In fact, the wave is reflected at that point. This phenomenon is the basis for an ionospheric sounding technique whereby one may determine the distance to layers of different plasma frequencies by measuring the time taken for a wave of given frequency to be reflected back to ground level. (See Problem 6.1.)

Let us now suppose that the wave is not parallel to the direction of the gradient of electron density. Then we must expect that the direction of propagation of the wave will change as the wave propagates in the medium. We may write

$$k^2 = k_\parallel^2 + k_\perp^2, \qquad (6.2.26)$$

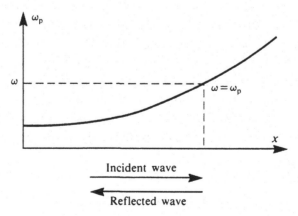

Fig. 6.2. Schematic representation of reflection of an electromagnetic wave in a plasma of slowly varying density.

where $k_\parallel$ and $k_\perp$ are the components of $\mathbf{k}$ parallel and normal, respectively, to the x direction, that is, to ∇n. Since the medium is assumed to be uniform in the directions transverse to x, $k_\perp$ will be constant. The component $k_\parallel$ will therefore be a function $k_\parallel(x)$ of x that is given in terms of $\omega_p(x)$ by

$$k_\parallel^2 = \frac{\omega^2 - \omega_p^2}{c^2} - k_\perp^2. \tag{6.2.27}$$

Clearly, the wave will now be reflected at the value of x for which $k_\parallel = 0$, for which $\omega_p(x)$ has the value

$$\omega_p^2 = \omega^2 = c^2 k_\perp^2. \tag{6.2.28}$$

This situation is represented schematically in Fig. 6.3.

Let us now consider a wave mode governed by a quite different dispersion relation, for which ω asymptotically approaches a limiting value ω_r as $k \to \infty$, as indicated schematically in Fig. 6.4. We shall meet such modes in Chapter 7. We now see that, as ω approaches ω_r, $k_\perp/k_\parallel$ becomes smaller and smaller, indicating that the wave is bent in the direction of the gradient. The behavior of the mode at $\omega = \omega_r$ is known as a 'resonance'. It is clear that the group velocity tends to zero at $\omega = \omega_r$. Hence the wave energy propagates into the location where $\omega = \omega_r$ and, as far as our current analysis goes, the energy remains in that location. In fact, of course, the energy would be absorbed by the medium. See Stix (1962) for further discussion.

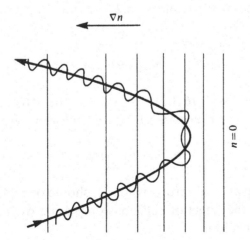

Fig. 6.3. Schematic representation of reflection of a wave obliquely incident on a gradient of refractive index n. The reflection occurs where $k_\parallel = 0$.

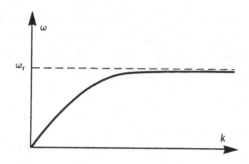

Fig. 6.4. Dispersion relation showing a resonance.

6.3 Effect of collisions

We now consider a simple treatment of the effect on the dispersion relation of collisions between electrons and either ions or neutral particles. A simple representation of the effect of collisions is to modify the equation of motion (6.2.1) as follows:

$$m_e\left(\frac{\partial \mathbf{v}}{\partial t}+\mathbf{v}\cdot\nabla\mathbf{v}\right) = -e\mathbf{E}-\nu m_e\mathbf{v}, \tag{6.3.1}$$

where ν is the effective 'collision frequency.' This equation may be regarded as representing either the mean effect of many small perturbations in the trajectory of an electron, or the average effect for a distribution of electrons. On making the change in notation indicated by (6.1.6), we find that if the stream velocity is zero, the relation (6.2.2) is now replaced by

$$\mathbf{v} = -\mathrm{i}\frac{e}{m_e\left(1+\mathrm{i}\dfrac{\nu}{\omega}\right)\omega}\mathbf{E}. \tag{6.3.2}$$

On comparing (6.2.2) and (6.3.2), we see that the effect of collisions has been to replace the electron mass m_e by the following expression

$$m_e \rightarrow \left(1+\mathrm{i}\frac{\nu}{\omega}\right)m_e. \tag{6.3.3}$$

(If the stream velocity is non-zero, ω in (6.3.3) should be replaced by $\omega-\mathbf{v}\cdot\mathbf{k}$.)

We now see that the effect of collisions on plasma oscillations, governed by the dispersion relation

$$\omega^2 = \omega_p{}^2 \equiv \frac{4\pi ne^2}{m_e}, \tag{6.3.4}$$

is given by the new dispersion relation

$$\omega^2 = \frac{\omega_p^2}{\left(1 + i\frac{\nu}{\omega}\right)}. \tag{6.3.5}$$

Alternatively, we may write

$$\omega = -\tfrac{1}{2}i\nu \pm (\omega_p^2 - \tfrac{1}{4}\nu^2)^{1/2}. \tag{6.3.6}$$

Hence the frequency is now complex:

$$\omega = \omega_r + i\omega_i. \tag{6.3.7}$$

Assuming that $\nu < 2\omega_p$, the real and imaginary parts are given by

$$\omega_r = \pm(\omega_p^2 - \tfrac{1}{4}\nu^2)^{1/2} \tag{6.3.8}$$

$$\omega_i = -\tfrac{1}{2}\nu. \tag{6.3.9}$$

Equation (6.3.9) shows that the amplitude of the electric field decays with time as follows:

$$E \propto e^{-\frac{1}{2}\nu t}, \tag{6.3.10}$$

so that the wave energy decays as $e^{-\nu t}$.

In the same way, we find that the dispersion relation for transverse electromagnetic waves takes the form

$$\omega^2 = \frac{\omega_p^2}{1 + i\frac{\nu}{\omega}} + c^2 k^2. \tag{6.3.11}$$

Although this is a cubic equation in ω, it is clear that ω will again be complex, and we may anticipate that the waves will once again be damped. On separating real and imaginary parts, we find that ω_r and ω_i satisfy the two equations

$$\omega_r^3 - 3\omega_r \omega_i^2 - 2\nu\omega_r\omega_i = (\omega_p^2 + c^2 k^2)\omega_r \tag{6.3.12}$$

$$3\omega_r^2\omega_i - \omega_i^3 + \nu(\omega_r^2 - \omega_i^2) = (\omega_p^2 + c^2 k^2)\omega_i + c^2 k^2 \nu. \tag{6.3.13}$$

If $\nu \ll \omega_r$, then ω_r satisfies approximately

$$\omega_r^2 = \omega_p^2 + c^2 k^2 \tag{6.3.14}$$

and ω_i is given approximately by

$$\omega_i = -\frac{1}{2}\frac{\omega_p^2 \nu}{\omega_r^2}. \tag{6.3.15}$$

Hence the damping of electromagnetic waves is less than that of plasma oscillations. In fact, the damping tends to zero as $\omega_r \to \infty$. The reason for this is that as $\omega_r \to \infty$, the medium behaves more and more like free space. In other words, most of the energy is in the electromagnetic field and very little in the plasma.

We may instead consider the propagation of a wave of given real frequency ω and write

$$k = k_r + i k_i. \tag{6.3.16}$$

For the case $\nu \ll \omega$, we find that k_r satisfies approximately the dispersion relation

$$\omega^2 = \omega_p^2 + c^2 k_r^2, \tag{6.3.17}$$

and the imaginary component k_i is given approximately by

$$k_i = \frac{\nu \omega_p^2}{2c\omega (\omega^2 - \omega_p^2)^{1/2}}. \tag{6.3.18}$$

Once again, we see that as $\omega \to \infty$ the damping tends to zero.

In the world of astrophysics, this damping is sometimes referred to as 'free–free absorption,' since it is due to 'free–free transitions' rather than 'free–bound transitions.'

It is interesting to compare (6.3.9) and (6.3.15) for the damping of plasma oscillations and electromagnetic waves, respectively. As $\omega_p \to 0$, the former remains finite whereas the latter tends to zero. The difference is to be attributed to the fact that the energy density in plasma oscillations tends to zero, for a given velocity amplitude, as the plasma density tends to zero. On the other hand, the energy density in electromagnetic waves remains finite, for given velocity amplitude, as the electron density tends to zero, since a vacuum still supports an electromagnetic wave that has nonzero energy density.

There is also an interesting difference between (6.3.15) and (6.3.18) for the damping of electromagnetic waves. For a given plasma frequency and collision frequency, the former is finite, no matter what the frequency. On the other hand, the latter expression tends to infinity as $\omega \to \omega_p$. This fact is to be understood by noting that the group velocity of the wave tends to zero in this limit. The wave may be decaying at a finite rate in time, but it is not propagating in space, so that its energy content drops off rapidly in space.

6.4 Electromagnetic waves in a cold magnetized electron plasma

We continue to consider wave propagation in a homogeneous cold plasma, and return to the assumption that collisions may be neglected. However, we now add to the system a static uniform magnetic field $\mathbf{B}_0$. The wave may generate both an electric field $\mathbf{E}$ and an additional component $\delta\mathbf{B}$ to the magnetic field $\mathbf{B}_0$. Then the equation of motion (6.2.1) must now be replaced by

$$\frac{d\mathbf{v}}{dt} = -\frac{e}{m_e}\mathbf{E} - \frac{e}{m_e c}\mathbf{v} \times \mathbf{B}_0. \tag{6.4.1}$$

Note that we neglect the second-order term $\mathbf{v} \cdot \Delta\mathbf{v}$, and we also neglect the second-order term involving $\mathbf{v} \times \delta\mathbf{B}$.

We may, for convenience, orient the rectangular coordinate system so that $\mathbf{B}_0$ is in the x_3 direction: $\mathbf{B}_0 = (0, 0, B_0)$. Then the equation of motion separates into the following equations:

$$\frac{dv_1}{dt} = -\frac{e}{m_e}E_1 - \frac{e}{m_e c}v_2 B_0, \tag{6.4.2}$$

$$\frac{dv_2}{dt} = -\frac{e}{m_e}E_2 + \frac{e}{m_e c}v_1 B_0, \tag{6.4.3}$$

and

$$\frac{dv_3}{dt} = -\frac{e}{m_e}E_3. \tag{6.4.4}$$

On considering Fourier components, according to the notation of (6.2.8), these equations become

$$i\omega v_1 = \frac{e}{m_e}E_1 + \omega_g v_2, \tag{6.4.5}$$

$$i\omega v_2 = \frac{e}{m_e}E_2 - \omega_g v_1, \tag{6.4.6}$$

and

$$i\omega v_3 = \frac{e}{m_e}E_3, \tag{6.4.7}$$

where ω_g is the gyrofrequency, now given by

$$\omega_g = \frac{eB_0}{m_e c}. \tag{6.4.8}$$

These equations lead to the following expressions for the velocity components in terms of the components of electric field:

$$v_1 = -\frac{ie}{m_e(\omega^2 - \omega_g^2)}(\omega E_1 - i\omega_g E_2), \tag{6.4.9}$$

$$v_2 = -\frac{ie}{m_e(\omega^2 - \omega_g^2)}(i\omega_g E_1 + \omega E_2), \tag{6.4.10}$$

and

$$v_3 = -\frac{ie}{m_e\omega}E_3. \tag{6.4.11}$$

We may now calculate the components of current density, in terms of the electric field, by using (6.2.3). In this way we obtain the matrix relation

$$\begin{pmatrix} j_1 \\ j_2 \\ j_3 \end{pmatrix} = -\frac{ne}{c} \begin{pmatrix} -\dfrac{ie\omega}{m_e(\omega^2-\omega_g^2)} & -\dfrac{e\omega_g}{m_e(\omega^2-\omega_g^2)} & 0 \\ \dfrac{e\omega_g}{m_e(\omega^2-\omega_g^2)} & -\dfrac{ie\omega_g}{m_e(\omega^2-\omega_g^2)} & 0 \\ 0 & 0 & \dfrac{ie}{m_e\omega} \end{pmatrix} \begin{pmatrix} E_1 \\ E_2 \\ E_3 \end{pmatrix}. \tag{6.4.12}$$

Hence we see that the components of the conductivity tensor, introduced in (6.1.3), are given by

$$\sigma = \begin{pmatrix} \dfrac{ine^2\omega}{m_e c(\omega^2-\omega_g^2)} & \dfrac{ne^2\omega_g}{m_e c(\omega^2-\omega_g^2)} & 0 \\ -\dfrac{ne^2\omega_g}{m_e c(\omega^2-\omega_g^2)} & \dfrac{ine^2\omega}{m_e c(\omega^2-\omega_g^2)} & 0 \\ 0 & 0 & \dfrac{ine^2}{m_e c\omega} \end{pmatrix}. \tag{6.4.13}$$

Equation (6.1.10) now gives the dielectric tensor:

$$K = \begin{pmatrix} 1 - \dfrac{\omega_p^2}{(\omega^2-\omega_g^2)} & \dfrac{i\omega_p^2\omega_g}{\omega(\omega^2-\omega_g^2)} & 0 \\ -\dfrac{i\omega_p^2\omega_g}{\omega(\omega^2-\omega_g^2)} & 1 - \dfrac{\omega_p^2}{(\omega^2-\omega_g^2)} & 0 \\ 0 & 0 & 1 - \dfrac{\omega_p^2}{\omega^2} \end{pmatrix}. \tag{6.4.14}$$

We may now study waves propagating in a uniform cold magnetized plasma by noting that, in Fourier transform notation, (6.1.11) takes the form

$$\mathbf{k} \times (\mathbf{k} \times \mathbf{E}) + \frac{\omega^2}{c^2} \mathbf{K} \cdot \mathbf{E} = 0. \qquad (6.4.15)$$

This equation is valid for an arbitrary direction of the wave vector $\mathbf{k}$ with respect to the DC magnetic field $\mathbf{B}_0$.

Before proceeding, it is interesting to note one or two special cases for the dielectric tensor given by (6.4.14). If ω_p is zero, that is if the electron density is zero, the tensor reduces to the unit diagonal tensor, as one would expect. Then (6.4.15) simply represents the dispersion relation for transverse electromagnetic waves in free space. In this case, (6.4.15) shows that $\mathbf{k} \cdot \mathbf{E} = 0$, so that there is no longitudinal mode for which $\mathbf{E}$ is parallel to $\mathbf{k}$.

If we now consider wave propagation in an extremely strong magnetic field, so that we can take the limit $\omega_g \to \infty$, we see that the dielectric tensor reduces to

$$\mathbf{K} = \begin{bmatrix} 1 & 0 & 0 \\ 0 & 1 & 0 \\ 0 & 0 & 1 - \dfrac{\omega_p^2}{\omega^2} \end{bmatrix} \qquad (6.4.16)$$

On taking the scalar product of (6.4.15) with the wave vector $\mathbf{k}$, we find that there is a longitudinal mode, with electric field parallel to $\mathbf{B}_0$, that still represents plasma oscillations for which the dispersion relation is (6.3.4). The magnetic field has no effect on the motion of electrons parallel to the field. On the other hand, on taking the vector product of $\mathbf{k}$ and (6.4.15), we find that the components of the electric field transverse to $\mathbf{B}_0$ satisfy the vacuum form of the dispersion relation (6.2.19) (the form with $\omega_p = 0$). The interpretation is that the very strong magnetic field effectively prevents electrons from moving transverse to the magnetic field so that they have no influence on wave propagation.

6.5 Wave propagation normal to the magnetic field

We now consider the special case of waves propagating perpendicular to the magnetic field. Without loss of generality, we may orient the x_1–x_2 axes so that the wave vector $\mathbf{k}$ is in the x_1 direction so that $\mathbf{k} = (k, 0, 0)$. On using the vector identity

$$\mathbf{k} \times (\mathbf{k} \times \mathbf{E}) = (\mathbf{k} \cdot \mathbf{E})\mathbf{k} - k^2 \mathbf{E} = 0, \qquad (6.5.1)$$

we find that the wave equation (6.4.15) becomes

$$-k^2 \begin{pmatrix} 0 \\ E_2 \\ E_3 \end{pmatrix} + \frac{\omega^2}{c^2} \mathbf{K} \cdot \begin{pmatrix} E_1 \\ E_2 \\ E_3 \end{pmatrix} = 0. \tag{6.5.2}$$

On introducing the notation

$$\left. \begin{array}{l} A = \dfrac{\omega^2}{c^2} \dfrac{\omega^2 - \omega_p^2 - \omega_g^2}{\omega^2 - \omega_g^2}, \\[4mm] B = \dfrac{\omega^2}{c^2} \dfrac{\omega_p^2 \omega_g}{\omega(\omega^2 - \omega_g^2)}, \end{array} \right\} \tag{6.5.3}$$

we find that (6.5.2) takes the form

$$\begin{pmatrix} A & iB & 0 \\ -iB & -k^2+A & 0 \\ 0 & 0 & -k^2+\dfrac{\omega^2-\omega_p^2}{c^2} \end{pmatrix} \begin{pmatrix} E_1 \\ E_2 \\ E_3 \end{pmatrix} \tag{6.5.4}$$

It is clear from the form of this matrix that the components of $\mathbf{E}$ perpendicular to $\mathbf{B}_0$ are decoupled from the component parallel to $\mathbf{B}_0$. The dispersion relation corresponding to $E_3 \neq 0$ is

$$\omega^2 = \omega_p^2 + c^2 k^2, \tag{6.5.5}$$

that is seen to be identical to (6.2.19), the dispersion relation for electromagnetic waves in a non-magnetized plasma. For the mode under consideration, the electric field is parallel to the DC magnetic field. Since electrons are free to move parallel to the magnetic field, we can see why the magnetic field has no effect on the dispersion relation.

The more interesting case involves E_1 and E_2. We see that they satisfy the homogeneous matrix equation

$$\begin{pmatrix} A & iB \\ -iB & -k^2+A \end{pmatrix} \begin{pmatrix} E_1 \\ E_2 \end{pmatrix} = 0. \tag{6.5.6}$$

The condition that the vector (E_1, E_2) be nonzero is that the determinant of the matrix in (6.5.6) be zero. That is,

$$A^2 - k^2 A - B^2 = 0. \tag{6.5.7}$$

This may be expressed as an equation for the magnitude of the wave vector:

$$k^2 = \frac{A^2 - B^2}{A}. \tag{6.5.8}$$

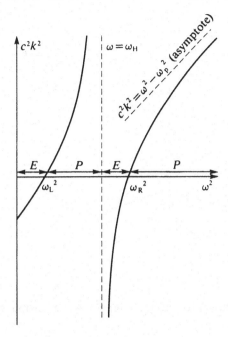

Fig. 6.5. Diagram giving the propagation properties of electromagnetic waves propagating transverse to a magnetic field.

Hence we obtain the dispersion relation for propagation of electromagnetic waves normal to a magnetic field as an expression for k^2 in terms of ω^2:

$$c^2 k^2 = \frac{(\omega^2 - \omega_p^2)^2 - \omega_g^2 \omega^2}{\omega^2 - \omega_p^2 - \omega_g^2}. \qquad (6.5.9)$$

The general form of this relationship is shown schematically in Fig. 6.5.

Note that regions with $c^2 k^2 < 0$ represent evanescent waves, while regions with $c^2 k^2 > 0$ represent propagating waves. The zeros of the numerator of (6.5.9) are the lower frequency limits (cutoff frequencies) of the two modes of propagation, and are given by

$$\omega_L = -\tfrac{1}{2}\omega_g + (\omega_p^2 + \tfrac{1}{4}\omega_g^2)^{1/2} \qquad (6.5.10)$$

and

$$\omega_R = \tfrac{1}{2}\omega_g + (\omega_p^2 + \tfrac{1}{4}\omega_g^2)^{1/2}. \qquad (6.5.11)$$

The denominator in (6.5.9) becomes zero at $\omega = \omega_H$ (a resonance frequency), known as the 'upper hybrid frequency,' where

$$\omega_H = (\omega_p^2 + \omega_g^2)^{1/2}. \qquad (6.5.12)$$

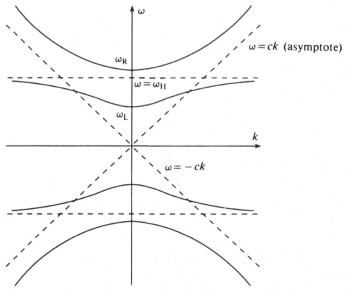

Fig. 6.6. Dispersion relation for electromagnetic waves propagating transverse to a magnetic field.

We see that electromagnetic waves can propagate transverse to the magnetic field in the lower branch for which $\omega_L < \omega < \omega_H$, and in the upper branch for which $\omega > \omega_R$. The dispersion relation is shown as an ω–k relation in Fig. 6.6.

6.6 Propagation parallel to the magnetic field

We next consider electromagnetic waves propagating in a cold, uniform magneto-plasma, for the case that the wave vector **k** is parallel to the direction of the imposed magnetic field $\mathbf{B}_0$. The procedure is very similar to that of Section 6.5, but the dispersion relation will be significantly different. We now adopt $\mathbf{B}_0 = (0, 0, B_0)$ and $\mathbf{k} = (0, 0, k)$. On using (6.4.14), we find that the wave equation (6.4.15) now takes the form

$$
\begin{bmatrix}
-k^2 + \dfrac{\omega^2}{c^2}\dfrac{\omega^2 - \omega_p^2 - \omega_g^2}{\omega^2 - \omega_g^2} & i\dfrac{\omega^2}{c^2}\dfrac{\omega_p^2 \omega_g}{\omega(\omega^2 - \omega_g^2)} & 0 \\[3ex]
-i\dfrac{\omega^2}{c^2}\dfrac{\omega_p^2 \omega_g}{\omega(\omega^2 - \omega_g^2)} & -k^2 + \dfrac{\omega^2}{c^2}\dfrac{\omega^2 - \omega_p^2 - \omega_g^2}{\omega^2 - \omega_g^2} & 0 \\[3ex]
0 & 0 & \dfrac{\omega^2}{c^2}\dfrac{\omega^2 - \omega_p^2}{\omega^2}
\end{bmatrix}
\begin{bmatrix}
E_1 \\[3ex] E_2 \\[3ex] E_3
\end{bmatrix} = 0.
$$

$$(6.6.1)$$

The terms involving E_3 lead once more to the dispersion relation (6.2.18) for longitudinal electrostatic waves, that is, plasma oscillations.

The terms involving E_1 and E_2, that will describe transverse waves, may be expressed in the form

$$\begin{pmatrix} A & iB \\ -iB & A \end{pmatrix} \begin{pmatrix} E_1 \\ E_2 \end{pmatrix} = 0, \tag{6.6.2}$$

where the symbols A, B are now used to represent the following functions of ω and k:

$$\left. \begin{aligned} A &= \omega^2 - c^2 k^2 - \frac{\omega_p^2 \omega^2}{\omega^2 - \omega_g^2}, \\ B &= \frac{\omega_p^2 \omega_g \omega}{\omega^2 - \omega_g^2}. \end{aligned} \right\} \tag{6.6.3}$$

By inspecting the determinant of the matrix in (6.6.2), we see that the dispersion relation is expressible as

$$A = \pm B. \tag{6.6.4}$$

We now see from (6.6.2) that the components of the transverse electric field, E_1 and E_2, have the same absolute magnitude but are in phase quadrature:

$$\frac{E_2}{E_1} = \pm i. \tag{6.6.5}$$

This shows that the normal modes of wave propagation parallel to the magnetic field are circularly polarized. With respect to the direction of the magnetic field,

$$\left. \begin{aligned} \frac{E_2}{E_1} &= +i \text{ corresponds to right-hand circular polarization,} \\ \frac{E_2}{E_1} &= -i \text{ corresponds to left-hand circular polarization.} \end{aligned} \right\} \tag{6.6.6}$$

We note that right-hand circular polarization leads to a possible synchronism between the rotation of the wave and the natural rotation of electrons in the magnetic field. We shall see that this synchronism leads to a singularity in the dispersion relation.

The dispersion relation (6.6.4) may be expressed as

$$c^2 k^2 = \omega^2 \left[1 - \frac{\omega_p^2}{\omega(\omega \mp \omega_g)} \right]. \tag{6.6.7}$$

On introducing the 'refractive index' defined by

$$n = \frac{ck}{\omega},$$ (6.6.8)

the dispersion relation may be expressed equivalently as

$$n^2 = 1 - \frac{\omega_p^2}{\omega(\omega \mp \omega_g)}.$$ (6.6.9)

We first examine the dispersion relation for the case that the plasma frequency is larger than the gyrofrequency. Specifically, we adopt $\omega_p = 2\omega_g$. The dependence of the refractive index upon frequency is then given by Fig. 6.7, and the corresponding $\omega-k$ diagram is shown in Fig. 6.8. As $\omega \to \infty$, $n \to 1$; that is, propagation is the same as in free space. As the frequency increases, the current contributed by the electrons decreases whereas the displacement current increases. In these diagrams, ω_R is the cutoff frequency for the R-mode (right-hand circularly polarized mode), and ω_L is the

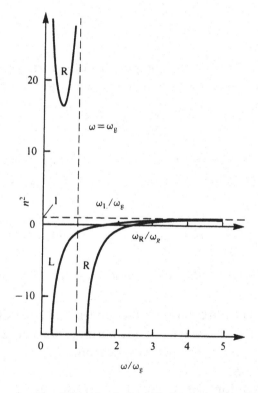

Fig. 6.7. Curve showing the refractive index of electromagnetic waves propagating parallel to the magnetic field for the case $\omega_p = 2\omega_g$. In this figure, R indicates right-hand polarization, corresponding to $E_2/E_1 = +i$, and L indicates left-hand polarization, corresponding to $E_2/E_1 = -i$.

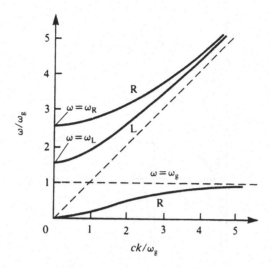

Fig. 6.8. Dispersion relation corresponding to the refractive index curve shown as Fig. 6.6.

cutoff frequency for the L-mode (left-hand circularly polarized mode). We see that there is a resonance at $\omega = \omega_g$ for the R-mode, since $k \to \infty$ as $\omega \to \omega_g$. This is the resonance that we anticipated, due to the synchronism of the electromagnetic wave and the electron gyromotion.

We now consider the form of the dispersion relation for the case that the gyrofrequency is larger than the plasma frequency. Specifically, we adopt $\omega_g = 2\omega_p$. The refractive index diagram and the ω–k diagram are now shown as Figs. 6.9 and 6.10, respectively. Once again, we see that there is a resonance at the frequency $\omega = \omega_g$ for the right-hand polarized mode.

We see there is an important difference between the cases $\omega_p > \omega_g$ and $\omega_p < \omega_g$. For the former case, there is a range of frequency between ω_g and ω_L for which no propagation is possible. In the latter case, at least one mode can propagate for any frequency.

The dispersion relation for the left-hand polarized mode is similar to that of an electromagnetic mode propagating in a plasma free from magnetic field, as shown in Fig. 6.1. For this reason, this is often referred to as the 'ordinary' mode. By contrast, the right-hand polarized mode differs substantially from that shown in Fig. 6.1, due to the resonance at $\omega = \omega_g$. Hence the right-hand circularly polarized mode is termed the 'extraordinary' mode.

6.7 Faraday rotation

Now consider that a linearly polarized, transverse wave is caused to propagate along the direction of the magnetic field. We may analyze this wave into two

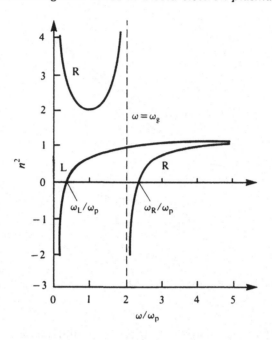

Fig. 6.9. Curve showing the refractive index of electromagnetic waves propagating parallel to the magnetic field for the case $\omega_g = 2\,\omega_p$.

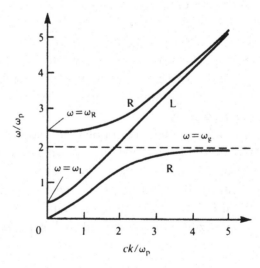

Fig. 6.10. Dispersion relation corresponding to the refractive index curve shown as Fig. 6.8.

circularly polarized waves that are characterized by different wave vectors given by

$$k_{\pm} = \frac{\omega}{c} \left[1 - \frac{\omega_p{}^2}{\omega(\omega \mp \omega_g)} \right]^{1/2}. \tag{6.7.1}$$

Since $k_+ \neq k_-$, there will be a progressive phase change between the two circularly polarized components as the wave propagates. The combined wave still exhibits linear polarization, but the direction of the electric field vector changes with distance. If the amplitude of each circularly polarized mode is E_0, we see from (6.2.8) and (6.6.5) that E_1 and E_2 vary with distance z as follows

$$
\begin{aligned}
E_1 &= E_0[\cos(k_+ z - \omega t) + \cos(k_- z - \omega t)], \\
E_2 &= E_0[\sin(k_+ z - \omega t) - \sin(k_- z - \omega t)].
\end{aligned}
\tag{6.7.2}
$$

This may be re-written alternatively as

$$
\begin{aligned}
E_1 &= 2E_0 \cos\left(\frac{k_+ + k_-}{2} z - \omega t \right) \cos\left(\frac{k_+ - k_-}{2} z \right), \\
E_2 &= 2E_0 \cos\left(\frac{k_+ + k_-}{2} z - \omega t \right) \sin\left(\frac{k_+ - k_-}{2} z \right).
\end{aligned}
\tag{6.7.3}
$$

If Ψ is the angle that the electric vector makes with respect to the x axis, so that

$$\frac{E_2}{E_1} = \tan \Psi, \tag{6.7.4}$$

we see that

$$\Psi = \frac{k_+ - k_-}{2} z. \tag{6.7.5}$$

That is,

$$\frac{d\Psi}{dz} = \frac{1}{2}(k_+ - k_-). \tag{6.7.6}$$

In problems of radio astronomy, we are typically concerned with the propagation of radio waves over large distances, over which the plasma frequency ω_p and the gyrofrequency ω_g are slowly varying. In such cases, the total angle of rotation of the electric vector, Ψ, is expressible as

$$\Psi = \int_0^d \frac{d\Psi}{ds}\, ds = \int_0^d \frac{1}{2}(k_+ - k_-)\, ds. \tag{6.7.7}$$

In interstellar space, the magnetic field strength is measured in μG, so that the gyrofrequency may be measured in Hz. Similarly, the plasma density is such that the plasma frequency will be at most a few kHz. Hence we may use the following approximate expressions for k_+, k_-:

$$k_{\pm} = \frac{\omega}{c}\left[1 - \frac{\omega_p^2}{2\omega^2}\left(1 \pm \frac{\omega_g}{\omega}\right)\right]. \qquad (6.7.8)$$

On using these expressions, we see that (6.7.6) becomes

$$\frac{d\Psi}{ds} = -\frac{\omega_p^2 \omega_g}{2c\omega^2}. \qquad (6.7.9)$$

This may be expressed alternatively as

$$\frac{d\Psi}{ds} = -\frac{2\pi n_e e^3 B}{m_e^2 c^2 \omega^2}. \qquad (6.7.10)$$

Hence the total rotation of the plane of polarization over the path of propagation of radio waves is expressible as

$$\Psi = -\frac{e^3}{2\pi m_e^2 c^2}\frac{1}{\nu^2}\int_0^d n_e B_{\parallel} \, ds, \qquad (6.7.11)$$

where we now represent the wave frequency by ν (Hz) rather than ω (radian s^{-1}).

In deriving an expression for Faraday rotation, we have assumed that the magnetic field is parallel (or anti-parallel) to the direction of propagation of the wave. However, if we were to consider a more general direction of propagation and still use the approximations ω_g, $\omega_p \ll \omega$, we would find that the rotation of the plane of polarization is still given by (6.7.11), where $B_{\parallel}$ now represents the component of magnetic field parallel to the wave vector **k**.

Equation (6.7.11) is conventionally expressed as

$$\Psi = -(RM)\nu^{-2}, \qquad (6.7.12)$$

where (RM) is termed the 'rotation measure' of the wave. Numerically,

$$(RM) = 10^{4.37}\int_0^d n_e B_{\parallel} \, ds. \qquad (6.7.13)$$

Since the direction of polarization is determined only modulo(π), one normally requires measurements at several frequencies to resolve this uncertainty in determining the rotation measure.

Note that we have defined rotation (and hence the terms 'left-hand' and 'right-hand') in terms of propagation along the magnetic field. In some applications – such as ionospheric propagation and radio astronomy – it is more common to use the convention that the sense of rotation of a wave is determined by an observer *receiving* the wave. If the latter convention is adopted, the terms 'right-hand' and 'left-hand' must be interchanged in (6.6.6).

6.8 Dispersion of radio waves

Electromagnetic waves of different frequencies propagate at different group velocities in a plasma. Hence measurement of the propagation time of a signal provides some information about the medium through which the signal propagates.

In radio astronomy, it is obviously not practical to measure the actual propagation time. What we can do is measure the relative arrival times of components of a signal of different frequencies. This technique is especially powerful when applied to the study of signals emitted by pulsars, since these signals are broad-band and have a duration of order 1 second, or sometimes much less.

In Section 6.7, we obtained the approximate relationship (6.7.8) between k_+, k_- and ω that is valid if ω_p, $\omega_g \ll \omega$. On introducing the group velocity, defined by

$$u = \frac{d\omega}{dk}, \qquad (6.8.1)$$

we see from (6.7.8) that, approximately,

$$u^{-1} = \frac{1}{c}\left[1 + \frac{\omega_p^2}{2\omega^2}\right]. \qquad (6.8.2)$$

Let us once again consider propagation over large distances, allowing for slow variation of the plasma frequency and gyrofrequency over the path of propagation. Then the total propagation time of the signal is given by

$$T = \int_0^d u^{-1}\, ds. \qquad (6.8.3)$$

Clearly this may be expressed as

$$T = \frac{d}{c} + \frac{1}{2c\omega^2}\int_0^d \omega_p^2\, ds. \qquad (6.8.4)$$

This may be expressed alternatively as

$$T = \frac{d}{c} + Dv^{-2}, \qquad (6.8.5)$$

where

$$D = \frac{e^2}{2\pi m_e c} \int_0^d n_e \, ds. \qquad (6.8.6)$$

The quantity D is known as the 'dispersion measure.' Numerically, (6.8.6) may be expressed as

$$D = 10^{-2.88} \int_0^d n_e \, ds. \qquad (6.8.7)$$

In practice, it is more usual to measure radio frequencies in MHz, and more usual to measure distances in parsec than in cm. (See Appendix A.) The appropriate form of the dispersion measure then becomes

$$D(\text{MHz}) = 10^{3.61} \int_0^d n_e (\text{cm}^{-3}) \, ds(\text{pc}), \qquad (6.8.8)$$

where the units of each term are as indicated. As implied earlier in this section, the dispersion measure is in fact determined by comparing the arrival times of a pulse at two (or more) different frequencies. That is, we use the relation

$$\Delta T \equiv T_2 - T_1 = D\left(\frac{1}{v_2^2} - \frac{1}{v_1^2}\right). \qquad (6.8.9)$$

In studying radio propagation from pulsars, one is able to get an estimate of the distance by studying the 21-centimeter neutral hydrogen absorption line. One may then use the measurement of the dispersion measure to estimate the electron density (see Manchester and Taylor, 1977). Using this method, the mean electron density is found to range from $0.01 \, \text{cm}^{-3}$ to $0.1 \, \text{cm}^{-3}$, with an average value of about $0.03 \, \text{cm}^{-3}$.

6.9 Whistlers

The term 'whistler' refers to very low frequency electromagnetic waves that propagate along magnetic field lines through the Earth's magnetosphere. Since the typical frequency of a whistler is in the audible range, whistlers can be detected (and were first detected) by listening to the signals picked up by a suitable antenna (Barkhausen, 1919; Eckersley, 1935).

It is now known that whistlers are commonly caused by lightning discharges. The lightning excites waves of all frequencies, that then travel along the earth's magnetic field lines to a 'conjugate point' on the Earth's surface. Sometimes the waves are reflected back and signals corresponding to several different 'bounces' may sometimes be distinguished. The typical sound is that of a whistle that begins at high frequency and then descends with a time-scale of order a few seconds to low frequency. Eckersley (1935) was the first person to demonstrate the low-frequency form of the dispersion law of whistlers, and Storey (1953) was the first to identify the magneto-spheric path of whistlers. Since that time, they have been extensively studied (see, for instance, Helliwell, 1965) and much has been learned about the physical conditions in the Earth's magnetosphere by whistler research.

In recent years, whistlers have also been identified in signals detected from Jupiter (Scarf *et al.*, 1979) and from Neptune (Gurnett *et al.*, 1990).

Consider the extraordinary wave propagating along the magnetic field. We see from (6.6.7) that the dispersion relation is expressible as

$$\frac{c^2 k^2}{\omega^2} = 1 - \frac{\omega_p{}^2}{\omega(\omega - \omega_g)}. \tag{6.9.1}$$

For most of the Earth's magnetosphere, we find that the electron plasma-frequency is substantially higher than the electron gyrofrequency. If we now consider waves of frequency comparable with the electron gyrofrequency, we may assume that ω, $\omega_g \ll \omega_p$, so that we may adopt the following approximate form of the dispersion relation:

$$\frac{c^2 k^2}{\omega^2} = \frac{\omega_p{}^2}{\omega(\omega_g - \omega)}. \tag{6.9.2}$$

Hence

$$k = \frac{\omega_p}{c} \left(\frac{\omega}{\omega_g - \omega} \right)^{1/2}. \tag{6.9.3}$$

We now see that the inverse of the group velocity is given by

$$u^{-1} = \frac{\omega_p \omega_g}{2c\omega^{1/2}(\omega_g - \omega)^{3/2}}, \tag{6.9.4}$$

so that

$$u = \frac{2c\omega^{1/2}(\omega_g - \omega)^{3/2}}{\omega_p \omega_g}. \tag{6.9.5}$$

We see from this equation that the group velocity vanishes at $\omega = 0$ and at $\omega = \omega_g$. One readily finds that the maximum value of the group velocity corresponds to the frequency $\omega = \frac{1}{4}\omega_g$, and is given by

$$u_M = \frac{3^{3/2}}{8} \frac{\omega_g}{\omega_p} c. \tag{6.9.6}$$

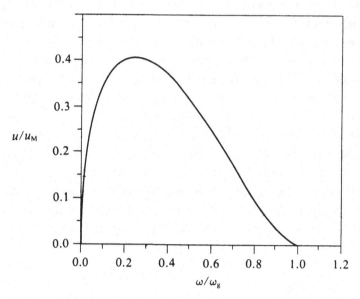

Fig. 6.11. Group velocity as a function of frequency, for the whistler mode.

Fig. 6.11 shows the dependence of the group velocity on frequency.

Modern studies of whistlers (Helliwell, 1965) take place by tape-recording signals received from large antennas and then spectrum-analyzing the signal. In this way, one can display the intensity of the signal as a function of both frequency and time, as shown schematically in Fig. 6.12. This shows that the familiar descending-tone signal is just part of a whistler. At higher frequencies (often too high to be heard), there is another branch for which the frequency increases as a function of time, as indicated by Fig. 6.12. When such a trace is recorded, it is called a 'nose whistler.' It is found that, for the earth's magnetosphere, the frequency of the 'nose' corresponds approximately to 0.4 times the minimum gyrofrequency along the propagation path, (Helliwell, 1965).

Research has shown that whistlers tend to propagate along discrete paths or 'ducts,' which are interpreted as flux tubes with higher than average electron density. On some occasions, the whistler reflects from the conjugate point, and it may in fact make several passages along the duct. In this case, there would be a sequence of signals with time delays in the ratio $1 : 3 : 5 : \ldots$ if the receiver is at the conjugate point from the lightning stroke, or in the ratio $1 : 2 : 3 : \ldots$ if the receiver is at the same location as the lightning stroke.

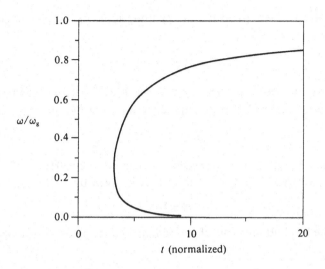

Fig. 6.12. Schematic representation of time delay (horizontal axis) as a function of frequency (vertical axis) for the whistler mode.

Problems

Problem 6.1. Consider the propagation of radio waves through the Earth's ionosphere, taking account only of the electron component. If z is the vertical coordinate, suppose that the base of the ionosphere is at height $z = h_0$, and that the plasma density increases linearly with height. Then the plasma frequency ν_p is expressible as

$$\nu_p{}^2 = K(z - h_0)$$

for $z > h_0$.

(a) Calculate the height $H(\nu)$ at which waves of frequency ν, transmitted vertically from a transmitter on the ground, are reflected from the ionosphere.
(b) Calculate the round-trip group travel time for a pulse of wave frequency ν which is transmitted vertically, reflected, and returns to its source.
(c) Express this time as the equivalent height $h'(\nu)$ at which, in the absence of the plasma, a horizontal plane reflector would have to be placed so as to produce an echo with the same delay time.
(d) Now suppose that, instead of increasing linearly with height indefinitely, the electron density passes through a maximum at some height $h_1 (> h_0)$ and then decreases linearly. Indicate qualitatively the form of h' as a function of ν for this case.

Problem 6.2. Type III solar radio bursts are caused by electron beams that travel upwards through the Sun's corona, exciting plasma oscillations. The plasma oscillations then produce electromagnetic radiation at frequencies ν_p and $2\nu_p$, where

$$\nu_p \approx 10^{4.0} \, n^{1/2}. \tag{1}$$

Consider a simple model of the corona as a plane-parallel electron–proton plasma of temperature T, the density varying with height z as

$$n = n_0 e^{-z/H}. \tag{2}$$

(a) Noting that the gravitational acceleration near the surface of the Sun is $10^{4.4} \, \text{cm s}^{-2}$, verify that the 'scale height' H is given by

$$H = 10^{3.8} T. \tag{3}$$

(b) Estimate the total attenuation of the radiation during radial propagation from the expression

$$A = \int_{z_0}^{\infty} k_i \, dz \tag{4}$$

and the formula

$$k_i = \frac{\nu_p^2 \nu_c}{2c\nu (\nu^2 - \nu_p^2)^{1/2}}. \tag{5}$$

Use the approximate expression

$$\nu_c = 10^{1.9} \, n T^{-3/2} \tag{6}$$

for the collision frequency. Hence show that the attenuation of radiation at the fundamental frequency may be expressed as

$$A = 10^{-8.9} T^{-3/2} H \int_0^{n_0} \frac{n \, dn}{n_0^{1/2} (n_0 - n)^{1/2}}, \tag{7}$$

where we define n_0 as the value of n at which $\nu_p = \nu$.

(c) Derive an expression for A in terms of ν and T, and evaluate the cutoff frequency at which $A = 1$ for the typical coronal temperature of $T = 10^{6.4} \, \text{K}$.

(d) Modify your response to parts (b) and (c) so as to obtain an expression for A, in terms of ν and T, that represents the attenuation of radiation emitted at the harmonic frequency $2\nu_p$. Hence find the cutoff frequency for harmonic radiation for an assumed coronal temperature $T = 10^{6.4} \, \text{K}$.

Problem 6.3. Radio signals from pulsars pass through the interstellar medium that contains free electrons.

(a) Show that the arrival time $t(\nu)$ of a signal will be a function of frequency of the form

$$t(\nu) = D\nu^{-2} + \text{constant}, \tag{1}$$

where ν is the frequency in Hz, and the 'dispersion coefficient' D is expressible as

$$D = C \int n_e \, ds \tag{2}$$

where the integral represents the path integral of the electron density along the propagation path of the radio signal.

(b) Find the coefficient C.

(c) For a particular pulsar it is found that the signal at 100 MHz arrives 2 s later than the signal at 200 MHz. What is the value of D for that pulsar?

(d) If the mean electron density is $10^{-1.5}$ cm^{-3}, what is the distance to the pulsar in centimeters and in parsecs?

(e) What complicating factors are neglected in deriving the above simple expression for time delay as a function of frequency?

Problem 6.4. The delay time (emission to reception) of a whistler-mode signal can be written as

$$\tau = \int_{\text{path}} u^{-1} \, ds, \tag{1}$$

where u is the group velocity of the mode. For high densities such that $\omega_p \gg \omega_g$,

$$u = \frac{2c\omega^{1/2}(\omega_g - \omega)^{3/2}}{\omega_p \omega_g}, \tag{2}$$

so that the delay time is expressible as

$$\tau = \int_{\text{path}} \frac{\omega_p \omega_g \, ds}{2c\omega^{1/2}(\omega_g - \omega)^{3/2}} \equiv D(\omega)\omega^{-1/2}. \tag{3}$$

$D(\omega)$ is called the 'dispersion' of the whistler and can be used to discover the electron plasma density in the magnetosphere. For low frequencies for which $\omega \gg \omega_g$,

$$D = \int_{\text{path}} \frac{\omega_p \, ds}{2c\omega_g^{1/2}} \tag{4}$$

which is a constant and can be measured from the time-trace of a whistler.

(a) Consider the magnetic field to be that of a dipole located at the center of the Earth, and derive the expression

$$\omega_g = \omega_{g,0} \left(\frac{R}{r}\right)^3 (1 + 3 \sin^2 \lambda)^{1/2} \tag{5}$$

for the gyrofrequency at radius r and latitude λ, where R is the Earth's radius and $\omega_{g,0}$ is the gyrofrequency at the surface at the geomagnetic equator.

(b) Show that a field line that intersects the surface of the Earth at geomagnetic latitude λ_0 is given by

$$r = R \cos^2 \lambda / \cos^2 \lambda_0. \tag{6}$$

(c) Integrate the integral for D along the path $s = s(r, \lambda)$ from $\lambda = 0$ to $\lambda = \lambda_0$, to obtain the expression

$$D(x_0) = \frac{R_0}{2c \cos^5 \theta \, \omega_{g,0}^{1/2}} \int_0^{x_0} \omega_p(x) (1 + 3x^2)^{1/4} (1 - x^2)^{3/2} \, dx \tag{7}$$

where $x = \sin \lambda$.

(d) If $\omega_p \propto \omega_g$, what is the form of the integral?

7

Electromagnetic waves in an electron–ion plasma

7.1 The dispersion relation

We now generalize calculations made in the previous chapter to include the presence of one or more species of ions and propagation at an arbitrary angle to the magnetic field. Following Stix (1962), we find it convenient to combine the components of electric field normal to the magnetic field as follows:

$$E^+ = E_1 + iE_2, \quad E^- = E_1 - iE_2. \tag{7.1.1}$$

Similarly, we introduce the following symbols

$$v^+ = v_1 + iv_2, \quad v^- = v_1 - iv_2. \tag{7.1.2}$$

We are once again assuming that $\mathbf{B}_0 = (0, 0, B_0)$.

Denoting the sense of the charge of species s by ε_s, the magnitude of the charge by $Z_s e$, and the mass by m_s, the equation of motion for each species becomes

$$m_s \frac{\mathrm{d}\mathbf{v}_s}{\mathrm{d}t} = \varepsilon_s Z_s e \left(\mathbf{E} + \frac{1}{c} \mathbf{v}_s \times \mathbf{B}_0 \right). \tag{7.1.3}$$

For electrons (s→e), $\varepsilon = -1$, $Z = 1$, $m = m_e$. Since we will be analyzing the properties of a wave with wave-vector $\mathbf{k}$ and frequency ω, we may replace (7.1.3) by

$$-i\omega \mathbf{v}_s = \frac{\varepsilon_s Z_s e}{m_s} \left(\mathbf{E} + \frac{1}{c} \mathbf{v}_s \times \mathbf{B}_0 \right). \tag{7.1.4}$$

The 1- and 2-components of this equation become

$$\left. \begin{aligned} v_1 &= i \frac{\varepsilon_s Z_s e}{\omega m_s} E_1 + i \frac{\varepsilon_s Z_s e}{\omega m_s c} B_0 v_2, \\[2mm] v_2 &= i \frac{\varepsilon_s Z_s e}{\omega m_s} E_2 - i \frac{\varepsilon_s Z_s e}{\omega m_s c} B_0 v_1, \end{aligned} \right\} \tag{7.1.5}$$

and the 3-component becomes

$$v_3 = i \frac{\varepsilon_s Z_s e}{\omega m_s} E_3.$$

(7.1.6)

On using the notation of (7.1.1) and (7.1.2), and on denoting the gyrofrequency of each species as follows,

$$\omega_{g,s} = \frac{Z_s e B_0}{m_s c},$$

(7.1.7)

we find that (7.1.5) may be expressed as

$$v^{\pm} = i \frac{\varepsilon_s Z_s e}{\omega m_s} E^{\pm} \pm \frac{\varepsilon_s \omega_{g,s}}{\omega} v^{\pm}.$$

(7.1.8)

Hence we find that

$$v^{\pm} = \frac{i \dfrac{\varepsilon_s \omega_{g,s}}{B_0} E^{\pm}}{1 (\mp) \dfrac{\varepsilon_s \omega_{g,s}}{\omega}}.$$

(7.1.9)

The total current density in the plasma may be expressed as

$$j = \frac{e}{c} \sum_s \varepsilon_s Z_s n_s v_s.$$

(7.1.10)

On using (7.1.9), we find that

$$j^{\pm} = \frac{i}{4\pi c} \sum_s \frac{\omega_{p,s}^2}{\omega(\mp)\varepsilon_s \omega_{g,s}} E^{\pm},$$

(7.1.11)

where

$$\omega_{p,s}^2 = \frac{4\pi n_s Z_s^2 e^2}{m_s}.$$

(7.1.12)

The 3-component of current is given by

$$j_3 = \frac{i}{4\pi c} \sum_s \frac{\omega_{p,s}^2}{\omega} E_3.$$

(7.1.13)

We can now find the dielectric tensor either by forming the conductivity tensor by means of (7.1.11) and (7.1.13) and using (6.1.10), or by noting that

$$D_r = K_{rs} E_s = E_r + \frac{4\pi i c}{\omega} j_r.$$

(7.1.14)

By either approach, we find the following relationship between the dielectric vector and the electric-field vector:

$$K = \begin{pmatrix} S & -iD & 0 \\ iD & S & 0 \\ 0 & 0 & P \end{pmatrix}. \qquad (7.1.15)$$

In this tensor,

$$P = 1 - \sum_s \frac{\omega_{p,s}^2}{\omega^2}, \qquad (7.1.16)$$

and

$$\left. \begin{aligned} S &= \tfrac{1}{2}(R+L), \\ D &= \tfrac{1}{2}(R-L), \end{aligned} \right\} \qquad (7.1.17)$$

where

$$R = 1 - \sum_s \frac{\omega_{p,s}^2}{\omega(\omega - \varepsilon_s \omega_{g,s})} \qquad (7.1.18)$$

and

$$L = 1 - \sum_s \frac{\omega_{p,s}^2}{\omega(\omega + \varepsilon_s \omega_{g,s})}.$$

We now consider the wave equation (6.2.9), i.e.

$$\mathbf{k} \times (\mathbf{k} \times \mathbf{E}) + \frac{\omega^2}{c^2} K \cdot \mathbf{E} = 0. \qquad (7.1.19)$$

It is convenient to introduce a vector defined by

$$\mathbf{n} = \frac{c\mathbf{k}}{\omega} \qquad (7.1.20)$$

that has the direction of the wave vector and the magnitude of the refractive index:

$$n = \frac{ck}{\omega}. \qquad (7.1.21)$$

With this notation, the wave equation becomes

$$\mathbf{n} \times (\mathbf{n} \times \mathbf{E}) + K \cdot \mathbf{E} = 0. \qquad (7.1.22)$$

We now choose to orient the x_1 and x_2 axes in such a way that the vector $\mathbf{n}$ is expressible as

$$\mathbf{n} = n(\sin \theta, 0, \cos \theta). \qquad (7.1.23)$$

Then we find that the first term on the left-hand side of (7.1.22) has the form

$$\mathbf{n} \times (\mathbf{n} \times \mathbf{E}) = n^2 \begin{pmatrix} -\cos^2\theta & 0 & \sin\theta\cos\theta \\ 0 & -1 & 0 \\ \sin\theta\cos\theta & 0 & -\sin^2\theta \end{pmatrix} \begin{pmatrix} E_1 \\ E_2 \\ E_3 \end{pmatrix}. \qquad (7.1.24)$$

On using (7.1.15) and (7.1.24), we find that the wave equation (7.1.22) takes the form

$$\begin{pmatrix} S - n^2\cos^2\theta & -iD & n^2\sin\theta\cos\theta \\ iD & S - n^2 & 0 \\ n^2\sin\theta\cos\theta & 0 & P - n^2\sin^2\theta \end{pmatrix} \begin{pmatrix} E_1 \\ E_2 \\ E_3 \end{pmatrix} = 0. \qquad (7.1.25)$$

We can now obtain the dispersion relation by setting the determinant of the matrix in (7.1.25) equal to zero. We find that the dispersion relation is expressible as

$$An^4 - Bn^2 + C = 0, \qquad (7.1.26)$$

where

$$\left.\begin{aligned} A &= S\,\sin^2\theta + P\,\cos^2\theta \\ B &= RL\,\sin^2\theta + PS(1 + \cos^2\theta). \\ C &= PRL, \end{aligned}\right\} \qquad (7.1.27)$$

where we have used the relation

$$S^2 - D^2 = RL. \qquad (7.1.28)$$

One may solve (7.1.26) to obtain the following expression for n^2:

$$n^2 = \frac{B \pm F}{2A}, \qquad (7.1.29)$$

where the expression for F may be reduced to the simple form

$$F^2 = (RL - PS)^2 \sin^4\theta + 4P^2D^2 \cos^2\theta. \qquad (7.1.30)$$

Alternatively, one may re-express (7.1.27) as

$$\left.\begin{aligned} A &= S\,\sin^2\theta + P\,\cos^2\theta \\ B &= (RL + PS)\sin^2\theta + 2PS\cos^2\theta, \\ C &= PRL\,\sin^2\theta + PRL\,\cos^2\theta \end{aligned}\right\} \qquad (7.1.31)$$

and so obtain from (7.1.26) an expression for $\tan^2\theta$:

$$\tan^2\theta = \frac{-P(n^2 - R)(n^2 - L)}{(Sn^2 - RL)(n^2 - P)}. \qquad (7.1.32)$$

We see at once that we may retrieve from (7.1.32) the dispersion relations for the various modes that propagate parallel to or normal to the magnetic field. For $\theta = 0$, the modes become

$$P=0, \quad n^2=R, \quad n^2=L. \tag{7.1.33}$$

We see from (7.1.16) that $P=0$ represents plasma oscillations, where we now include the effects of all electron and ion species. We see from the second line of (7.1.25) that

$$\frac{E_2}{E_1} = \frac{iD}{n^2-S}. \tag{7.1.34}$$

Hence we find that $n^2=R$ leads to $E_2/E_1=i$, and $n^2=L$ leads to $E_2/E_1=-i$. That is, $n^2=R$ corresponds to right-hand rotating circularly polarized wave, and $n^2=L$ corresponds to a left-hand rotating circularly polarized wave.

For propagation normal to the magnetic field, $\theta = \pi/2$, we see that (7.1.32) leads to the two modes

$$n^2=P, \quad n^2=\frac{RL}{S}. \tag{7.1.35}$$

From an electron plasma, the first term corresponds to the mode previously found as (6.5.5), and one finds that the second mode corresponds to the mode represented by (6.5.9).

7.2 Wave propagation in an electron plasma

We now, for simplicity, ignore ion motion and consider only the response of the electrons. It is convenient to introduce the parameters

$$\alpha = \frac{\omega_p}{\omega}, \quad \beta = \frac{\omega_g}{\omega}. \tag{7.2.1}$$

Then we find that the coefficients of the matrix (7.1.15) may be expressed as

$$\left. \begin{array}{c} S = 1 - \dfrac{\alpha^2}{1-\beta^2}, \\[2mm] P = 1 - \alpha^2, \\[2mm] D = \dfrac{\alpha^2 \beta}{1-\beta^2}. \end{array} \right\} \tag{7.2.2}$$

As explained in Chapter 6, we used the term 'cutoff' to refer to the condition $n=0$. We see from (7.1.26) that this requires that $C=0$ and that either A or B be nonzero. The condition $C=0$ is found to be expressible as

$$\left(1-\frac{\alpha^2}{1-\beta}\right)\left(1-\frac{\alpha^2}{1+\beta}\right)(1-\alpha^2)=0 \qquad (7.2.3)$$

showing that cutoffs occur for the following conditions:

$$\text{or} \quad \left.\begin{array}{r}(1-\alpha^2)^2-\beta^2=0,\\ 1-\alpha^2=0.\end{array}\right\} \qquad (7.2.4)$$

The first of these conditions is referred to as the 'cyclotron cutoff' since it involves the cyclotron frequency through β. The cutoff indicated by the second condition of (7.2.4) is termed the 'plasma cutoff.' It is to be noted that the angle θ does not appear in the above conditions, since the term C does not involve the angle θ. Hence, when a cutoff occurs, it occurs for all directions of propagation simultaneously.

We now investigate the resonances defined by $n=\infty$. We see from (7.1.29) that these are determined by the condition $A=0$. Since A depends upon θ, there will be a critical cone defined by $\theta=\theta_{\text{res}}$, the 'resonance cone.' For some modes, propagation is allowed only *inside* this cone, whereas for other modes propagation is allowed only *outside* the cone. In either case, the topological form of the phase-velocity surface changes in crossing the cone.

For propagation along the magnetic field, $\theta=0$, we see from (7.1.31) that the condition $A=0$ reduces to $P=0$ that describes plasma oscillations. However, closer inspection of (7.1.26) shows that, for $\theta=0$, P factors out of the equation completely leaving

$$n^4-2Sn^2+RL=0. \qquad (7.2.5)$$

We see from (7.1.17) that this may be expressed alternatively as

$$(n^2-R)(n^2-L)=0. \qquad (7.2.6)$$

We see from (7.1.18) that the resonance at $n^2=R$, may alternatively be expressed as

$$n^2=1-\frac{\alpha^2}{1-\beta} \qquad (7.2.7)$$

so that the resonance occurs at $\beta=1$, i.e. at $\omega=\omega_{g,e}$. Similarly, we find that the resonance associated with the mode $n^2=L$ corresponds to $\beta=-1$. However, this condition cannot be satisfied since β is, by definition, non-negative. Hence only the right-hand circularly polarized wave has a resonance.

For propagation perpendicular to the magnetic field, $\theta = \pi/2$, the condition $A = 0$ yields the relation

$$S \equiv 1 - \frac{\alpha^2}{1 - \beta^2} = 0, \qquad (7.2.8)$$

that is

$$\alpha^2 + \beta^2 = 1 \qquad (7.2.9)$$

or

$$\omega^2 = \omega_p^2 + \omega_g^2. \qquad (7.2.10)$$

This agrees with the result found in Chapter 6, that a resonance occurs at the upper hybrid frequency for propagation normal to the magnetic field.

It is now possible to divide the (α^2)–(β^2) plane into distinct regions, using the information we have obtained about cutoffs and resonances. In Fig. 7.1,

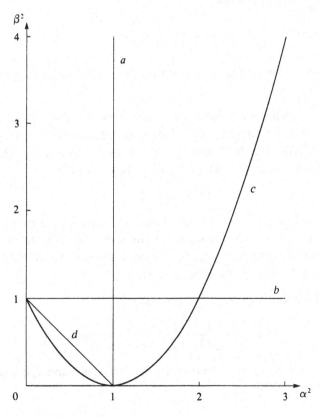

Fig. 7.1. Division of the α^2–β^2 plane according to the cutoffs and resonances.

we denote by a the cutoff condition $\alpha^2 = 1$ that we have termed the 'plasma cutoff.' Line b is $\beta = 1$, the electron cyclotron resonance. Line c is the cyclotron cutoff defined by the first of equations (7.2.4), and the line d is the upper hybrid resonance defined by (7.2.9).

Problems

Problem 7.1. Consider the electron gas as a fluid and use the equation of motion

$$mn\frac{\partial v}{\partial t} = qnE - \frac{\partial p}{\partial x}. \tag{1}$$

Assume that the pressure p and the density n are related by an equation of state of the form $p \propto n^\gamma$.

(a) Find the dispersion relation for longitudinal (electrostatic) waves.
(b) What type of gas produces the value of γ that reconciles the dispersion relation with that for a 'warm plasma'

$$\omega^2 = \omega_p{}^2 + 3\frac{k_B T}{m_e}k^2, \tag{2}$$

where we use k_B for Boltzmann's constant to avoid confusion with the wave number k.

Problem 7.2. Consider the polarization, given by (7.1.34), as a function of the angle θ between the direction of the propagation vector and the direction of the applied magnetic-field. Show that the sense of rotation of the electric-field vector changes sign at a critical value of θ given by

$$\sin^2 \theta_{cr} = P/S. \tag{1}$$

Problem 7.3. In Problem 6.3, we discussed the dispersion of radio waves emitted by a pulsar due to propagation through the interstellar medium, ignoring the role of magnetic field. Consider the propagation of electromagnetic waves parallel to the magnetic field.

(a) Show that the group velocity u is expressible as

$$u^{-1} = \frac{1}{c}\left[\frac{1 \mp [\omega_p{}^2\omega_g/2\omega(\omega \pm \omega_g)^2]}{[1 - (\omega_p{}^2/\omega(\omega \pm \omega_g))]^{1/2}}\right] \tag{1}$$

(b) Hence, using $\nu = \omega/2\pi$, etc., and assuming that $\nu_p \ll \nu$ and $\nu_g \ll \nu$, obtain the approximate expression

$$u^{-1} = c^{-1}\left[1 + \frac{\nu_p{}^2}{2\nu^2}\left(1 + \frac{3\nu_p{}^2}{4\nu^2} \pm \frac{2\nu_g}{\nu}\right)\right], \tag{2}$$

(c) Show that the time delay between the reception of signals at frequencies ν_1 and ν_2 may be expressed as

$$\Delta t = D\left(\frac{1}{\nu_1{}^2} - \frac{1}{\nu_2{}^2}\right)(1 + T_1 \mp T_2), \tag{3}$$

where

$$T_1 = \frac{\langle n_e^2 \rangle}{\langle n_e \rangle^2} \frac{3Dc(\nu_1{}^2 + \nu_2{}^2)}{2d\nu_1{}^2\nu_2{}^2} \tag{4}$$

and

$$T_2 = 2\langle \nu_g \rangle \frac{(\nu_2{}^3 - \nu_1{}^3)}{\nu_1 \nu_2 (\nu_2{}^2 - \nu_1{}^2)}. \tag{5}$$

As before,

$$D = \frac{e^2}{2\pi m_e c} \int_0^d n_e \, ds \tag{6}$$

and the terms in brackets are averages of the different quantities over the path of propagation:

$$\langle f \rangle \equiv d^{-1} \int_0^d f \, ds. \tag{7}$$

(d) For $\nu_1 = 40\,\text{MHz}$ and $\nu_2 = 430\,\text{MHz}$, observations have shown the corrections T_1 and T_2 to be each less than 3×10^{-4}. From this information, obtain upper limits on D and on the magnetic field strength. How does the upper limit on D compare with the value found in Problem 6.3? Are the values of ν_p and ν_g at these limits much less than ν_1 and ν_2?

8

Two-stream instability

We have met the concept of plasma oscillations in Chapter 2 and again in Chapter 6. In these analyses, we considered oscillations in a simple homogeneous plasma at zero temperature. In Chapter 9, we shall adopt a more general approach to the study of plasma oscillations that will allow us, among other things, to investigate the effect of nonzero temperature of the plasma. In this chapter, we focus on a more specific modification of the plasma: that is, we consider the possibility that it is composed of two or more streams of particles with different velocities. For simplicity, we begin by considering that each stream is of zero temperature. This leads to the important concept of 'two-stream instability' in its simplest form.

8.1 Particle streams of zero temperature

We now consider a collection of cold interpenetrating plasma streams with the goal of considering the instabilities that can arise in such systems. As before, we consider different 'species' of particles denoted by $s = e$ (electrons), 1, 2, ... Particles of species s have mass m_s and charge q_s. We also assume that, in the unperturbed state, the density and velocity of species s are n_s and $\mathbf{v}_s$. We consider the perturbations indicated in

$$
\begin{aligned}
n_s &\rightarrow n_s + \delta n_s, \\
\mathbf{v}_s &\rightarrow \mathbf{v}_s + \delta \mathbf{v}_s.
\end{aligned}
\tag{8.1.1}
$$

We assume that

$$
\zeta \equiv \sum_s n_s q_s = 0,
\tag{8.1.2}
$$

so that we may assume that there is no electric field in the unperturbed state. Then the only electric field is $\delta \mathbf{E}$, that satisfies the equation

$$\nabla \cdot \delta \mathbf{E} = 4\pi \sum_{s} q_s \delta n_s. \tag{8.1.3}$$

We assume that there is no magnetic field in the unperturbed state, and we consider only nonrelativistic velocities, so that the effects of the magnetic field produced by the particle streams may be neglected. Then the equation of motion is

$$\frac{\partial \mathbf{v}_s}{\partial t} + \mathbf{v}_s \cdot \nabla \mathbf{v}_s = \frac{q_s}{m_s} \mathbf{E}. \tag{8.1.4}$$

Hence we obtain the following equation for the perturbations $\delta \mathbf{v}_s$:

$$\frac{\partial \delta \mathbf{v}_s}{\partial t} + \mathbf{v}_s \cdot \nabla \delta \mathbf{v}_s = \frac{q_s}{m_s} \delta \mathbf{E}. \tag{8.1.5}$$

In order to complete our system of equations, we need to include also the continuity equation,

$$\frac{\partial n_s}{\partial t} + \nabla \cdot (n_s \mathbf{v}_s) = 0. \tag{8.1.6}$$

The linearized form of this equation is

$$\frac{\partial \delta n_s}{\partial t} + \mathbf{v}_s \cdot \nabla \delta n_s + n_s \nabla \cdot \delta \mathbf{v}_s = 0. \tag{8.1.7}$$

Since we are assuming that the unperturbed state is homogeneous and steady, we may analyze the properties of this system by Fourier analysis. Within the context of linear theory, each mode may be considered separately; hence we may consider a single wave, that we take to have wave vector $\mathbf{k}$ and frequency ω. Then, with the conventions adopted in previous chapters, we find that (8.1.3), (8.1.5) and (8.1.7) become

$$i\mathbf{k} \cdot \mathbf{E} = 4\pi \sum_{s} q_s \delta n_s, \tag{8.1.8}$$

$$-i\omega \delta \mathbf{v}_s + i(\mathbf{k} \cdot \mathbf{v}_s) \delta \mathbf{v}_s = \frac{q_s}{m_s} \mathbf{E}, \tag{8.1.9}$$

and

$$-i\omega \delta n_s + i(\mathbf{k} \cdot \mathbf{v}_s) \delta n_s + i n_s \mathbf{k} \cdot \delta \mathbf{v}_s = 0. \tag{8.1.10}$$

Equation (8.1.9) leads to the following expression for the velocity perturbation in terms of the electric-field perturbation:

$$\delta \mathbf{v}_s = \frac{\dfrac{q_s}{m_s} \mathbf{E}}{-i(\omega - \mathbf{k} \cdot \mathbf{v}_s)} . \qquad (8.1.11)$$

Equation (8.1.10) leads to

$$\delta n_s = \frac{n_s \mathbf{k} \cdot \delta \mathbf{v}_s}{\omega - \mathbf{k} \cdot \mathbf{v}_s}, \qquad (8.1.12)$$

so that, using (8.1.11), we obtain

$$\delta n_s = \frac{n_s \dfrac{q_s}{m_s} \mathbf{k} \cdot \mathbf{E}}{-i(\omega - \mathbf{k} \cdot \mathbf{v}_s)^2} . \qquad (8.1.13)$$

On substituting (8.1.13) into (8.1.8), we obtain

$$i\mathbf{k} \cdot \mathbf{E} = i \left\{ \sum_s \frac{4\pi q_s^2 n_s}{m_s} \frac{1}{(\omega - \mathbf{k} \cdot \mathbf{v}_s)^2} \right\} \mathbf{k} \cdot \mathbf{E}, \qquad (8.1.14)$$

that is

$$\left\{ \sum_s \frac{\omega_{ps}^2}{(\omega - \mathbf{k} \cdot \mathbf{v}_s)^2} - 1 \right\} \mathbf{k} \cdot \mathbf{E} = 0. \qquad (8.1.15)$$

Hence, in order for there to be a longitudinal mode of nonzero amplitude, it is necessary that the following dispersion relation be satisfied:

$$\sum_s \frac{\omega_{ps}^2}{(\omega - \mathbf{k} \cdot \mathbf{v}_s)^2} = 1. \qquad (8.1.16)$$

If we consider that all streams have zero velocity, we recover once more from (8.1.16) the following dispersion relation

$$\omega^2 = \sum_s \omega_{ps}^2. \qquad (8.1.17)$$

If we consider only a single stream, say an electron stream, then the relation becomes

$$(\omega - \mathbf{k} \cdot \mathbf{v}_e)^2 = \omega_{pe}^2. \qquad (8.1.18)$$

In this case, it is clear that the group velocity is equal to $\mathbf{v}_e$, the unperturbed velocity of the electron stream. Hence we find, once more, that simple plasma oscillations propagate with the electron plasma.

8.2 Two-stream instability

Now suppose that we need consider only two streams of particles. These may be a stream of electrons and a stream of ions, of equal but opposite charge density, with different velocities. On the other hand, we might consider two streams of electrons, limiting our attention to sufficiently high frequencies that the motion of the background ions may be neglected. If the velocity vectors are not parallel, we may adopt a moving coordinate frame such that the velocities become parallel. We also consider only waves with wave vectors parallel to the velocities of the streams. Hence the problem has become one-dimensional, and the dispersion relation (8.1.16) now becomes

$$\frac{\omega_{p1}^2}{(\omega - v_1 k)^2} + \frac{\omega_{p2}^2}{(\omega - v_2 k)^2} = 1. \tag{8.2.1}$$

Since this is a quartic equation in ω, there is no simple general expression for ω in terms of k. The form of the dispersion relation (8.2.1) will be examined, for a particular case, in the next section.

If we introduce the phase velocity v_ϕ defined by

$$v_\phi = \frac{\omega}{k}, \tag{8.2.2}$$

the dispersion relation becomes

$$\frac{\omega_{p1}^2}{k^2(v_\phi - v_1)^2} + \frac{\omega_{p2}^2}{k^2(v_\phi - v_2)^2} = 1. \tag{8.2.3}$$

This may be expressed alternatively as

$$G(v_\phi) \equiv \frac{\omega_{p1}^2}{(v_\phi - v_1)^2} + \frac{\omega_{p2}^2}{(v_\phi - v_2)^2} = k^2. \tag{8.2.4}$$

Since this equation is of fourth order and has real coefficients, there will be four solutions for v_ϕ that can be arranged in two pairs, each pair being a complex-conjugate pair.

The general form of the function $G(v_\phi)$ is shown in Fig. 8.1. We see that there is a critical value of k, that we label k_c. For $k^2 > k_c^2$, (8.2.4) has four real roots so that the system supports four waves that propagate without growth or decay. On the other hand, if $k^2 < k_c^2$, (8.2.4) has only two real roots. Hence the other two roots must be complex, corresponding to complex frequencies. Since these two frequencies are complex conjugates, one of them will correspond to a mode that grows exponentially in time. Hence any small perturbation of the system will grow to arbitrarily large values (within the

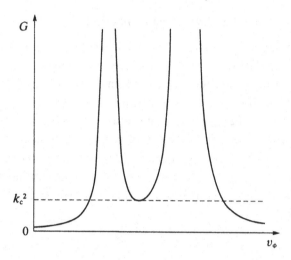

Fig. 8.1. Schematic representation of the function $G(v_\phi)$ defined by (8.2.4).

framework of linear theory). That is, the system is unstable. The instability we have found is called the 'two-stream instability.'

We may find the value k_c by finding the value of v_ϕ for which the derivative dG/dv_ϕ is zero. This is found to be

$$v_{\phi,c} = \frac{\omega_{p1}^{2/3} v_2 + \omega_{p2}^{2/3} v_1}{\omega_{p1}^{2/3} + \omega_{p2}^{2/3}}. \tag{8.2.5}$$

Hence we find that

$$k_c^2 = \frac{(\omega_{p1}^{2/3} + \omega_{p2}^{2/3})^3}{(v_1 - v_2)^2}. \tag{8.2.6}$$

We see that (8.2.6) has a curious implication. Our physical intuition tells us that as $v_1 - v_2 \to 0$, the system reverts to a one-stream system that we know to be stable. On the other hand, we see from (8.2.6) that as $(v_1 - v_2)$ becomes arbitrarily small, k_c becomes arbitrarily large. This indicates that the system becomes unstable for a larger and larger range of wave number. This paradox will be resolved when we consider the properties of streams of finite temperature.

For the one-dimensional problem now being considered, it is simple to extend the results to relativistic motion. For one-dimensional motion, the equation of motion (8.1.4) is changed only in the requirement that the mass m should be replaced by the 'longitudinal' mass m_L given by

$$m_L = \gamma^3 m, \tag{8.2.7}$$

where γ is defined by (3.1.2). We would therefore obtain the same form (8.2.1) for the dispersion relation, except that the plasma frequency is now given by

$$\omega_p^2 = \frac{4\pi ne^2}{\gamma^3 m}. \tag{8.2.8}$$

8.3 Two identical but opposing streams

If the streams have the same density and equal but opposite velocities, we may adopt

$$v_1 = -v_2 = v, \quad \omega_{p1} = \omega_{p2} = \omega_p, \tag{8.3.1}$$

so that (8.2.1) becomes

$$\frac{\omega_p^2}{(\omega - vk)^2} + \frac{\omega_p^2}{(\omega + vk)^2} = 1. \tag{8.3.2}$$

This is now a quadratic equation in ω^2:

$$2\omega_p^2(\omega^2 + v^2 k^2) = (\omega^2 - v^2 k^2)^2, \tag{8.3.3}$$

that is

$$\omega^4 - 2(\omega_p^2 + v^2 k^2)\omega^2 - 2\omega_p^2 v^2 k^2 + v^4 k^4 = 0. \tag{8.3.4}$$

Hence we obtain the following expression for ω^2:

$$\omega^2 = \omega_p^2 + v^2 k^2 \pm \omega_p(\omega_p^2 + 4v^2 k^2)^{1/2}. \tag{8.3.5}$$

This relationship between ω^2 and k^2 is shown graphically in Fig. 8.2. We see that, for $k^2 > k_c^2$, all values of ω are real. However, for $k^2 < k_c^2$, two values of ω will be complex, and one of these will represent a growing wave, that is, an instability. We find that

$$k_c = \frac{2^{1/2}\omega_p}{v}. \tag{8.3.6}$$

Once again, we see that the range of unstable wave numbers becomes increasingly large as $v \to 0$.

We may find the wave number k_m that corresponds to the most unstable mode by finding the value of k^2 for which $d\omega^2/dk^2 = 0$. Hence we find from (8.3.5) that

$$k_m = \frac{3^{1/2}}{2} \frac{\omega_p}{v}. \tag{8.3.7}$$

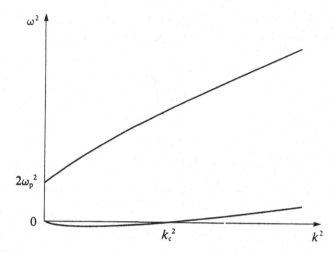

Fig. 8.2. Representation of relationship between ω^2 and k^2 given by (8.3.5).

We find that the maximum imaginary value of the frequency ω_{im} is given by

$$\omega_{im} = \tfrac{1}{2}\omega_p. \tag{8.3.8}$$

Hence the most unstable mode grows exponentially at a rate that is comparable with the plasma frequency.

We may also examine the properties of waves of given real frequency by inspecting Fig. 8.2. We see that for $\omega^2 > \omega_c^2$, all four waves have real wave number. However for $\omega^2 < \omega_c^2$, two values of k will be real and the other two will be imaginary. The value of ω_c is found to be

$$\omega_c = 2^{1/2}\omega_p. \tag{8.3.9}$$

Waves that grow in space divide into two categories. One category is that of 'evanescent' waves, such as waves in a wave guide at frequencies below the cutoff frequency. These waves inevitably decay away from the source. The other waves are similar to waves in a traveling-wave tube, that grow in amplitude as they propagate away from the source and hence can be used for amplification purposes (see, for instance, Pierce, 1950). There is also a similar classification of instabilities into 'convective' and 'absolute.' The basic ideas underlying this classification are presented briefly in Appendix C. The spatially growing modes in the current problem are found to represent amplifying waves. The instability is found to be a convective instability if v_1 and v_2 have the same sign, and an absolute instability if they have opposite signs.

The ω–k diagram for the case of two equal streams of opposite velocities

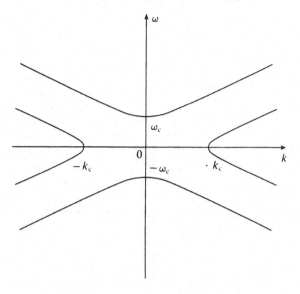

Fig. 8.3. Dispersion relation for two identical but opposing streams.

is found to have the form shown in Fig. 8.3. A schematic representation of the form of the dispersion relation for the case of two streams traveling in the same direction is given in Fig. 8.4.

8.4 Stream moving through a stationary plasma

We now consider a stationary plasma with plasma frequency ω_p and a second stream of plasma frequency ω_{p1} moving with velocity v through the stationary stream. Then the dispersion relation (8.2.1) becomes

$$\frac{\omega_p^2}{\omega^2} + \frac{\omega_{p1}^2}{(\omega - vk)^2} = 1. \tag{8.4.1}$$

This may be expressed alternatively as

$$G(v_\phi) \equiv \frac{\omega_p^2}{v_\phi^2} + \frac{\omega_{p1}^2}{(v_\phi - v)^2} = k^2. \tag{8.4.2}$$

Proceeding as before, we find that the condition for instability is that $k^2 < k_c^2$ where

$$k_c = \frac{\omega_p}{v} \left[1 + \left(\frac{\omega_{p1}}{\omega_p} \right)^{2/3} \right]^{3/2}. \tag{8.4.3}$$

If we assume that $\omega_{p1} \ll \omega_p$ this becomes, approximately,

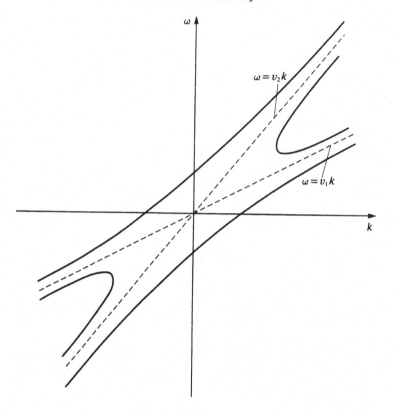

Fig. 8.4. Schematic representation of the form of the dispersion relation for two streams traveling in the same direction.

$$k_c = \frac{\omega_p}{v}\left[1 + \frac{3}{2}\left(\frac{\omega_{pl}}{\omega_p}\right)^{2/3}\right]. \tag{8.4.4}$$

We may obtain an estimate of the maximum growth rate for the case $\omega_{pl} \ll \omega_p$. If we write

$$\omega = \omega_p + \Delta\omega, \quad k = v^{-1}\omega_p + \Delta k, \tag{8.4.5}$$

and substitute these expressions in (8.4.1), we find that, to lowest order in the small quantity ω_{pl}/ω_p, the dispersion relation takes the form

$$\Delta\omega(\Delta\omega - v\Delta k)^2 = \tfrac{1}{2}\omega_p\omega_{pl}^2. \tag{8.4.6}$$

We need to find the value of the real quantity Δk that makes the imaginary part of $\Delta\omega$ a maximum. In order to do this, we may introduce the dimensionless quantities defined by

$$\xi = (\tfrac{1}{2}\omega_p\omega_{pl}{}^2)^{-1/3}\Delta\omega_r, \quad \eta = (\tfrac{1}{2}\omega_p\omega_{pl}{}^2)^{-1/3}\Delta\omega_i, \quad \kappa = (\tfrac{1}{2}\omega_p\omega_{pl}{}^2)^{-1/3}v\Delta k.$$
$$(8.4.7)$$

Then (8.4.6) takes the form

$$(\xi + i\eta)(\xi + i\eta - \kappa)^2 = 1. \tag{8.4.8}$$

The imaginary term yields

$$\eta^2 = 3\xi^2 - 4\xi\kappa + \kappa^2 \tag{8.4.9}$$

so that the real part of (8.4.8) may be expressed as

$$-8\xi^3 + 16\xi^2\kappa - 10\xi\kappa^2 + 2\kappa^3 = 1. \tag{8.4.10}$$

We can now consider small perturbations $\delta\xi$, $\delta\eta$ and $\delta\kappa$, but note that we should set $\delta\eta = 0$ since we are looking for a maximum in ω_i. Hence from (8.4.9) and (8.4.10) we find two linear relationships between $d\xi$ and $d\kappa$. The condition that these two equations be satisfied for all $d\xi$ and $d\kappa$ yields three possible relationships between ξ and κ.

$$\frac{\xi}{\kappa} = \frac{1}{2}.$$

$$\frac{\xi}{\kappa} = 1. \tag{8.4.11}$$

$$\kappa = 0.$$

We find that the first two choices yield η^2 negative or $\eta^2 = 0$ respectively, so that only the last choice is acceptable. From (8.4.9) and (8.4.10) we find:

$$\xi = -\frac{1}{2}, \quad \eta = \frac{\sqrt{3}}{2}, \quad \text{and } \kappa = 0. \tag{8.4.12}$$

Hence we obtain the following expressions for ω and k that correspond to the maximum growth rate:

$$\omega = \omega_p - 0.396\,\omega_p{}^{1/3}\omega_{pl}{}^{2/3} + i\,0.687\,\omega_p{}^{1/3}\omega_{pl}{}^{2/3},$$

$$k = v^{-1}\omega_p. \tag{8.4.13}$$

The maximum growth rate may be expressed as

$$\frac{\omega_{i,m}}{\omega_p} = 0.687\left(\frac{\omega_{pl}}{\omega_p}\right)^{2/3}. \tag{8.4.14}$$

This theory can be applied to an electron–ion plasma in which the electron and ion streams have a relative speed v. Since the ion plasma frequency is smaller than the electron plasma frequency, we may adopt

$$\omega_p{}^2 = \omega_{pe}{}^2 \equiv \frac{4\pi n_e e^2}{m_e}, \qquad \omega_{p1}{}^2 = \omega_{pi}{}^2 \equiv \frac{4\pi n_i Z_i e^2}{m_i}. \qquad (8.4.15)$$

For the particular case that the ions are protons, we see that

$$\frac{\omega_{pp}}{\omega_{pe}} = \left(\frac{m_e}{m_p}\right)^{1/2} = 10^{-1.63}. \qquad (8.4.16)$$

Hence we find from (8.4.14) that the maximum growth rate is given by

$$\omega_{i,m} = 10^{-1.25}\omega_{pe}. \qquad (8.4.17)$$

The two-stream instability of an electron–ion plasma is known as the 'Buneman instability' (Buneman, 1959).

Problems

Problem 8.1. Consider two streams of electrons, of equal density so that each stream has plasma frequency ω_p, moving in a given direction with velocities V and $-V$ with respect to a neutralizing fixed background of positive charge. Take into account collisions between the electrons of each stream and the positive charges by including the term $-m_e \nu \mathbf{v}$ in the equation of motion of each electron stream, as indicated in Section 6.3.

(a) Derive the dispersion relation

$$\frac{\omega_p{}^2}{(\omega - Vk + i\nu)^2} + \frac{\omega_p{}^2}{(\omega + Vk + i\nu)^2} = 1.$$

(b) Find the minimum value of ν required to suppress the two-stream instability.

Problem 8.2. Consider two streams of ions, each of density $\frac{1}{2}n_0$, that move in the x-direction with velocities V and $-V$ with respect to a fixed background distribution of electrons of density n_0. Consider also that there is a magnetic field of strength B_0 oriented in the z-direction.

(a) Consider that the ion gyroradius is so large that we may neglect the effect of the magnetic field on the ion orbits, and so obtain the dispersion relation

$$\frac{\omega_{P,i}{}^2}{2(\omega - Vk)^2} + \frac{\omega_{P,i}{}^2}{2(\omega + Vk)^2} + \frac{\omega_{P,e}{}^2}{\omega^2 - \omega_{g,e}{}^2} = 1,$$

where $\omega_{P,e}$ is the plasma frequency of the electron distribution, and $\omega_{P,i}$ is the plasma frequency of an ion distribution of density n_0.

(b) Now consider wave frequencies in the range

$$\omega_{g,i} \ll \omega \ll \omega_{g,e}.$$

Find the range of wave-numbers k for which the system is unstable, and calculate the real and imaginary parts of ω as functions of k.

9

Electrostatic oscillations in a plasma of nonzero temperature

We recall from the previous chapter the curious result that, as measured by the range of unstable wave numbers, a two-stream plasma becomes more unstable as the relative velocity of the two species becomes smaller. This seems paradoxical, since we know that the system becomes stable as the relative speed tends to zero.

In order to resolve this paradox, we need to set up a more comprehensive theory that can take account of the thermal motion of the electrons and ions in a plasma. In order to do this, we need to introduce the concept of a 'distribution function' as representing the state of a particle species in a plasma.

9.1 Distribution functions

The distribution function $f(\mathbf{x}, \mathbf{v}, t)$ is defined by the requirement that the number of particles in the volume element x_1 to $x_1 + dx_1$, x_2 to $x_2 + dx_2$, x_3 to $x_3 + dx_3$ and with velocities in the range v_1 to $v_1 + dv_1$, v_2 to $v_2 + dv_2$, v_3 to $v_3 + dv_3$, is given by

$$d^6 N = f(\mathbf{x}, \mathbf{v}, t) \, dx_1 \, dx_2 \, dx_3 \, dv_1 \, dv_2 \, dv_3. \qquad (9.1.1)$$

Note that there is some subtlety in this definition. If we imagine that f is defined in the limit as $d^3\mathbf{x}$ and $d^3\mathbf{v}$ become arbitrarily small, then we find that either $f \to \infty$ (in the case that the volume elements are centered on a particle) or $f \to 0$ otherwise.

One way around this difficulty is to imagine that we are dealing with velocity distributions for which the spatial gradients and the velocity gradients are not too large. For instance, if the spatial gradients are characterized by a spatial dimension Δx and the velocity gradients by a velocity range Δv, and if

$$f(\Delta x)^3 (\Delta v)^3 \gg 1, \tag{9.1.2}$$

then we may estimate the distribution function f by considering spatial volumes that are small compared with $(\Delta x)^3$ and velocity volumes that are small compared with $(\Delta v)^3$, for which the corresponding value of $\mathrm{d}^3 N$ is nevertheless large compared with unity. In what follows, we shall assume that this is the case.

There are alternative interpretations of (9.1.1). For instance, we can consider an 'ensemble' of systems and then imagine that we are discussing an average over that ensemble. Alternatively, we can imagine that (9.1.1) represents the probability of finding a particle in the specified element of space and volume. Such assumptions certainly extend the range of applicability of the concept of distribution function. However, one then needs to be cautious about interpreting the results of calculations that involve distribution functions.

We now set these subtleties on one side and proceed with our calculations. It is clear from (9.1.1) that the particle density n is related to the distribution function f by

$$n(\mathbf{x}, t) = \int \mathrm{d}^3 v f(\mathbf{x}, \mathbf{v}, t). \tag{9.1.3}$$

We may also introduce a particle flux $\mathbf{F}$ defined by

$$\mathbf{F} = \int \mathrm{d}^3 v f(\mathbf{x}, \mathbf{v}, t) \mathbf{v}. \tag{9.1.4}$$

However, the flux may be represented alternatively in terms of an average velocity $\langle \mathbf{v} \rangle$ by

$$\mathbf{F} = n \langle \mathbf{v} \rangle \tag{9.1.5}$$

so that

$$\langle \mathbf{v} \rangle = n^{-1} \int \mathrm{d}^3 v f(\mathbf{x}, \mathbf{v}, t) \mathbf{v}. \tag{9.1.6}$$

The average values of higher moments of the velocity may be defined analogously. For instance

$$\langle v^2 \rangle = n^{-1} \int \mathrm{d}^3 v f(\mathbf{x}, \mathbf{v}, t) v^2. \tag{9.1.7}$$

In order to analyze the properties of such a plasma, we need to find expressions for the charge density and current density in terms of the distribution function. If we now consider that the plasma is composed of a number of species $\sigma = e, 1, 2, \ldots$, then the electric charge density is given by

$$\zeta = \sum_\sigma q_\sigma \int d^3v \, f_\sigma(\mathbf{x}, \mathbf{v}, t) \qquad (9.1.8)$$

and the current density is given by

$$\mathbf{j} = \sum_\sigma \frac{q_\sigma}{c} \int d^3v \, f_\sigma(\mathbf{x}, \mathbf{v}, t)\mathbf{v}. \qquad (9.1.9)$$

We now need to find a way to express the equation of motion as an equation for the distribution function f.

It is convenient to define the acceleration vector field $\mathbf{a}(\mathbf{x}, \mathbf{v}, t)$. Considering, for the moment, only nonrelativistic motion, we see that, for any particular species,

$$\mathbf{a}(\mathbf{x}, \mathbf{v}, t) \equiv \frac{d\mathbf{v}}{dt} = \frac{q}{m}\,\mathbf{E} + \frac{q}{mc}\,\mathbf{v} \times \mathbf{B}. \qquad (9.1.10)$$

Let us now consider the motion of particles of this species in the time interval dt. Then the position and velocity change as follows

$$\mathbf{x} \to \mathbf{x}' = \mathbf{x} + \mathbf{v}\,dt, \qquad (9.1.11)$$

and

$$\mathbf{v} \to \mathbf{v}' = \mathbf{v} + \mathbf{a}\,dt. \qquad (9.1.12)$$

The distribution function will change in time, and we can write

$$f \to f' = f + \frac{\partial f}{\partial t}\,dt. \qquad (9.1.13)$$

However, (9.1.11) and (9.1.12) are written from the Lagrangian viewpoint of following particles in their motion, whereas (9.1.13) is written from the Eulerian viewpoint of following the change in a quantity at a fixed position and a fixed value of velocity.

In order to relate these quantities, we note that the number of particles in the original volume $d^3x\,d^3v$ will be the same as the number of particles in the later volume $d^3x'\,d^3v'$, so that

$$f'(\mathbf{x}', \mathbf{v}', t + dt)d^3x'\,d^3v' = f(\mathbf{x}, \mathbf{v}, t)d^3x\,d^3v. \qquad (9.1.14)$$

The volume elements are related by the Jacobian quantity:

$$d^3x'\,d^3v' = \frac{\partial(x_1', x_2', x_3', v_1', v_2', v_3')}{\partial(x_1, x_2, x_3, v_1, v_2, v_3)}\,d^3x\,d^3v. \qquad (9.1.15)$$

On using (9.1.11) and (9.1.12) we find that, to first order in dt,

$$\mathrm{d}^3x'\,\mathrm{d}^3v' = \left(1 + \frac{\partial a_r}{\partial v_r}\,\mathrm{d}t\right)\mathrm{d}^3x\,\mathrm{d}^3v. \tag{9.1.16}$$

We may now expand (9.1.14) and obtain, to first order in $\mathrm{d}t$,

$$\left[f + v_r\,\mathrm{d}t\,\frac{\partial f}{\partial x_r} + a_r\,\mathrm{d}t\,\frac{\partial f}{\partial v_r} + \frac{\partial f}{\partial t}\,\mathrm{d}t\right]\left[1 + \frac{\partial a_s}{\partial v_s}\,\mathrm{d}t\right] = f + \left(\frac{\partial f}{\partial t}\right)_c\,\mathrm{d}t, \tag{9.1.17}$$

where we have included on the right-hand side a 'collision term' $(\partial f/\partial t)_c$ that represents the effects of collisions. The form of this term will be investigated in a later chapter.

By gathering together terms that are linear in $\mathrm{d}t$, we finally obtain the equation

$$\frac{\partial f}{\partial t} + v_r\frac{\partial f}{\partial x_r} + \frac{\partial}{\partial v_r}(a_r f) = \left(\frac{\partial f}{\partial t}\right)_c. \tag{9.1.18}$$

We may easily verify from (9.1.10) that $\partial a_r/\partial v_r = 0$, so that the equation finally takes the form

$$\frac{\partial f}{\partial t} + v_r\frac{\partial f}{\partial x_r} + a_r\frac{\partial f}{\partial v_r} = \left(\frac{\partial f}{\partial t}\right)_c. \tag{9.1.19}$$

If we were considering a relativistic gas, according to the present formalism, we would find that $\partial a_r/\partial v_r \neq 0$. We might then decide to represent the particle distribution function in terms of $\mathbf{x}$ and $\mathbf{p}$, rather than $\mathbf{x}$ and $\mathbf{v}$, since we would find that $\partial \dot{p}_r/\partial p_r = 0$.

In the case that the collision term represents the effects of two-body large-angle collisions, (9.1.19) would be known as the 'Boltzmann equation.' For the case that the collision term represents the accumulated effects of small-angle collisions, we will find another form for the collision term and the equation will then be known as the 'Fokker–Planck equation.' For the case that the effects of collisions may be neglected so that $(\partial f/\partial t)_c = 0$, (9.1.19) is known as the 'Vlasov equation.' The equation may then be expressed as follows

$$\frac{\partial f}{\partial t} + v_r\frac{\partial f}{\partial x_r} + \left[\frac{q}{m}E_r + \frac{q}{mc}\varepsilon_{rst}v_s B_t\right]\frac{\partial f}{\partial v_r} = 0. \tag{9.1.20}$$

Equation (9.1.20) is sometimes referred to as the 'collision-less Boltzmann equation.' However, as Marshall Rosenbluth once remarked, it is perhaps more appropriate to call it the 'Boltzmann-less collision equation.'

9.2 Linear perturbation analysis of the Vlasov equation

We here consider the conceptually simple problem of a uniform plasma with infinitely massive (i.e. fixed) ions, no magnetic field, and no electric field. We consider small perturbations from the equilibrium condition by writing

$$f = f_0 + f_1, \quad \mathbf{E} = \mathbf{E}_1, \tag{9.2.1}$$

where f_0 is the unperturbed value of the distribution function f. For a non-relativistic plasma, for which the particle velocities are small compared with c, we may neglect the effect of the magnetic field associated with the pertubation.

On keeping terms to first order in the perturbation, we obtained from (9.1.20) the equation

$$\frac{\partial f_1}{\partial t} + v_r \frac{\partial f_1}{\partial x_r} - \frac{e}{m_e} E_{1r} \frac{\partial f_0}{\partial v_r} = 0. \tag{9.2.2}$$

We now further simplify the problem by supposing that the system is one-dimensional, that is, $\mathbf{E}_1$ has only one nonzero component in the x_1 direction, and $\mathbf{E}_1$ and f_1 are independent of x_2 and x_3. Writing x for x_1, the equation then becomes

$$\frac{\partial f_1}{\partial t} + v \frac{\partial f_1}{\partial x} - \frac{e}{m_e} E_1 \frac{\partial f_0}{\partial v} = 0. \tag{9.2.3}$$

Poisson's equation (2.1.3) now becomes

$$\frac{\partial E_1}{\partial x} = -4\pi e \int dv f_1. \tag{9.2.4}$$

It is convenient to introduce the normalized distribution function defined by

$$g(x, v, t) = n_0^{-1} f(x, v, t). \tag{9.2.5}$$

Equations (9.2.3) and (9.2.4) then become

$$\frac{\partial g_1}{\partial t} + v \frac{\partial g_1}{\partial x} - \frac{e}{m_e} E_1 \frac{dg_0}{dv} = 0 \tag{9.2.6}$$

and

$$\frac{\partial E_1}{\partial x} = -4\pi n_0 e \int dv\, g_1. \tag{9.2.7}$$

Since we are assuming that n_0 and g_0 are uniform in space and time, we may conveniently apply Fourier analysis to the present problem. We use the notation that g_1 and its transform $\tilde{g}_1$ are related by

$$g_1(x, v, t) = \int\int dk \, d\omega \, e^{i(kx - \omega t)} \tilde{g}_1(k, v, \omega) \qquad (9.2.8)$$

and

$$\tilde{g}_1(k, v, \omega) = (2\pi)^{-2} \int\int dx \, dt \, e^{-i(kx - \omega t)} g_1(x, v, t). \qquad (9.2.9)$$

With this notation, (9.2.6) and (9.2.7) become

$$-i\omega\tilde{g}_1 + ivk\tilde{g}_1 - \frac{e}{m}\tilde{E}_1\frac{dg_0}{dv} = 0 \qquad (9.2.10)$$

and

$$ik\tilde{E}_1 = -4\pi n_0 e \int dv \, \tilde{g}_1. \qquad (9.2.11)$$

Except for values of k and ω for which $\omega = vk$ for v in the range of values for which $g_0 \neq 0$, we may obtain from (9.2.10) the following expression for $\tilde{g}_1$:

$$\tilde{g}_1 = \frac{i\frac{e}{m}\tilde{E}_1\frac{dg_0}{dv}}{\omega - vk}. \qquad (9.2.12)$$

On substituting this expression in (9.2.11), we obtain

$$\left[1 - \frac{\omega_p^2}{k^2}\int dv \frac{\frac{dg_0}{dv}}{v - \frac{\omega}{k}}\right]\tilde{E}_1 = 0. \qquad (9.2.13)$$

Hence we find, as the condition that $\tilde{E}_1$ be nonzero, the following dispersion relation

$$\varepsilon(k, \omega) \equiv 1 - \frac{\omega_p^2}{k^2}\int dv \frac{\frac{dg_0}{dv}}{v - \frac{\omega}{k}} = 0. \qquad (9.2.14)$$

Since, in writing (9.2.12), we have already assumed that $\omega - vk$ does not vanish (except perhaps where $dg_0/dv = 0$), we may integrate by parts to obtain, in place of (9.2.14), the alternative form of the dispersion relation

$$\varepsilon(k, \omega) \equiv 1 - \frac{\omega_p^2}{k^2}\int \frac{g_0 dv}{\left(v - \frac{\omega}{k}\right)^2} = 0. \qquad (9.2.15)$$

The term on the left-hand side of (9.2.14). or (9.2.15) may be shown to be the dielectric function, that we now write as $\varepsilon(k, \omega)$, that effectively contains all the physics of the problem.

It is obvious that if we replace g_0 by $\delta(v)$, so that we once again consider a cold plasma, the dispersion relation reduces to

$$1 - \frac{\omega_p^2}{\omega^2} = 0 \qquad (9.2.16)$$

as we would expect. Hence our new dispersion relation describes the generalization of plasma oscillations for the case that the plasma must be described by a distribution function rather than as a collection of particle streams of zero temperature.

9.3 Dispersion relation for a warm plasma

Let us examine the form of the dispersion relation for the case that g_0 has the form sketched in Fig. 9.1, for which

$$g_0 = 0 \quad \text{for} \quad |v| > v_0. \qquad (9.3.1)$$

Then, for $|\omega/k| \gg v_0$, we may expand the denominator in the integral (9.2.15) and so obtain

$$1 - \frac{\omega_p^2}{\omega^2} \int g_0 \, dv \left[1 + \frac{2vk}{\omega} + \frac{3v^2k^2}{\omega^2} + \ldots \right] = 0. \qquad (9.3.2)$$

If we consider the case that g_0 is symmetric in v, the second term in the integral vanishes, and we obtain

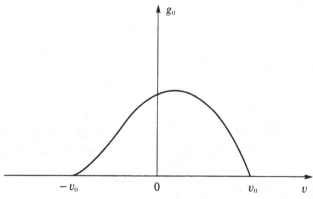

Fig. 9.1. The normalized distribution function $g_0(v)$ is zero for $|v| > v_0$.

$$1 - \frac{\omega_p^2}{\omega^2}\left[1 + \frac{3k^2}{\omega^2}\langle v^2 \rangle\right] = 0. \qquad (9.3.3)$$

In the limit that $v^2 k^2 \ll \omega^2$, the dispersion relation becomes

$$\omega^2 = \omega_p^2 + 3\langle v^2 \rangle k^2. \qquad (9.3.4)$$

This is referred to as the dispersion relation for Langmuir waves in a 'warm' plasma.

If the electron distribution is Maxwellian, the above derivation is approximately correct. On using the relation

$$\tfrac{1}{2} m_e \langle v^2 \rangle = \tfrac{1}{2} k_B T, \qquad (9.3.5)$$

where k_B is Boltzmann's constant, we obtain

$$\omega^2 = \omega_p^2 + 3\frac{k_B T}{m_e} k^2. \qquad (9.3.6)$$

In deriving this simple result, we have assumed that we are considering electrostatic waves with phase velocities much higher than the thermal velocities of the electrons so that none of the electrons is in resonance with the wave. In the next section, we investigate the changes that occur when this restriction is relaxed.

9.4 The Landau initial-value problem

The dispersion relation (9.3.6) gives two values of the frequency for each value of k, the same as if we were dealing with a cold single-stream plasma. On the other hand, we found in Section 8.2 that a two-stream situation gave four values of ω for each value of k, and we would have found that n streams give $2n$ modes. Hence, since we are now dealing with a distribution of electron velocities, that is to say an infinite number of streams, we should expect there to be an infinite number of modes.

The Case–Van-Kampen approach to this problem does indeed yield a solution of the equations of Section 9.2 for each pair of real values of ω and k. As a result, any initial disturbance with a single value of k excites a range of values of frequency, and these components will in the course of time get out of phase, leading to damping of the original disturbance, as measured for instance by the electric field E_1.

However, there is another approach to this problem that was originally developed by Landau (1946), that treats the plasma behavior as an initial-value problem. As we shall see, this leads naturally to complex values of the frequency or damping. When it represents damping, it is equivalent to the phase-mixing procedure that arises in the Case–Van-Kampen approach.

Once again, we restrict our attention to one-dimensional electrostatic oscillations of the electron component of a plasma. We assume that the ions may be regarded as infinitely massive so that their motion may be neglected, that the unperturbed state of the plasma is homogeneous and time-independent, that there are no electric or magnetic fields in the unperturbed state, and that collisions may be neglected. Hence g_0 is a function of v only.

We suppose that at time $t=0$, the initial value of $g_1(x, v, t)$ is $g_{10}(x, v)$. Then the evolution of the system is determined by the Vlasov equation

$$\frac{\partial g_1}{\partial t} + v\frac{\partial g_1}{\partial x} - \frac{e}{m_e}E_1\frac{dg_0}{dv} = 0, \quad t>0 \tag{9.4.1}$$

subject to the initial condition

$$g_1(x, v, 0) = g_{10}(x, v). \tag{9.4.2}$$

As before, the electric field is related to the perturbation of the distribution function by

$$\frac{\partial E_1}{\partial x} = -4\pi n_0 e\int dv\, g_1. \tag{9.4.3}$$

The above initial-value problem may be solved by the Laplace-transform procedure. However, an equivalent procedure, that is more closely related to the techniques we have previously introduced, is to use Fourier-transform analysis. In order to use Fourier transforms, it is convenient to modify the problem so that we can consider the complete range of t, $-\infty < t < \infty$, rather than the restricted range denoted in (9.4.1). This may be achieved by replacing (9.4.1) with the equation

$$\frac{\partial g_1}{\partial t} + v\frac{\partial g_1}{\partial x} - \frac{e}{m_e}E_1\frac{dg_0}{dv} = g_{10}\delta(t) \tag{9.4.4}$$

with the understanding that

$$g_1 = 0, \quad E_1 = 0, \quad \text{for} \quad t<0. \tag{9.4.5}$$

This is equivalent to introducing an impulsive 'forcing term' that suddenly changes a plasma from a quiescent state into the initial state given by (9.4.2).

Since the unperturbed state is homogeneous and steady, and since we are now considering a complete range of both x and t, we may now introduce the Fourier transforms defined by (9.2.8) and (9.2.9). Then (9.4.4) and (9.4.3) become

$$-\mathrm{i}\omega\tilde{g}_1 + \mathrm{i}vk\tilde{g}_1 - \frac{e}{m_e}\tilde{E}_1\frac{\mathrm{d}g_0}{\mathrm{d}v} = \frac{1}{2\pi}\tilde{g}_{10} \qquad (9.4.6)$$

and

$$\mathrm{i}k\tilde{E}_1 = -4\pi n_0 e \int \mathrm{d}v\,\tilde{g}_1. \qquad (9.4.7)$$

Since g_{10} is independent of t, we are of course using the notation

$$g_{10}(x, v) = \int \mathrm{d}k\,\mathrm{e}^{\mathrm{i}kx}\tilde{g}_{10}(k, v), \quad \tilde{g}_{10}(k, v) = (2\pi)^{-1}\int \mathrm{d}x\,\mathrm{e}^{-\mathrm{i}kx}g_{10}(x, v).$$

$$(9.4.8)$$

We note from (9.2.9) that the integral defining $\tilde{g}_1(k, v, \omega)$ and the corresponding integral defining $\tilde{E}_1(k, \omega)$ are well defined for $\omega_\mathrm{i} > 0$ since the integrand is well behaved for $t > 0$ and the integrand is zero if $t < 0$. Hence the quantities $\tilde{g}_1(k, v, \omega)$ and $\tilde{E}_1(k, \omega)$ are well defined and analytic in the upper half of the complex ω-plane. If these functions are known in the upper-half plane, they may be determined in the lower-half plane by analytic continuation.

On using (9.4.7), we may re-write (9.4.6) in the form

$$\tilde{g}_1(v) = \frac{\omega_\mathrm{p}^2}{k^2\left[v - \dfrac{\omega}{k}\right]}\,g_0'(v)\int \mathrm{d}v'\tilde{g}_1(v') + \frac{1}{2\pi\mathrm{i}k}\frac{\tilde{g}_{10}(v)}{v - \dfrac{\omega}{k}} \qquad (9.4.9)$$

where, for present purposes, it is convenient to show explicitly the argument v, and it is therefore convenient to write the derivative of $g_0(v)$ as $g_0'(v)$. On integrating over velocity, we obtain

$$\int \mathrm{d}v\,\tilde{g}_1(v) = -\frac{\mathrm{i}}{2\pi k\varepsilon(k, \omega)}\int_{\Gamma_v}\frac{g_{10}\mathrm{d}v}{v - \dfrac{\omega}{k}}, \qquad (9.4.10)$$

where the 'dielectric function' $\varepsilon(k, \omega)$ is defined by (9.2.14) and Γ_v is a contour, yet to be determined, in the complex v-plane.

Once (9.4.10) has been evaluated, we can determine $\tilde{E}_1$ from (9.4.7) and $\tilde{g}_1$ from (9.4.9). We can then determine the evolution in time from the equations

$$g_1(k, v, t) = \int_{\Gamma_\omega}\mathrm{d}\omega\,\mathrm{e}^{-\mathrm{i}\omega t}\tilde{g}_1(k, v, \omega) \qquad (9.4.11)$$

and

$$E_1(k, t) = \int_{\Gamma_\omega}\mathrm{d}\omega\,\mathrm{e}^{-\mathrm{i}\omega t}\tilde{E}_1(k, \omega) \qquad (9.4.12)$$

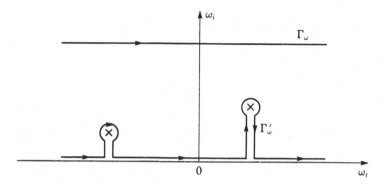

Fig. 9.2. Causality considerations indicate that the line denoted by Γ_ω is an acceptable contour of integration in (9.4.11) and (9.4.12). This contour may be moved lower in the complex plane, provided it does not cross any singularities, for instance to yield the contour Γ'_ω. Singularities in the ω_r–ω_i plane are denoted by Xs.

where the contour Γ_ω in the complex ω-plane is yet to be determined. For convenience, we consider only a wave with a given value of wave number k.

An appropriate form for the contour Γ_ω can be identified from the causal requirements (9.4.5) for g_1 and E_1. If the contour Γ_ω lies above all singularities, then it can be displaced in the direction $\omega = +i\infty$. Then clearly (9.4.11) and (9.4.12) yield $g_1 = 0$ and $E_1 = 0$ for $t < 0$. An equivalent contour is one that runs along the real ω axis except for indentations that run above all singularities in the upper-half complex ω-plane. Such a contour is shown in Fig. 9.2. This contour may be deformed into the contour shown in Fig. 9.3 so that (9.4.11) becomes

$$g_1(k, v, t) = \lim\left\{\int_A^B + \int_B^C + \int_C^D\right\} d\omega\, e^{-i\omega t}\tilde{g}_1(k, v, \omega). \quad (9.4.13)$$

In the limit that the rectangle $ABCD$ becomes arbitrarily large, the contributions along the segments AB and CD are zero since the integrand oscillates with an arbitrarily high frequency. The contribution from the element

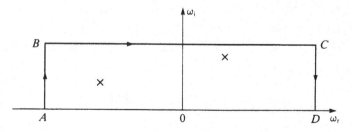

Fig. 9.3. The contour Γ_ω is here deformed into the contour $ABCD$, where the sections AB and CD are supposed to be located at infinite distances left and right.

$BC \rightarrow 0$ because the integrand now involves the factor $e^{\omega_i t}$ where $t < 0$ and $\omega_i \rightarrow \infty$. The same choice of the contour shows that $E_1(k, t)$ is zero for $t < 0$. The same result follows from (9.4.7).

For $t > 0$, one can evaluate the integrals by deforming the contour as shown in Fig. 9.4. We now see that the only nonzero contributions to the integral (9.4.11) arise from the poles in the complex ω-plane.

In order to change the path of integration from Γ_ω or Γ'_ω as in Fig. 9.2 and 9.3 to Γ''_ω as in Fig. 9.4, it is necessary that the function ε, that enters (9.4.10), should be extended analytically from the upper half plane to the lower half plane. We assume that k is real and now assume also that it is positive. We also assume that $g_0(v)$ is well behaved so that $g'_0(v)$ has no poles on the real v-axis. The path of integration in the integral in (9.2.14) was originally the real v-axis. The integral is then well defined for $\omega_i > 0$, but it cannot be used to evaluate $\varepsilon(k, \omega)$ if ω is real. However, we may extend the definition of $\varepsilon(k, \omega)$ to real values of ω and into the lower half ω-plane by the process of analytic continuation. This is achieved by beginning with the choice of the contour Γ_v as the real axis for the case that $\omega_i > 0$ and then deforming the contour of integration in the v-plane, as ω_i takes on negative values, so that no singularity is allowed to cross the contour in the v-plane.

It is convenient to consider separately the three cases that arise:

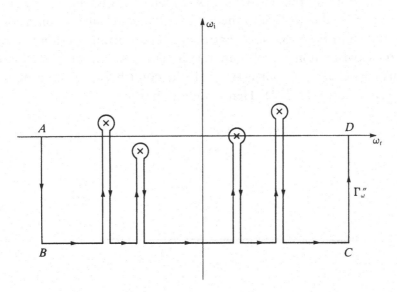

Fig. 9.4. For $t > 0$, we may adopt the contour Γ''_ω, where the sections AB and CD are supposed to move to be located at infinite distances left and right, and the horizontal section BC is supposed to be located at infinite negative values of ω_i.

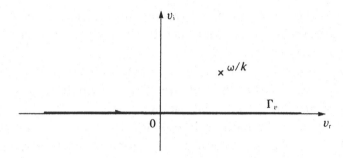

Fig. 9.5. The appropriate contour of integration in the v_r-v_i plane for case (a), that $\omega_i > 0$. The contour Γ_v runs along the v_r-axis.

(a) $\omega_i > 0$. This case has been discussed earlier in this section. We are assuming that k is real and positive. The integration may now be performed with the original path of integration Γ_v that runs along the real v-axis as shown in Fig. 9.5. The resulting formula for $\varepsilon(k, \omega)$ is

$$\varepsilon(k, \omega) = 1 - \frac{\omega_p^2}{k^2} \int_{-\infty}^{\infty} \frac{g_0' \, dv}{v - \dfrac{\omega}{k}}, \quad \text{for } \omega_i > 0. \tag{9.4.14}$$

(b) $\omega_i = 0$. In this case, a pole at $v = \omega/k$ appears on the real v-axis. Since we need to continue the function $\varepsilon(k, \omega)$ analytically, it is necessary to deform the original path Γ_v into the path Γ_v' that continues to run below the pole, as shown in Fig. 9.6. Note that the direction of integration is positive (counterclockwise) along the semi-circle indentation, hence a contribution from this part of the contour is one half the contribution from a path that completely encloses the pole. Hence we see that

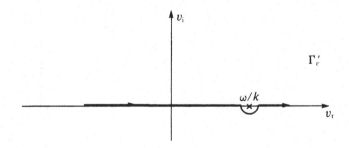

Fig. 9.6. The appropriate contour of integration in the v_r-v_i plane for case (b), that $\omega_i = 0$. The contour Γ_v' runs along the v_r-axis, except for a semicircular detour to run below the singularity at $v = \omega/k$.

$$\varepsilon(k,\omega) = 1 - \frac{\omega_p^2}{k^2}\, P \int \frac{g_0'\, dv}{v - \frac{\omega}{k}} - i\pi \frac{\omega_p^2}{k^2}\, g_0'(\omega/k), \quad \text{for } \omega_i > 0.$$

$$(9.4.15)$$

In this expression, the integral refers to the Cauchy principal value, defined by

$$P \int f(v)\, dv = \lim_{\delta \to 0} \left[\int_{-\infty}^{v_0 - \delta} f(v)\, dv + \int_{v_0 + \delta}^{\infty} f(v)\, dv \right], \quad (9.4.16)$$

where $v = v_0$ is the location of the singularity on the real v-axis.

(c) $\omega_i < 0$. As ω_i takes on negative values, the singularity is depressed into the lower half v-plane. Hence it is now necessary to adopt the contour Γ_v'' shown in Fig. 9.7. The deformed contour now completely encircles the pole so that we obtain a full contribution from the circular integration around the pole. Hence we now obtain

$$\varepsilon(k,\omega) = 1 - \frac{\omega_p^2}{k^2} \int \frac{g_0'\, dv}{v - \frac{\omega}{k}} - 2i\pi \frac{\omega_p^2}{k^2}\, g_0'(\omega/k), \quad \text{for } \omega_i < 0.$$

$$(9.4.17)$$

We may now find the dispersion relation for these three cases simply by setting $\varepsilon(k,\omega) = 0$:

$$\varepsilon(k,\omega) \equiv 1 - \frac{\omega_p^2}{k^2} \int \frac{g_0'\, dv}{v - \frac{\omega}{k}} = 0, \quad \text{for } \omega_i > 0, \qquad (9.4.18)$$

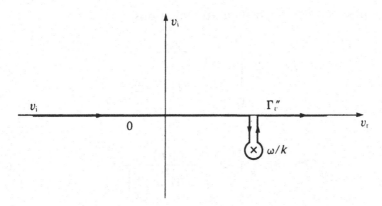

Fig. 9.7. The appropriate contour of integration in the v_r-v_i plane for case (c), that $\omega_i < 0$. The contour Γ_v'' runs along the v_r-axis, except for a detour to encircle the singularity at $v = \omega/k$.

$$\varepsilon(k,\omega) \equiv 1 - \frac{\omega_p^2}{k^2} P \int \frac{g_0' \, dv}{v - \frac{\omega}{k}} - i\pi \frac{\omega_p^2}{k^2} g_0'(\omega/k) = 0, \quad \text{for } \omega_i = 0,$$

(9.4.19)

and

$$\varepsilon(k,\omega) \equiv 1 - \frac{\omega_p^2}{k^2} \int \frac{g_0' \, dv}{v - \frac{\omega}{k}} - 2i\pi \frac{\omega_p^2}{k^2} g_0'(\omega/k) = 0, \quad \text{for } \omega_i < 0.$$

(9.4.20)

If ω lies in the upper-half plane, the wave grows with time and the mode is unstable. If ω lies on the real ω-axis, the wave has a steady amplitude and the mode is stable. If ω lies in the lower-half ω-plane, the wave decays with time, so that the mode is once again stable.

9.5 Gardner's theorem

We can now prove Gardner's theorem that 'a single-hump distribution cannot support growing waves and is therefore stable.' We may prove this by assuming that the contrary is true, that there is a growing wave with $\omega_i > 0$. In terms of ω_r and ω_i, we may write (9.4.18) explicitly as

$$1 - \frac{\omega_p^2}{k^2} \int dv \frac{\left(v - \frac{\omega_r}{k} + i\frac{\omega_i}{k}\right) g_0'(v)}{\left(v - \frac{\omega_r}{k}\right)^2 + \left(\frac{\omega_i}{k}\right)^2} = 0.$$

(9.5.1)

This therefore yields the two separate equations

$$\varepsilon_r \equiv 1 - \frac{\omega_p^2}{k^2} \int dv \frac{\left(v - \frac{\omega_r}{k}\right) g_0'(v)}{\left(v - \frac{\omega_r}{k}\right)^2 + \left(\frac{\omega_i}{k}\right)^2} = 0$$

(9.5.2)

and

$$\varepsilon_i \equiv -\frac{\omega_p^2}{k^2} \frac{\omega_i}{k} \int dv \frac{g_0'(v)}{\left(v - \frac{\omega_r}{k}\right)^2 + \left(\frac{\omega_i}{k}\right)^2} = 0.$$

(9.5.3)

We now assume that $g_0(v)$ is single-humped with a maximum at $v = v_0$, as shown in Fig. 9.8. Then

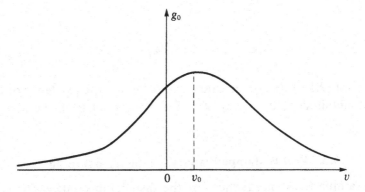

Fig. 9.8. The distribution function $g_0(v)$ has a single maximum at $v=v_0$.

$$g_0'(v)>0 \quad \text{for} \quad v<v_0, \quad g_0'(v)<0 \quad \text{for} \quad v>v_0, \qquad (9.5.4)$$

so that

$$(v_0-v)g_0'(v)\geq 0 \quad \text{for} \quad -\infty<v<\infty. \qquad (9.5.5)$$

Since $\varepsilon_r=0$ and $\varepsilon_i=0$, clearly

$$\varepsilon_r+\left(\frac{v_0 k-\omega_r}{\omega_i}\right)\varepsilon_i=0. \qquad (9.5.6)$$

On using (9.5.2) and (9.5.3), we see that this becomes

$$1+\frac{\omega_p^2}{k^2}\int dv\,\frac{(v_0-v)g_0'(v)}{\left(v-\frac{\omega_r}{k}\right)^2+\left(\frac{\omega_i}{k}\right)^2}=0. \qquad (9.5.7)$$

However, we see from (9.5.5) that the integrand is non-negative for all values of v. Hence the left-hand side of the (9.5.7) is greater than unity so that the equation cannot be satisfied. This proves that our original assumption must be incorrect. That is to say, a single-humped distribution cannot support a growing wave. Hence a single-humped distribution is necessarily stable.

We may also examine the condition on the distribution function $g_0(v)$ that must be satisfied in order that the system may support a wave of constant amplitude. In this case, it is clear from (9.4.19) that we must satisfy both of the following equations:

$$\varepsilon_r\equiv 1-\frac{\omega_p^2}{k^2}\,P\int\frac{dv g_0'(v)}{v-\frac{\omega}{k}}=0 \qquad (9.5.8)$$

and

$$\varepsilon_{i} \equiv -\pi \frac{\omega_{p}^{2}}{k^{2}} g_{0}'(\omega/k) = 0. \tag{9.5.9}$$

It is clear from (9.5.9) that a necessary condition for the existence of a wave of constant amplitude is that $g_{0}'(v) = 0$ at the phase velocity of the wave.

9.6 Weakly damped waves – Landau damping

We now examine the consequences of the dispersion relation for the case that $\omega_{i} \ll \omega_{r}$. We may now express (9.2.14) as

$$\varepsilon \equiv 1 - \frac{\omega_{p}^{2}}{k^{2}} \int_{\Gamma_{v}} dv \frac{g_{0}'(v)}{v - \frac{\omega_{r}}{k}} \left[1 - i \frac{\frac{\omega_{i}}{k}}{v - \frac{\omega_{r}}{k}} \right]^{-1} = 0. \tag{9.6.1}$$

On expanding to lowest order in ω_{i}/k, we obtain

$$\varepsilon \equiv 1 - \frac{\omega_{p}^{2}}{k^{2}} \int dv \frac{g_{0}'(v)}{v - \frac{\omega_{r}}{k}} \left[1 + i \frac{\frac{\omega_{i}}{k}}{v - \frac{\omega_{r}}{k}} \right] = 0. \tag{9.6.2}$$

On integrating by parts, this may be re-expressed as

$$\varepsilon \equiv 1 - \frac{\omega_{p}^{2}}{k^{2}} \int dv \frac{g_{0}'(v)}{v - \frac{\omega_{r}}{k}} - i \frac{\omega_{p}^{2}}{k^{2}} \frac{\omega_{i}}{k} \int dv \frac{g_{0}''(v)}{v - \frac{\omega_{r}}{k}} = 0. \tag{9.6.3}$$

In these integrals, the pole is on the real axis so that it is necessary to adopt the contour Γ_{v}' indicated in Fig. 9.6. Hence, we obtain from (9.6.3) the following two equations:

$$\varepsilon_{r} \equiv 1 - \frac{\omega_{p}^{2}}{k^{2}} P \int dv \frac{g_{0}'(v)}{v - \frac{\omega_{r}}{k}} + \pi \frac{\omega_{p}^{2}}{k^{2}} \frac{\omega_{i}}{k} g_{0}''(\omega_{r}/k) = 0 \tag{9.6.4}$$

and

$$\varepsilon_{i} \equiv -\pi \frac{\omega_{p}^{2}}{k^{2}} g_{0}'(\omega_{r}/k) - \frac{\omega_{p}^{2}}{k^{2}} \frac{\omega_{i}}{k} P \int dv \frac{g_{0}''(v)}{v - \frac{\omega_{r}}{k}} = 0. \tag{9.6.5}$$

We see from (9.6.5) that, to first order, ω_{i} is given by

$$\omega_i \equiv -\frac{\pi k g_0'(\omega_r/k)}{P \displaystyle\int dv \, \frac{g_0''(v)}{v - \dfrac{\omega_r}{k}}} \qquad (9.6.6)$$

We find that it is possible to evaluate the integral in this expression.

To lowest order in the quantity ω_i/ω_r, (9.6.4) reduces to a dispersion relation between ω_r and k:

$$\varepsilon_r \equiv 1 - \frac{\omega_p^2}{k^2} P \int dv \, \frac{g_0'(v)}{v - \dfrac{\omega_r}{k}} = 0. \qquad (9.6.7)$$

A variation δk in k will lead a variation $\delta\omega_r$ in ω_r, the two increments being related by

$$\frac{2\omega_p^2}{k^2} \delta k \, P \int dv \, \frac{g_0'(v)}{v - \dfrac{\omega_r}{k}} - \frac{\omega_p^2}{k^2} P \int dv \, \frac{g_0'(v)}{\left(v - \dfrac{\omega_r}{k}\right)^2} \left\{ \frac{\delta\omega_r}{k} - \frac{\omega_r \delta k}{k^2} \right\}. \qquad (9.6.8)$$

On multiplying through by $k/\delta k$, using (9.6.7), and noting that, $\delta\omega_r$ and δk are related by the group velocity

$$\frac{\delta\omega_r}{\delta k} = u \equiv \frac{d\omega_r}{dk}, \qquad (9.6.9)$$

we obtain

$$2 - \frac{\omega_p^2}{k^2} \left(u - \frac{\omega_r}{k} \right) P \int dv \, \frac{g_0'(v)}{\left(v - \dfrac{\omega_r}{k}\right)^2} = 0. \qquad (9.6.10)$$

We may now integrate by parts to obtain

$$2 - \frac{\omega_p^2}{k^2} \left(u - \frac{\omega_r}{k} \right) P \int dv \, \frac{g_0''(v)}{\left(v - \dfrac{\omega_r}{k}\right)} = 0. \qquad (9.6.11)$$

Hence we find that the integral that appears in (9.6.6) has the value

$$P \int dv \, \frac{g_0''(v)}{\left(v - \dfrac{\omega_r}{k}\right)} = -\frac{2k^3}{\omega_p^2 (\omega_r - uk)}. \qquad (9.6.12)$$

On substituting this value into (9.6.6), we finally obtain the expression

$$\omega_i = \frac{\pi}{2} \frac{\omega_p^2}{k^2} (\omega_r - uk) g_0'(\omega_r/k). \qquad (9.6.13)$$

On examining the dispersion relation (9.6.7) for typical distributions, we find that the group velocity is less than the phase velocity, i.e. $u < \omega_r/k$. Hence we see from (9.6.13) that $\omega_i < 0$, that is the wave is damped, if $g_0'(\omega_r/k) < 0$. This is known as 'Landau damping.'

In the case that we are dealing with a Maxwellian distribution, and that we consider waves of phase speed much larger than the mean thermal speed, we may use the approximate dispersion relation (9.3.4). We see from this that

$$u = 3 \frac{\langle v^2 \rangle k}{\omega_r}, \qquad (9.6.14)$$

so that

$$\omega_r - uk = \omega_r - \frac{3\langle v^2 \rangle k^2}{\omega_r} = \frac{\omega_p^2}{\omega_r}. \qquad (9.6.15)$$

This confirms that $\omega_r/k > u$. Hence the expression (9.6.13) becomes

$$\omega_i = \frac{3\pi}{2} \frac{\omega_p^4 \langle v^2 \rangle}{\omega_r(\omega_r^2 - \omega_p^2)} g_0'(\omega_r/k). \qquad (9.6.16)$$

In order to examine the significance of this equation, we remember that we are considering a wave for which $k > 0$. Suppose that g_0 is single-humped, with a maximum value at $v = 0$. Then the relationship between ω_r and k is given by (9.3.4) and leads to two equal and opposite values of ω_r. If we adopt the positive value of ω_r, then the phase velocity is positive and the sign ω_i is the same as the sign of g_0' at that phase velocity. However, for positive v, $g_0' < 0$. Hence $\omega_i < 0$ and the process represented by equation (9.6.16) represents (Landau) damping.

If there is a slight bump on the tail of the distribution, as shown in Fig. 9.9, then the relationship between ω_r and k is not greatly affected. However, for waves of the appropriate phase velocity, $g_0 > 0$. Hence such distributions would exhibit two-stream instability in the range of phase velocities for which $g_0' > 0$. We note that these results are all in accordance with Gardner's theorem discussed in Section 9.4.

9.7 The Penrose criterion for stability

Gardner's theorem shows that it is necessary that a distribution have at least two peaks for the two-stream instability to occur. The analysis of the previous

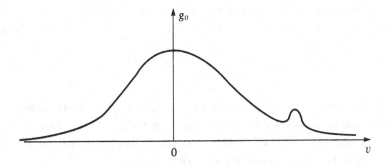

Fig. 9.9. The case that there is a slight bump on the tail of the distribution function.

section is applicable to the case that one of the peaks is on the tail of the principal distribution. In this section, we will develop a more robust criterion for the stability of a plasma against two-stream instability. This is due to work by Penrose (1960) based on earlier work Nyquist (1932).

The wave properties of the system we are discussing are determined by the dispersion relation (9.2.14), that is by

$$\varepsilon(k, \omega) \equiv 1 - \frac{\omega_p^2}{k^2} \int dv \frac{g_0'(v)}{v - \frac{\omega}{k}} = 0. \tag{9.7.1}$$

The problem is to determine the location in the complex ω-plane of the zeros of the function $\varepsilon(k, \omega)$ for given real values of k. Nyquist's theorem states that the number of zeros of the function ε within a closed contour Γ_ω is given by

$$N = \frac{1}{2\pi i} \int_{\Gamma_\omega} \frac{1}{\varepsilon} \frac{\partial \varepsilon}{\partial \omega} d\omega, \tag{9.7.2}$$

where the path of integration is in the counterclockwise direction, and we assume that ε has no poles in the region indicated. For instance, we may consider a small contour embracing a zero at $\omega = \omega_0$. Then we may expand ε in the neighborhood of ω_0 as follows:

$$\varepsilon = \left(\frac{\partial \varepsilon}{\partial \omega}\right)_{\omega_0} (\omega - \omega_0) + \dots, \tag{9.7.3}$$

so that

$$\frac{\partial \varepsilon}{\partial \omega} = \left(\frac{\partial \varepsilon}{\partial \omega}\right)_{\omega_0} + \left(\frac{\partial^2 \varepsilon}{\partial \omega^2}\right)_{\omega_0} (\omega - \omega_0) + \dots. \tag{9.7.4}$$

On substituting these expressions in (9.7.2) and contracting the contour to a small circle enclosing $\omega = \omega_0$, we see that, provided $(\partial \varepsilon / \partial \omega)_{\omega_0} \neq 0$,

$$N = \frac{1}{2\pi i} \int_{\Gamma_\omega} \frac{d\omega}{\omega - \omega_0} = 1. \tag{9.7.5}$$

Since we are interested in determining whether there are unstable modes of a system, we are interested only in positive values of ω_i. Hence we can determine whether the system is stable or unstable by determining the number of roots of (9.7.1) in the upper half of the ω-plane.

This quantity, that we write as N_u, is given by (9.7.2) if the contour Γ_ω now comprises the real axis and a semi-circle in the upper-half plane of infinite radius, as indicated schematically in Fig. 9.10.

The contour Γ_ω in the ω-plane may alternatively be mapped into a contour Γ_ε in the ε-plane. If there is one root of the equation $\varepsilon = 0$ in the upper-half plane, then Γ_ε will encircle the origin once, as indicated in Fig. 9.11(a). If there are no zeros of $\varepsilon = 0$ in the upper half plane, then the contour Γ_ε will not encircle the origin, as indicated in Fig. 9.11(b), and the system is stable. If the contour embraces three zeros of $\varepsilon = 0$, as indicated in Fig. 9.11(c), then the contour Γ_ε will encircle the origin three times.

Since the contour Γ_ω shown schematically in Fig. 9.10 is essentially an integration along the real ω-axis, we may use (9.4.19), that is:

$$\varepsilon(k, \omega) \equiv 1 - \frac{\omega_p^2}{k^2} P \int \frac{g_0'(v) dv}{v - \frac{\omega}{k}} - i\pi \frac{\omega_p^2}{k^2} g_0'(\omega/k). \tag{9.7.6}$$

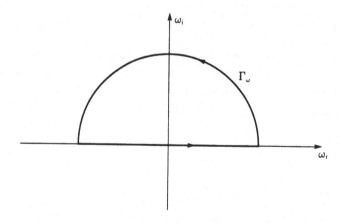

Fig. 9.10. A schematic representation of the contour Γ_ω comprising the real axis and a semicircle of infinite radius in the upper half plane.

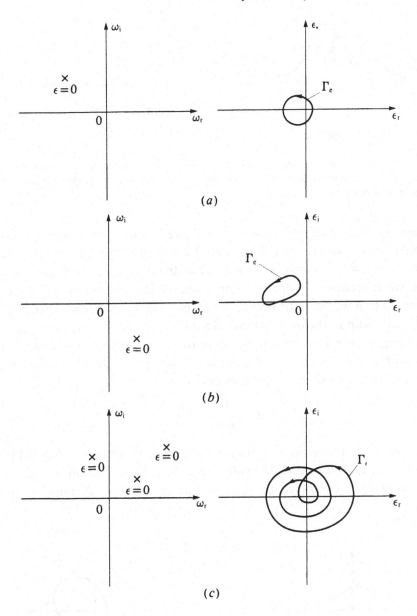

Fig. 9.11. In case (*a*), there is one root of $\varepsilon=0$ in the upper half plane; then Γ_ε encircles the origin once. In case (*b*), there are no roots of $\varepsilon=0$ in the upper half plane; then Γ_ε will not encircle the origin. In case (*c*), there are three roots of $\varepsilon=0$ in the upper half plane; then Γ_ε will encircle the origin three times.

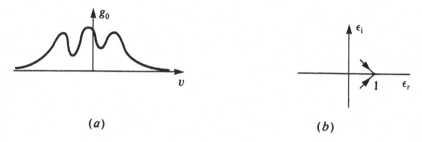

(a)   (b)

Fig. 9.12. If, as we expect, g_0 decreases to zero as $v \to \pm\infty$, $\varepsilon(k, \omega)$ approaches the point $\varepsilon = 1$ from above as $\omega/k \to \infty$, and leaves the point $\varepsilon = 1$ in the downward direction as ω/k increases from $-\infty$.

When $\omega = \infty$, the second term in the above equation becomes zero, and the third term also vanishes if g_0 is well behaved so that $g_0' \to 0$ as $v \to \infty$. Hence $\varepsilon(k, \infty) = 1$. For any well behaved distribution, $g_0'(v) < 0$ as $v \to +\infty$, so that we approach $\varepsilon = 1$ from the upper-half plane. Similarly, $g_0'(v) > 0$ as $v \to -\infty$, so that we leave the point $\varepsilon = 1$ by moving into the lower-half plane. This relationship is shown schematically in Fig. 9.12.

The simplest form of the contour Γ_ε in the ε-plane would seem to be that shown in Fig. 9.13(a), that embraces the origin once in the clockwise sense. However, (9.7.2) leads to the equivalent formula

$$N = \frac{1}{2\pi i} \int_{\Gamma_\varepsilon} \frac{d\varepsilon}{\varepsilon} \qquad (9.7.7)$$

and we see from this equation that, for the contour shown in Fig. 9.13(a), $N = -1$. This is an unacceptable value since N is, by definition, the number of zeros of $\varepsilon = 0$ embraced by the contour Γ_ω, so that N must be non-negative.

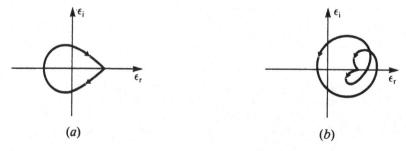

(a)   (b)

Fig. 9.13. The simplest contour compatible with Fig. 9.12 that embraces the origin (a) is found to be unacceptable since it circles the origin in the clockwise direction. The contour shown as (b) is acceptable since it is compatible with Fig. 9.12 and embraces the origin in the anti-clockwise direction.

A contour that is consistent with Fig. 9.12(*b*) but embraces the origin $\varepsilon = 0$ once in the anti-clockwise direction is that shown in Fig. 9.13(*b*). This leads to $N = +1$.

On inspecting Fig. 9.13(*b*), we see that the number of times that the contour crosses the real axis restricts the possible values of N. If the contour Γ_ε crosses the real axis only once, as shown in 9.13(*a*), then the system will be stable. In order for there to be instability, the contour must cross the real ε-axis at least three times. The crossing of the real axis corresponds to $\varepsilon_i = 0$. From (9.7.6), we see that this corresponds to a zero of $g_0'(\omega/k)$. Hence we have once again confirmed Gardner's theorem: if $g_0'(v) = 0$ for only one value of v, the system is stable.

However, the number of values of ω for which $\varepsilon_i = 0$ is not a sufficient criterion for instability. Let us now consider a two-humped velocity distribution shown schematically in Fig. 9.14. There are two maxima of g_0 at v_1 and v_2 and one minimum at v_0. For definiteness, we assume that $g_0(v_1) > g_0(v_2)$.

There are two possible forms of the contour Γ_ε, consistent with the restrictions noted above, and these are shown in Fig. 9.15. In case (*a*), the contour does not embrace the origin, and in case (*b*) it embraces the origin once. Hence case (*a*) corresponds to a stable situation, and case (*b*) corresponds to an unstable situation. We can distinguish between these two cases by determining the sign of $\varepsilon_r(k, \omega)$ at $\omega = v_0 k$, that is by determining the value of $\varepsilon_r(k, v_0 k)$. We know that $\varepsilon_i(k, v_0 k) = 0$ since $g_0'(v_0) = 0$. We see from (9.7.6) that

$$\varepsilon_r(k, v_0 k) = 1 - \frac{\omega_p^2}{k^2} \, \mathrm{P} \int \frac{g_0'(v)\,\mathrm{d}v}{v - v_0} . \qquad (9.7.8)$$

The condition that the system be unstable is that $\varepsilon_r(k, v_0 k) < 0$. This requires that

$$k^2 < k_c^2 \equiv \omega_p^2 \, \mathrm{P} \int \frac{g_0'(v)\,\mathrm{d}v}{v - v_0} . \qquad (9.7.9)$$

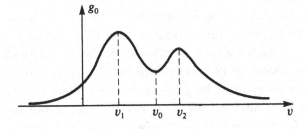

Fig. 9.14. A two-peak velocity distribution.

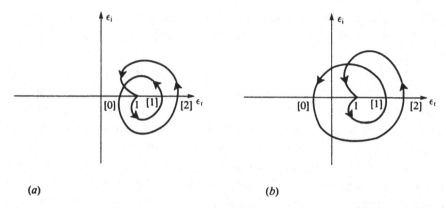

Fig. 9.15. Two possible Nyquist diagrams corresponding to the velocity distribution shown in Fig. 9.14. The points [0], [1] and [2] correspond to the conditions $\omega/k = v_0$, $\omega/k = v_1$ and $\omega/k = v_2$, respectively. In (a) the contour does not embrace the origin, so the system is stable. In (b) the contour embraces the origin, and the system is unstable.

Clearly, the condition that there should exist an unstable range of wave numbers is that the integral in (9.7.9) should be positive. Clearly, this condition can also be written as

$$\mathrm{P} \int dv \frac{\dfrac{d}{dv}(g_0(v) - g_0(v_0))}{v - v_0} > 0. \tag{9.7.10}$$

On integrating by parts, we see that this condition becomes

$$\mathrm{P} \int \frac{dv(g_0(v) - g_0(v_0))}{(v - v_0)^2} > 0. \tag{9.7.11}$$

This is called the 'Penrose criterion' for instability, and it is a necessary and sufficient condition for linear instability.

The significance of this criterion may be understood by referring to Fig. 9.16. In region B of this figure, $g_0(v) - g_0(v_0) > 0$, while in regions A and C, $g_0(v) - g_0(v_0) < 0$. Hence condition (9.7.11) requires that the contribution to the integral coming from region B should outweigh the sum of the contributions coming from regions A and C. If we consider the case that $g_0(v_1) = g_0(v_2)$, and if $g_0(v_0)$ is only slightly smaller than the maximum value of g_0, then condition (9.7.11) would clearly not be satisfied. In other words, the dip at $v = v_0$ must be deep enough for the Penrose criterion to indicate that an instability occurs. For the case considered in Section 9.5, that of a small bump on the tail of a Maxwellian distribution, we may verify from Fig. 9.16 that it is reasonable to expect the system to be unstable.

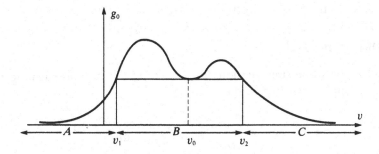

Fig. 9.16. The condition for instability of the two-peaked velocity distribution shown in Fig. 9.14 is that the contribution to the integral (9.7.11) from regions A and C should outweigh that from region B.

Problems

Problem 9.1. Consider a plasma for which (neglecting the role of the ions) the normalized distribution of electron velocity v_x has the form

$$g_0(v_x) = \begin{cases} 0, & \text{if } |v_x| > V, \\ 1/(2V), & \text{if } |v_x| < V \end{cases} \tag{1}$$

(a) Find the dispersion relation.
(b) Define an equivalent one-dimensional temperature by the equation

$$\tfrac{1}{2}k_B T = \tfrac{1}{2}m\langle v_x^2\rangle \tag{2}$$

and express the dispersion relation in terms of T rather than V. How does this compare with the dispersion relation of Problem 7.1?
(c) Can you find an isotropic three-dimensional velocity distribution $G_0(v_x, v_y, v_z)$ such that g_0 is the integral of G_0 over v_y and v_z?

Problem 9.2. By using Gardner's theorem, show that electrostatic modes are stable for any isotropic velocity distribution.

Problem 9.3. Consider a plasma comprising two streams of electrons, each with a Cauchy distribution, neutralized by infinitely massive ions. The normalized electron distribution function may be written as

$$g(v) = \frac{v_0^2}{2\pi}\left[\frac{1}{(v-v_1)^2 + v_0^2} + \frac{1}{(v+v_1)^2 + v_0^2}\right].$$

(a) Sketch the form of $g(v)$ for the three cases $v_0 \ll v_1$, $v_0 = v_1$ and $v_0 \gg v_1$.
(b) Show that, according to Gardner's theorem, the system should be stable if $(v_1/v_0) < 3^{-1/2}$.

Problem 9.4. Consider an electron–proton plasma with normalized three-dimensional velocity distribution functions $G_e(\mathbf{v})$ and $G_p(\mathbf{v})$ for the electrons and protons, respectively.

(a) Show that the dispersion relation for electrostatic waves propagating in the x-direction may be written in the form

$$1 + \frac{\omega_{p,e}^2}{k^2} \int \frac{dv_x}{\frac{\omega}{k} - v_x} \left\{ \frac{dg_e(v_x)}{dv_x} + \frac{m_e}{m_p} \frac{dg_p(v_x)}{dv_x} \right\} = 0 \tag{1}$$

where

$$g(v_x) = \int\int dv_y\, dv_z\, G(\mathbf{v}). \tag{2}$$

(b) Assume that the electrons and protons have Maxwellian distributions with temperatures T_e and T_p, respectively, and that $T_p \ll T_e$. Find the dispersion relation describing ion-acoustic waves in this system by examining the dispersion relation for waves with phase velocities v_ϕ larger than the ion thermal velocity $v_{th,p}$ and smaller than the electron thermal velocity $v_{th,e}$, and adopting the approximations that

$$v_\phi \ll v_{th,e} \text{ and } v_\phi \gg v_{th,p}.$$

10

Collision theory

We have already considered the Vlasov equation that describes the evolution of a distribution function when the effects of collisions may be ignored. In this chapter, we consider the effects of collisions on the distribution functions of the various species that comprise a plasma. If the plasma contained neutral particles, it would be necessary to consider the effects of large-angle collisions. This is normally carried out by a procedure due originally to Boltzmann. However, if neutral particles do not play a significant role in the plasma, the collision processes are due to the long-range Coulomb force. As we found in Chapter 2, Coulomb interactions give rise mainly to many small-angle collisions. The appropriate equation to use in this case is the Fokker–Planck equation, that we shall now introduce.

10.1 Lagrange expansion

The Fokker–Planck equation may be conveniently approached through a generalization of the Lagrange expansion (Sturrock, 1960a). We begin by considering a three-dimensional version of this expansion, but this is readily generalized to n dimensions.

We consider a distribution of particles in three-dimensional space that is described by a density distribution $n(\mathbf{x})$. We then suppose that the particle that was at position $\mathbf{x}$ in the original state is displaced to the position $\mathbf{x}'$ given by

$$x'(\mathbf{x}, \lambda) = \mathbf{x} + \lambda \xi(\mathbf{x}) \tag{10.1.1}$$

where ξ is a scalar. That is to say, the particle originally at position $\mathbf{x}$ is displaced by an amount $\lambda \xi(\mathbf{x})$. As a result of the perturbation, the original density distribution $n_0(\mathbf{x})$ will be changed to a new distribution that we write as $n'(\mathbf{x})$. Our goal is to calculate $n'(\mathbf{x})$ in terms of $n(\mathbf{x})$, $\xi(\mathbf{x})$ and λ.

It is convenient to introduce the Fourier transform of $n(\mathbf{x})$:

$$n(\mathbf{x}) = \int d^3k \, e^{i\mathbf{k}\cdot\mathbf{x}} \tilde{n}(\mathbf{k}) \qquad (10.1.2)$$

and

$$\tilde{n}(\mathbf{k}) = (2\pi)^{-3} \int d^3x \, e^{-i\mathbf{k}\cdot\mathbf{x}} n(\mathbf{x}). \qquad (10.1.3)$$

Assuming that particles are conserved, the number of particles contained in the volume d^3x before the perturbation is the same as the number of particles contained in the volume d^3x' after the perturbation, that is

$$n'(\mathbf{x}') \, d^3x' = n(\mathbf{x}) \, d^3x. \qquad (10.1.4)$$

On combining this equation with (10.1.3), we see that

$$\tilde{n}'(\mathbf{k}) = (2\pi)^{-3} \int d^3x \, n(\mathbf{x}) \exp[-i\mathbf{k}\cdot(\mathbf{x}+\lambda\boldsymbol{\xi}(\mathbf{x}))]. \qquad (10.1.5)$$

On substituting this expression in (10.1.2), we obtain

$$n'(\mathbf{x}) = (2\pi)^{-3} \int d^3k \, e^{i\mathbf{k}\cdot\mathbf{x}} \int d^3x' \, n(\mathbf{x}') \exp[-i\mathbf{k}\cdot(\mathbf{x}'+\lambda\boldsymbol{\xi}(\mathbf{x}'))]. \qquad (10.1.6)$$

We may invert the order of integration and also expand in powers of λ to obtain

$$n'(\mathbf{x}) = (2\pi)^{-3} \int d^3x' \, n(\mathbf{x}') \int d^3k \, e^{i\mathbf{k}\cdot(\mathbf{x}-\mathbf{x}')}$$

$$\{1 - i\lambda\mathbf{k}\cdot\boldsymbol{\xi}(\mathbf{x}') - \tfrac{1}{2}\lambda^2(\mathbf{k}\cdot\boldsymbol{\xi}(\mathbf{x}'))^2 + \ldots\}. \qquad (10.1.7)$$

The integration over $\mathbf{k}$ now yields

$$n'(\mathbf{x}) = \int d^3x' \, n(\mathbf{x}')\Big\{\delta^3(\mathbf{x}-\mathbf{x}') - \lambda\xi_r(\mathbf{x}')\frac{\partial}{\partial x_r}\delta^3(\mathbf{x}-\mathbf{x}')$$

$$+ \tfrac{1}{2}\lambda^2\xi_r(\mathbf{x}')\xi_s(\mathbf{x}')\frac{\partial^2}{\partial x_r \partial x_s}\delta^3(\mathbf{x}-\mathbf{x}') - \ldots\Big\}. \qquad (10.1.8)$$

On noting that

$$\frac{\partial}{\partial x_r}\delta^3(\mathbf{x}-\mathbf{x}') = -\frac{\partial}{\partial x_r'}\delta^3(\mathbf{x}-\mathbf{x}'), \text{ etc.,} \qquad (10.1.9)$$

and integrating by parts, (10.1.8) becomes

$$n'(x) = \int d^3x' \, \delta^3(x - x') \left\{ n(x') - \lambda \frac{\partial}{\partial x'_r} (n(x') \xi_r(x')) \right.$$

$$\left. + \frac{1}{2} \lambda^2 \frac{\partial^2}{\partial x'_r \partial x'_s} (n(x') \xi_r(x') \xi_s(x')) - \dots \right\}. \qquad (10.1.10)$$

Hence we finally obtain the generalized Lagrange expansion

$$n'(x) = n(x) - \lambda \frac{\partial}{\partial x_r} (n(x) \xi_r(x)) + \frac{1}{2} \lambda^2 \frac{\partial^2}{\partial x_r \partial x_s} (n(x) \xi_r(x) \xi_s(x)) - \dots$$

$$(10.1.11)$$

10.2 The Fokker–Planck equation

We now wish to study the evolution of the distribution function for a particle species. Since the instantaneous state of a particle is now defined by the six-component vector $(x_1, x_2, x_3, v_1, v_2, v_3)$, we need to consider the six-dimensional form of the Lagrange expansion (10.1.10). We consider the increments

$$t \to t + \Delta t, \quad x \to x + \Delta x, \quad v \to v + \Delta v. \qquad (10.2.1)$$

Then (10.1.10) yields the equation

$$f(x, v, t + \Delta t) = f(x, v, t) - \frac{\partial}{\partial x_r} (f \Delta x_r) - \frac{\partial}{\partial v_r} (f \Delta v_r) + \frac{1}{2} \frac{\partial^2}{\partial x_r \partial x_s} (f \Delta x_r \Delta x_s)$$

$$+ \frac{\partial^2}{\partial x_r \partial v_s} (f \Delta x_r \Delta v_s) + \frac{1}{2} \frac{\partial^2}{\partial v_r \partial v_s} (f \Delta v_r \Delta v_s) - \dots,$$

$$(10.2.2)$$

where all the terms on the right-hand side of the equation are evaluated at time t.

The term Δx consists of first-order variations due to the instantaneous velocity plus terms that are of second order in Δt due to accelerations. There are no instantaneous changes of position due to collisions. Hence, to first order in Δt, we may write

$$\Delta x_r = v_r \Delta t. \qquad (10.2.3)$$

The term Δv comprises two contributions, one due to the macroscopic electromagnetic field, and the other due to collisions. We therefore write

$$\Delta v = a \Delta t + (\Delta v)_c, \qquad (10.2.4)$$

where **a** is the particle acceleration due to the macroscopic fields, and $(\Delta \mathbf{v})_c$ is the contribution to $\Delta \mathbf{v}$ due to collisions.

Since collisions represent a random process, the effect of collisions must be treated statistically. In considering the statistical averages of the quantities $(\Delta v_r)_c$, $(\Delta v_r)_c (\Delta v_s)_c$, etc., we find that the first and second moments of $(\Delta \mathbf{v})_c$ contain contributions that scale as Δt, but all higher moments scale as higher powers of Δt. This is consistent with our understanding of the random-walk process, according to which the mean square displacement increases linearly with time. We therefore introduce the expressions

$$Av\left(\frac{(\Delta v_r)_c}{\Delta t}\right) = \left\langle \frac{\Delta v_r}{\Delta t}\right\rangle, \quad Av\left(\frac{(\Delta v_r)_c (\Delta v_s)_c}{\Delta t}\right) = \left\langle \frac{\Delta v_r \Delta v_s}{\Delta t}\right\rangle. \quad (10.2.5)$$

On substituting from the above equations into (10.2.2), we obtain

$$f + \frac{\partial f}{\partial t}\Delta t = f - \frac{\partial}{\partial x_r}(f v_r \Delta t) - \frac{\partial}{\partial v_r}(f a_r \Delta t) - \frac{\partial}{\partial v_r}\left[f\left\langle\frac{\Delta v_r}{\Delta}\right\rangle \Delta t\right]$$

$$+ \frac{1}{2}\frac{\partial^2}{\partial v_r \partial v_s}\left[f\left\langle\frac{\Delta v_r \Delta v_s}{\Delta t}\right\rangle \Delta t\right], \quad (10.2.6)$$

Since $\partial v_r / \partial x_r = 0$, since v_r and x_r are independent variables, (10.2.6) yields

$$\frac{\partial f}{\partial t} + v_r\frac{\partial f}{\partial x_r} + \frac{\partial f}{\partial v_r}(a_r f) = -\frac{\partial}{\partial v_r}\left(\left\langle\frac{\Delta v_r}{\Delta t}\right\rangle f\right) + \frac{1}{2}\frac{\partial^2}{\partial v_r \partial v_s}\left(\left\langle\frac{\Delta v_r \Delta v_s}{\Delta t}\right\rangle f\right),$$

$$(10.2.7)$$

that is known as the Fokker–Planck equation.

For nonrelativistic motion of charged particles in electric and magnetic fields,

$$a_r = \frac{q}{m}\left(E_r + \frac{1}{c}\varepsilon_{rst}v_s B_t\right). \quad (10.2.8)$$

We see that $\partial a_r / \partial v_r = 0$, so that the Fokker–Planck equation takes the following form for the nonrelativistic motion of charged particles in electromagnetic fields, taking account of Coulomb collisions in the small-angle-deflection approximation:

$$\frac{\partial f}{\partial t} + v_r\frac{\partial f}{\partial x_r} + \frac{q}{m}\left(E_r + \frac{1}{c}\varepsilon_{rst}v_s B_t\right)\frac{\partial f}{\partial v_r} =$$

$$-\frac{\partial}{\partial v_r}\left(\left\langle\frac{\Delta v_r}{\Delta t}\right\rangle f\right) + \frac{1}{2}\frac{\partial^2}{\partial v_r \partial v_s}\left(\left\langle\frac{\Delta v_r \Delta v_s}{\Delta t}\right\rangle f\right). \quad (10.2.9)$$

10.3 Coulomb collisions

We now wish to derive the differential cross section for Coulomb scattering. Consider the interaction of two particles, one with mass m and position x, and the other with mass m_1 and position x_1. They interact through a mutual force $\mathbf{F}$ such that

$$\frac{d^2 x}{dt^2} = \frac{1}{m}\mathbf{F}, \quad \frac{d^2 x_1}{dt^2} = -\frac{1}{m_1}\mathbf{F}. \tag{10.3.1}$$

We introduce the following expression for the position of the center of mass,

$$\mathbf{X} = \frac{m x + m_1 x_1}{m + m_1}, \tag{10.3.2}$$

and also introduce the following expression for the relative position of the two particles:

$$\boldsymbol{\xi} = x - x_1. \tag{10.3.3}$$

We may verify that

$$\frac{d^2 \mathbf{X}}{dt^2} = 0 \tag{10.3.4}$$

showing that the center of mass moves without acceleration. We also find that

$$\frac{d^2 \boldsymbol{\xi}}{dt^2} = \frac{1}{\mu}\mathbf{F}, \tag{10.3.5}$$

where μ is the 'reduced mass' defined by

$$\mu = \frac{m m_1}{m + m_1}. \tag{10.3.6}$$

Since the Coulomb force is parallel to $x - x_1$, we see that

$$\boldsymbol{\xi} \times \mathbf{F} = 0 \tag{10.3.7}$$

so that

$$\frac{d}{dt}(\boldsymbol{\xi} \times \dot{\boldsymbol{\xi}}) = \boldsymbol{\xi} \times \ddot{\boldsymbol{\xi}} = \frac{1}{\mu}\boldsymbol{\xi} \times \mathbf{F} = 0. \tag{10.3.8}$$

Hence

$$h \equiv \boldsymbol{\xi} \times \dot{\boldsymbol{\xi}} = \text{const.} \tag{10.3.9}$$

This may be interpreted as a statement that the fictitious particle of mass μ

and position ξ has constant angular momentum. We see that $\boldsymbol{h} \cdot \xi(t) = 0$, so that the relative motion is contained in a plane. It follows that the problem of Coulomb interaction reduces to a one-body problem with only two degrees of freedom.

The geometry of the equivalent one-body problem is shown in Fig. 10.1. In this equivalent problem, we consider a particle of mass μ, initial velocity v_0, and impact parameter b. We adopt polar coordinates r, ϕ, as shown. The scattering angle θ will be related to the final value of ϕ by

$$\theta = \pi - \phi_f. \tag{10.3.10}$$

The sum of the kinetic and potential energies of the particle is constant, so that

$$\frac{1}{2} \mu (\dot{r}^2 + r^2 \dot{\phi}^2) + \frac{qq_1}{r} = \frac{1}{2} \mu v_0^2. \tag{10.3.11}$$

Since the angular momentum is constant, we see that

$$r^2 \dot{\phi} = b v_0. \tag{10.3.12}$$

Equation (10.3.11) may be rearranged as

$$\frac{\dot{r}^2}{\dot{\phi}^2} + r^2 = \frac{v_0^2 - \frac{2}{\mu} \frac{qq_1}{r}}{\dot{\phi}^2}. \tag{10.3.13}$$

On using (10.3.12), this becomes

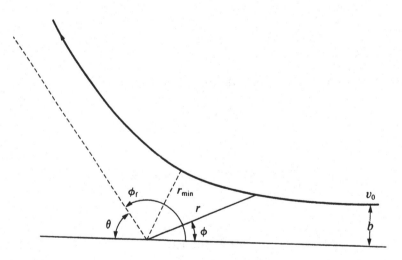

Fig. 10.1. Geometry of one-body equivalent of two particle scattering.

$$\left(\frac{dr}{d\phi}\right)^2 = \frac{\left(v_0^2 - \frac{2qq_1}{\mu r}\right)r^4}{b^2 v_0^2} - r^2 \qquad (10.3.14)$$

or, equivalently,

$$\frac{1}{r^2}\frac{dr}{d\phi} = \pm \left[\frac{1}{b^2} - \frac{2qq_1}{\mu b^2 v_0^2 r} - \frac{1}{r^2}\right]^{1/2}. \qquad (10.3.15)$$

For the time being, we consider the case with positive sign. By making the substitution

$$u = \frac{1}{r}, \qquad (10.3.16)$$

we obtain

$$\frac{du}{d\phi} = -\left[\frac{1}{b^2} - \frac{2qq_1}{\mu b^2 v_0^2}u - u^2\right]^{1/2}. \qquad (10.3.17)$$

If we define η by

$$\eta = bu = \frac{b}{r} \qquad (10.3.18)$$

and Q by

$$Q = \frac{qq_1}{\mu b v_0^2}, \qquad (10.3.19)$$

(10.3.17) may be rewritten as

$$\frac{d\eta}{d\phi} = -[1 + Q^2 - (Q+\eta)^2]^{1/2}. \qquad (10.3.20)$$

Since the trajectory is symmetric on either side of the position at which $r = r_{min}$, we may integrate (10.3.20) to obtain

$$\phi_f = 2\int_0^{\eta_m} \frac{d\eta}{(1 + Q^2 - (Q+\eta)^2)^{1/2}}. \qquad (10.3.21)$$

The limits of integration correspond to integrating from $r = \infty$ to $r = r_{min}$.

If we write

$$\sin \Psi = \frac{Q+\eta}{(1+Q^2)^{1/2}}, \qquad (10.3.22)$$

we see that

$$\cos \Psi \, d\Psi = \frac{d\eta}{(1+Q^2)^{1/2}} \qquad (10.3.23)$$

and, as a consequence, the integral (10.3.21) reduces to

$$\phi_f = 2 \int_{\Psi_0}^{\Psi_m} d\Psi = 2(\Psi_m - \Psi_0), \tag{10.3.24}$$

where the limits of integration are given by

$$\sin \Psi_0 = \frac{Q}{(1+Q^2)^{1/2}} \tag{10.3.25}$$

and

$$\sin \Psi_m = \frac{Q + \eta_m}{(1+Q^2)^{1/2}}. \tag{10.3.26}$$

We see from (10.3.25) that

$$\Psi_0 = \arctan Q. \tag{10.3.27}$$

Since $du/d\phi = 0$ at $\eta = \eta_m$, we see from (10.3.20) that

$$(Q + \eta_m)^2 = 1 + Q^2. \tag{10.3.28}$$

On combining this expression with (10.3.26), we find that

$$\Psi_m = \frac{\pi}{2}. \tag{10.3.29}$$

Hence (10.3.24) becomes

$$\phi_f = \pi - 2 \arctan Q. \tag{10.3.30}$$

On referring back to (10.3.10), we now see that the deflection angle θ is given by

$$\theta = 2 \arctan Q. \tag{10.3.31}$$

That is,

$$\theta = 2 \arctan\left(\frac{qq_1}{\mu b v_0^2}\right). \tag{10.3.32}$$

We can now calculate the differential scattering cross section. We consider the increments

$$b \to b + db, \quad \theta \to \theta + d\theta. \tag{10.3.33}$$

The differential cross section is defined by

$$\frac{d\sigma}{d\Omega} d\Omega = |2\pi b \, db|, \tag{10.3.34}$$

where the element of solid angle $d\Omega$ is given by

$$d\Omega = |2\pi \sin\theta \, d\theta|. \tag{10.3.35}$$

Hence

$$\frac{d\sigma}{d\Omega} = \left| \frac{b \, db}{\sin\theta \, d\theta} \right|. \tag{10.3.36}$$

From (10.3.32), we see that

$$\frac{qq_1}{\mu v_0^2} \frac{1}{b} = \tan\left(\frac{1}{2}\theta\right) \tag{10.3.37}$$

so that

$$\frac{qq_1}{\mu v_0^2} \left| \frac{db}{b^2} \right| = \frac{1}{2} \sec^2\left(\frac{1}{2}\theta\right) |d\theta|. \tag{10.3.38}$$

On using this relation between db and $d\theta$ in (10.3.36), and expressing b in terms of θ by means of (10.3.37), we finally obtain the following for the differential cross section for Coulomb scattering:

$$\frac{d\sigma}{d\Omega} = \frac{1}{4}\left(\frac{qq_1}{\mu v_0^2}\right)^2 \mathrm{cosec}^4\left(\frac{1}{2}\theta\right). \tag{10.3.30}$$

This formula is known as the 'Rutherford cross-section.'

10.4 The Fokker–Planck equation for Coulomb collisions

In Section 10.2, we derived the Fokker–Planck equation in the form (10.2.7). In Section 10.3, we have expressed the effects of Coulomb collisions in terms of the Rutherford cross section (10.3.39). In this section, we use the Rutherford cross section to evaluate the quantities $\langle \Delta v_r / \Delta t \rangle$, $\langle \Delta v_r \Delta v_s / \Delta t \rangle$ that appear on the right-hand side of the Fokker–Planck equation. In doing so, we take advantage of the fact that, in a plasma for which the plasma parameter is large compared with unity, small-angle collisions play the dominant role, and large-angle collisions may be neglected (as discussed in Chapter 2).

We now consider a collision between two particles of mass m and m_1 and charge q and q_1. We denote by $\mathbf{v}$ and $\mathbf{v}_1$ the velocities of the particles prior to the collision, and by $\mathbf{v}'$ and $\mathbf{v}_1'$ the velocities after the collision. The velocity of the center of mass is given by

$$\mathbf{V} = \frac{m\mathbf{v} + m_1\mathbf{v}_1}{m + m_1}. \tag{10.4.1}$$

We denote by $\mathbf{u}$ and $\mathbf{u}'$ the relative velocity of the particles before and after the collision:

$$\mathbf{u} = \mathbf{v} - \mathbf{v}_1, \qquad (10.4.2)$$

and

$$\mathbf{u}' = \mathbf{v}' - \mathbf{v}_1'. \qquad (10.4.3)$$

We also introduce the following symbol for the change in the relative velocity due to the collision:

$$\delta\mathbf{u} = \mathbf{u}' - \mathbf{u}. \qquad (10.4.4)$$

Since

$$\mathbf{v} = \mathbf{V} + \frac{\mu}{m}\mathbf{u} \quad \text{and} \quad \mathbf{v}_1 = \mathbf{V} - \frac{\mu}{m_1}\mathbf{u}, \qquad (10.4.5)$$

the velocity change of the two particles may be expressed in terms of $\delta\mathbf{u}$ as follows,

$$\delta\mathbf{v} = \frac{\mu}{m}\delta\mathbf{u} \quad \text{and} \quad \delta\mathbf{v}_1 = -\frac{\mu}{m_1}\delta\mathbf{u}, \qquad (10.4.6)$$

where μ is the reduced mass that was defined by (10.3.6). The Rutherford cross section gives us information about $\delta\mathbf{u}$, but we need to know $\delta\mathbf{v}$ and $\delta\mathbf{v}_1$ in order to calculate the effect of the collision process on the particle distribution functions.

The collision terms in the Fokker–Planck equation (10.2.7) may now be expressed as

$$\left\langle \frac{\Delta v_r}{\Delta t} \right\rangle = \int d^3 v_1 \, f_1 \frac{\mu}{m} \{\delta u_r\} \qquad (10.4.7)$$

and

$$\left\langle \frac{\Delta v_r \Delta v_s}{\Delta t} \right\rangle = \int d^3 v_1 \, f_1 \frac{\mu^2}{m^2} \{\delta u_r \, \delta u_s\} \qquad (10.4.8)$$

where we have introduced the following notation:

$$\{\delta u_r\} = \int d\Omega \, \frac{d\sigma}{d\Omega} u \, \delta u_r \qquad (10.4.9)$$

and

$$\{\delta u_r \, \delta u_s\} = \int d\Omega \, \frac{d\sigma}{d\Omega} u \, \delta u_r \, \delta u_s, \qquad (10.4.10)$$

where $u = |\mathbf{u}|$. The integrals in (10.4.9) and (10.4.10) represent an averaging over initial impact parameters.

The integration in (10.4.9) and (10.4.10) are simplified by choosing a coordinate frame in which $\mathbf{u}$ is along one of the axes. We assume that

$$\mathbf{u} = (0, 0, u) \qquad (10.4.11)$$

and then express the value of $\mathbf{u}$ after the collision in the form

$$\mathbf{u}' = u(\sin\theta\cos\phi, \sin\theta\sin\phi, \cos\theta). \qquad (10.4.12)$$

Hence

$$\delta\mathbf{u} = u(\sin\theta\cos\phi, \sin\theta\sin\phi, -2\sin^2\tfrac{1}{2}\theta). \qquad (10.4.13)$$

In terms of our current notation, the differential cross section for Coulomb scattering takes the form

$$\frac{d\sigma}{d\Omega} = \frac{1}{4}\left(\frac{qq_1}{\mu u^2}\right)^2 \operatorname{cosec}^4 \tfrac{1}{2}\theta. \qquad (10.4.14)$$

Equation (10.4.9) may now be expressed as

$$\{\delta\mathbf{u}\} = \int d\Omega \frac{d\sigma}{d\Omega} u^2 \left(\sin\theta\cos\phi, \sin\theta\sin\phi, -2\sin^2\tfrac{1}{2}\theta\right). \qquad (10.4.15)$$

The first and second terms integrate to zero, leaving

$$\{\delta u_1\} = 0, \quad \{\delta u_2\} = 0, \quad \{\delta u_3\} = -\pi\left(\frac{qq_1}{\mu u}\right)^2 \int_0^\pi d\theta \sin\theta \operatorname{cosec}^2\tfrac{1}{2}\theta. \qquad (10.4.16)$$

Similarly, (10.4.10) may be written in dyadic form as

$$\{\delta\mathbf{u}\,\delta\mathbf{u}\} = \int d\Omega \frac{d\sigma}{d\Omega} u^3 \begin{pmatrix} \sin\theta\cos\phi \\ \sin\theta\sin\phi \\ -2\sin^2\tfrac{1}{2}\theta \end{pmatrix} \begin{pmatrix} \sin\theta\cos\phi \\ \sin\theta\sin\phi \\ -2\sin^2\tfrac{1}{2}\theta \end{pmatrix}. \qquad (10.4.17)$$

We find that the off-diagonal components integrate to zero, leaving

$$\{\delta\mathbf{u}\,\delta\mathbf{u}\} = \begin{pmatrix} \tfrac{1}{2}\{(\delta u_\perp)^2\} & 0 & 0 \\ 0 & \tfrac{1}{2}\{(\delta u_\perp)^2\} & 0 \\ 0 & 0 & \{(\delta u_\parallel)^2\} \end{pmatrix}, \qquad (10.4.18)$$

where

$$\{(\delta u_\perp)\}^2 = \frac{\pi q^2 q_1^2}{2\mu^2 u} \int_0^\pi d\theta \sin^3\theta \operatorname{cosec}^4\tfrac{1}{2}\theta \qquad (10.4.19)$$

and

$$\{(\delta u_\parallel)\}^2 = \frac{2\pi q^2 q_1^2}{\mu^2 u} \int_0^\pi d\theta \sin\theta. \tag{10.4.20}$$

Our simple treatment of the scattering process fails at this point, as the integrals in (10.4.16) and (10.4.19) diverge as the scattering angle θ tends to zero. This divergence is due to the fact that the Coulomb force falls off only as the inverse square of the distance between two particles, and the number of particles with impact parameters less than a certain value of b increases as b^2. It is this combination of factors that leads to the logarithmic divergence in the above integrals.

However, we learned in Chapter 2 that the effect of one particle on a neighboring particle is modified by the presence of the plasma. We found that the potential of a stationary test charge in a plasma varies as shown in (2.3.8). At short distances, the potential is the same as in free space, but at large distances the potential decreases exponentially. The transition occurs for particle separations comparable with the Debye length. This result is modified if one takes account of the velocity with which the test particle moves with respect to the plasma. However, for present purposes, we adopt the simplifying assumption that the Debye length represents the maximum distance over which the Coulomb force is operative. We see from (10.3.37) that the maximum value λ_D of the impact parameter b leads to the following equation for the minimum value θ_m of the scattering angle θ:

$$\tan \tfrac{1}{2}\theta_m = \frac{q q_1}{\mu u^2 \lambda_D}. \tag{10.4.21}$$

We shall find that if the plasma parameter $\Lambda \gg 1$, θ_m may be taken to be

$$\theta_m = \frac{2 q q_1}{\mu u^2 \lambda_D}. \tag{10.4.22}$$

Let us first consider scattering between similar singly charged particles, for instance electrons. The mass is related to the reduced mass by $\mu = \tfrac{1}{2} m_e$ and the mean relative velocity is related to the mean square electron velocity by $\langle u^2 \rangle = 2\langle v_e^2 \rangle$. For a thermal distribution, the mean square electron velocity is given by

$$\tfrac{1}{2} m_e \langle v_e^2 \rangle = \tfrac{3}{2} k T_e. \tag{10.4.23}$$

Hence

$$\langle u^2 \rangle = 6 \frac{k T_e}{m_e}. \tag{10.4.24}$$

On using this expression and noting that Λ_e is defined by (2.4.1), (10.4.22) becomes

$$\theta_m = \frac{1}{6\pi} n_e^{-1} \lambda_{De}^{-3} \equiv \frac{1}{6\pi} \Lambda_e^{-1} \quad \text{for } e\text{-}e \text{ collisions.} \quad (10.4.25)$$

For scattering between electrons and protons, the reduced mass is given approximately by $\mu = m_e$ and the mean square relative velocity is approximately the mean square velocity of electrons: $\langle u^2 \rangle = \langle v_e^2 \rangle$. Hence we obtain, in place of (10.4.25),

$$\theta_m = \frac{1}{6\pi} \Lambda_e^{-1} \quad \text{for } e\text{-}p \text{ collisions.} \quad (10.4.26)$$

In writing this expression, we are including only the screening effects of the electrons, since the proton motion is too slow to significantly affect the Coulomb field of electrons.

We need to evaluate the integrals (10.4.16), (10.4.19) and (10.4.20), replacing the lower limit of integration by θ_m. We find that

$$\{\delta u_3\} = -4\pi \frac{q^2 q_1^2}{\mu^2 u^2} [\ln(\sin \tfrac{1}{2}\theta)]_{\theta_m}^\pi, \quad (10.4.27)$$

$$\{(\delta u_\perp)\}^2 = \frac{2\pi q^2 q_1^2}{\mu^2 u} \{4[\ln(\sin \tfrac{1}{2}\theta)]_{\theta_m}^\pi + [\cos\theta]_{\theta_m}^\pi\} \quad (10.4.28)$$

and

$$\{(\delta u_\parallel)\}^2 = \frac{2\pi q^2 q_1^2}{\mu^2 u} [-\cos\theta]_{\theta_m}^\pi. \quad (10.4.29)$$

Clearly, the dominant terms will be those that arise from the divergence of the integral at small values of θ_m. Retaining only these terms, we may write the above expressions as

$$\{\delta u_3\} = -\Gamma \frac{q_1^2}{q^2} \frac{m^2}{\mu^2} \frac{1}{u^2} \quad (10.4.30)$$

$$\{(\delta u_\perp)^2\} = 2\Gamma \frac{q_1^2}{q^2} \frac{m^2}{\mu^2} \frac{1}{u} \quad (10.4.31)$$

and

$$\{(\delta u_\parallel)\}^2 = 0, \quad (10.4.32)$$

where

$$\Gamma = \frac{4\pi q^4}{m^2} \ln\left(\frac{2}{\theta_m}\right). \quad (10.4.33)$$

The above calculations were made for a particular coordinate system for which **u** can be expressed in the form (10.4.11). We now express the results in the following form for a general coordinate system:

$$\{\delta u_r\} = -\Gamma \frac{q_1^2}{q^2} \frac{m^2}{\mu^2} \frac{u_r}{u^3} \tag{10.4.34}$$

and

$$\{\delta u_r \, \delta u_s\} = \Gamma \frac{q_1^2}{q^2} \frac{m^2}{\mu^2} \left(\frac{\delta_{rs}}{u} - \frac{u_r u_s}{u^3} \right). \tag{10.4.35}$$

On substituting these expressions in (10.4.7) and (10.4.8), we obtain the following expressions for the coefficients in the Fokker-Planck equation:

$$\left\langle \frac{\Delta v_r}{\Delta t} \right\rangle = -\Gamma \frac{q_1^2}{q^2} \frac{m + m_1}{m_1} \int d^3 v_1 \, f_1 \frac{u_r}{u^3} \tag{10.4.36}$$

and

$$\left\langle \frac{\Delta v_r \Delta v_s}{\Delta t} \right\rangle = \Gamma \frac{q_1^2}{q^2} \int d^3 v_1 \, f_1 \left(\frac{\delta_{rs}}{u} - \frac{u_r u_s}{u^3} \right). \tag{10.4.37}$$

These expressions can be simplified by recognizing that the terms involving the relative velocity are expressible in the following form:

$$\frac{u_r}{u^3} = -\frac{\partial}{\partial v_r} \left(\frac{1}{u} \right) \tag{10.4.38}$$

and

$$\frac{\delta_{rs}}{u} - \frac{u_r u_s}{u^3} = \frac{\partial^2 u}{\partial v_r \partial v_s}. \tag{10.4.39}$$

We may therefore express the Fokker-Planck coefficients in terms of two scalar potentials G and H, termed the Rosenbluth potentials:

$$G = \frac{q_1^2}{q^2} \int d^3 v_1 \, f_1 u \tag{10.4.40}$$

and

$$H = \frac{q_1^2}{q^2} \frac{m + m_1}{m_1} \int d^3 v_1 \, f_1 \frac{1}{u}. \tag{10.4.41}$$

In terms of these potentials, expressions (10.4.36) and (10.4.37) become

$$\left\langle \frac{\Delta v_r}{\Delta t} \right\rangle = \Gamma \frac{\partial H}{\partial v_r} \tag{10.4.42}$$

and

$$\left\langle \frac{\Delta v_r \Delta v_s}{\Delta t} \right\rangle = \Gamma \frac{\partial^2 G}{\partial v_r \partial v_s}. \qquad (10.4.43)$$

In this way, we finally arrive at the following form of the Fokker–Planck equation for Coulomb collisions:

$$\frac{\partial f}{\partial t} + v_r \frac{\partial f}{\partial x_r} + \frac{q}{m}\left[E_r + \frac{1}{c}\,\varepsilon_{rst}\,v_s B_t\right]\frac{\partial f}{\partial v_r} = -\Gamma \frac{\partial}{\partial v_r}\left(f\frac{\partial H}{\partial v_r}\right) + \tfrac{1}{2}\Gamma \frac{\partial^2}{\partial v_r \partial v_s}\left(f\frac{\partial^2 G}{\partial v_r \partial v_s}\right).$$

$$(10.4.44)$$

10.5 Relaxation times

In this section, we examine the characteristic times in which a velocity distribution function changes in time. For instance, we shall examine the evolution of particles with momentum different from the mean momentum of the plasma, and examine the time-scale by which the momentum relaxes to the mean value. Other characteristic times that are of interest are the time-scales in which a distribution becomes isotropic; the time-scale in which a distribution becomes Maxwellian; and the time-scale in which electrons and ions approach a common temperature. We shall deal with the momentum relaxation time in detail, and simply quote results for the other relaxation times.

One way to study the evolution of a plasma containing a background distribution of particles and another group of particles that, for instance, have a specified momentum, would be that of integrating the dynamical equation for the distribution function f. However, it does not appear to be possible to carry out such an integration in analytical terms. An alternative is to use numerical integration, but each computer run then applies only to a specific situation. In this section, we use a third approach which is to some extent approximate, but yields results that are in good agreement with numerical calculations for specific cases.

We consider a field-free Maxwellian fully ionized hydrogen plasma. For convenience, we here represent the distribution functions for the electrons and protons by f_e and f_p, respectively:

$$f_e \equiv n_0 g_e = n_0 \left(\frac{m_e}{2\pi k T_e}\right)^{3/2} e^{-\alpha_e^2 v^2}$$

$$f_p \equiv n_0 g_p = n_0 \left(\frac{m_p}{2\pi k T_p}\right)^{3/2} e^{-\alpha_p^2 v^2} \qquad (10.5.1)$$

where

$$\alpha_e^2 = \left(\frac{m_e}{2kT_e}\right), \quad \alpha_p^2 = \frac{m_p}{2kT_p}. \tag{10.5.2}$$

We assume that, in addition to the Maxwellian component, the plasma also contains a component comprised of 'test' particles of mass m_t and charge $Z_t e$. We assume that the distribution function of these particles is uniform in space and that

$$g_t(\mathbf{v}, t) = \delta^3(\mathbf{v} - \mathbf{V}) \quad \text{at } t = 0. \tag{10.5.3}$$

We assume that the number density of test particles (that according to (10.5.3) is unity) is small compared with the number density of electrons and protons of the Maxwellian plasma, so that we may neglect the effect of the test particles on the background plasma. Hence we need consider only the effect of the background plasma on the distribution of test particles. *A fortiori*, we may neglect the effect of interaction among test particles.

As we see, the test particles begin as a single stream with a unique velocity. In the course of time, the mean velocity of this distribution will change, due to the first term on the right-hand side of (10.4.44), and the distribution will acquire a spread in velocity, due to the second term on the right-hand side (10.4.44). In the calculation that follows, we focus on the rate of change of the velocity of the stream and make the further simplification that we calculate the rate of change of this velocity only at the time $t = 0$.

Introducing the quantity

$$\Gamma_t = \frac{4\pi e^4 Z_t^2}{m_t^2} \ln\Lambda, \tag{10.5.4}$$

we find that the Fokker–Planck equation (10.4.44) takes the form

$$\frac{1}{n_0\Gamma_t}\frac{\partial g_t}{\partial t} = -\frac{\partial}{\partial v_r}\left[g_t\frac{\partial}{\partial v_r}\left\{\left(\frac{m_t+m_e}{m_e}\right)\int d^3v_1\frac{g_e}{u} + \left(\frac{m_t+m_p}{m_p}\right)\int d^3v_1\frac{g_p}{u}\right\}\right]$$
$$+ \frac{1}{2}\frac{\partial^2}{\partial v_r\partial v_s}\left[g_t\frac{\partial^2}{\partial v_r\partial v_s}\left\{\int d^3v_1(g_e+g_p)u\right\}\right]. \tag{10.5.5}$$

We now multiply the equation by $\mathbf{v}$ and integrate over $\mathbf{v}$. We are assuming that the principal change in the test particle distribution function is the change in the velocity $\mathbf{V}$, so that the term $\partial g_t/\partial t$ on the left-hand side of (10.5.5) leads to

$$\int d^3v\, v_r\frac{\partial g_t}{\partial t} \to \int d^3v\, v_r\frac{\partial}{\partial t}(\delta^3(\mathbf{v}-\mathbf{V}(t))) = \frac{\partial}{\partial t}\int d^3v\, v_r\delta^3(\mathbf{v}-\mathbf{V}(t)) = \frac{dV_r(t)}{dt}. \tag{10.5.6}$$

This approximate relationship is acceptable at $t=0$. Hence (10.5.5) leads to the following equation for the rate of change of the mean velocity $\mathbf{V}$ of the stream:

$$\frac{dV_r}{dt} = -n_0\Gamma_t \int d^3v\, v_r \frac{\partial}{\partial v_s} \left\{ \delta^3(\mathbf{v}-\mathbf{V}) \frac{\partial}{\partial v_s} \left[\left(\frac{m_t+m_e}{m_e} \right) \int d^3v_1 \frac{g_e}{u} \right. \right.$$

$$\left. \left. + \left(\frac{m_t+m_p}{m_p} \right) \int d^3v_1 \frac{g_p}{u} \right] \right\}$$

$$+ \frac{1}{2} n_0\Gamma_t \int d^3v\, v_r \frac{\partial^2}{\partial v_s \partial v_t} \left\{ \delta^3(\mathbf{v}-\mathbf{V}) \frac{\partial^2}{\partial v_s \partial v_t} \left[\int d^3v_1 (g_e+g_p) u \right] \right\}.$$

$$(10.5.7)$$

Introducing the temporary notation

$$Q_{st} = \delta^3(\mathbf{v}-\mathbf{V}) \frac{\partial^2}{\partial v_s \partial v_t} \left[\int d^3v_1 (g_e+g_p) u \right], \qquad (10.5.8)$$

we see that the second term on the right-hand side of (10.5.7) may be integrated by parts to yield

$$\int d^3v\, v_r \frac{\partial^2 Q_{st}}{\partial v_s \partial v_t} = -\int d^3v \frac{\partial v_r}{\partial v_s} \frac{\partial Q_t}{\partial v_t} + \int dS_s v_r \frac{\partial Q_{st}}{\partial v_t}, \qquad (10.5.9)$$

where the second term represents an integration over the relevant surface in velocity space. Since this surface should be taken to be a sphere of infinite radius, and since g_e and g_p decrease exponentially with v^2, the surface integral vanishes. Carrying out a further integration by parts, we see that

$$-\int d^3v\, \delta_{rs} \frac{\partial Q_t}{\partial v_t} = \int d^3v \frac{\partial}{\partial v_t} (\delta_{rs}) Q_t = 0. \qquad (10.5.10)$$

Hence the second term on the right-hand side of (10.5.7) gives no contribution to the rate of change of the mean velocity of the test-particle stream. This is to be expected, since the second term represents diffusion rather than a frictional deceleration.

Turning now to the first term on the right-hand side of (10.5.7), we may again carry out an integration by parts and so obtain

$$\frac{dV_r}{dt} = n_0\Gamma_t \int d^3v \frac{\partial v_r}{\partial v_s} \left\{ \delta^3(\mathbf{v}-\mathbf{V}) \frac{\partial}{\partial v_s} \left[\left(\frac{m_t+m_e}{m_e} \right) \int d^3v_1 \frac{g_e}{u} \right. \right.$$

$$\left. \left. + \left(\frac{m_t+m_p}{m_p} \right) \int d^3v_1 \frac{g_p}{u} \right] \right\}. \qquad (10.5.11)$$

This becomes

$$\frac{\mathrm{d}V_r}{\mathrm{d}t}=n_0\Gamma_t\frac{\partial}{\partial V_r}\left[\left(\frac{m_t+m_e}{m_e}\right)\int\mathrm{d}^3v\,\frac{g_e(\mathbf{v})}{|\mathbf{v}-\mathbf{V}|}+\left(\frac{m_t+m_p}{m_p}\right)\int\mathrm{d}^3v\,\frac{g_p(\mathbf{v})}{|\mathbf{v}-\mathbf{V}|}\right]$$

(10.5.12)

where we have expressed $\mathbf{u}$ explicitly as $\mathbf{v}-\mathbf{V}$.

We now substitute for $g_e(\mathbf{v})$ and $g_p(\mathbf{v})$ from (10.5.1) and introduce the error function defined by

$$\Phi(y)=2\pi^{-1/2}\int_0^y\mathrm{d}x\,\mathrm{e}^{-x^2}.$$

(10.5.13)

On noting that

$$\int\mathrm{d}^3v\,\frac{\mathrm{e}^{-\alpha^2v^2}}{|\mathbf{v}-\mathbf{V}|}=\frac{\pi^{3/2}}{\alpha^3}\frac{1}{V}\Phi(\alpha V),$$

(10.5.14)

we see that (10.5.12) may be rewritten as

$$\frac{\mathrm{d}V_r}{\mathrm{d}t}=n_0\Gamma_t\left[\left(1+\frac{m_t}{m_e}\right)\frac{\partial}{\partial V_r}\left\{\frac{1}{V}\Phi(\alpha_e V)\right\}+\left(1+\frac{m_t}{m_p}\right)\frac{\partial}{\partial V_r}\left\{\frac{1}{V}\Phi(\alpha_p V)\right\}\right].$$

(10.5.15)

Since the distribution functions are isotropic in velocity, the velocity gradient terms on the right-hand side of (10.5.15) will yield vectors parallel to $\mathbf{V}$. Hence the equation will have the form

$$\frac{\mathrm{d}V_r}{\mathrm{d}t}=-\frac{V_r}{\tau_s},$$

(10.5.16)

where τ_s represents the characteristic 'slowing-down time'. The expression for τ_s is, in general, complicated but the form becomes manageable in certain limiting cases. For instance,

$$\text{if }V\gg\alpha_e^{-1},\alpha_p^{-1}\qquad\tau_s=\frac{m_t^2V^3}{4\pi e^4n_0\ln\Lambda\left(2+\dfrac{m_t}{m_e}+\dfrac{m_t}{m_p}\right)Z_t^2},$$

(10.5.17)

since in this limit $\Phi\to1$. If we take the test particles to be electrons, so that $m_t=m_e$ and $Z_t=1$, we obtain

$$\tau_{s,e}=\frac{m_e^2V^3}{12\pi e^4n_0\ln\Lambda}\quad\text{for electrons.}$$

(10.5.18)

Similarly, if we consider that the test particles are protons, so that $m_t=m_p$ and $Z_t=1$, we obtain

$$\tau_{s,p} = \frac{m_p m_e V^3}{4\pi e^4 n_0 \ln\Lambda} \quad \text{for protons.} \quad (10.5.19)$$

As another special case, suppose that the test particles move slowly with respect to the electrons and ions. Since

$$\Phi(y) \approx 2\pi^{-1/2}(y - \tfrac{1}{3}y^3 + \ldots) \quad \text{for } y \ll 1, \quad (10.5.20)$$

we see that

$$\text{if } V \ll \alpha_e^{-1}, \alpha_p^{-1}, \quad \tau_s^{-1} = \frac{16\pi^{1/2}e^4 Z_t^2 n_0 \ln\Lambda}{3m_t^2}\left[\left(1 + \frac{m_t}{m_e}\right)\alpha_e^3 + \left(1 + \frac{m_t}{m_p}\right)\alpha_p^3\right].$$
$$(10.5.21)$$

If we now consider that the test particles are electrons and note that $m_e/m_p \ll 1$, we obtain

$$\tau_{s,e} = \frac{3m_e^2 m_p^{-3/2}(kT_p)^{3/2}}{2^{5/2}\pi^{1/2}e^4 n_0 \ln\Lambda} \quad \text{for electrons.} \quad (10.5.22)$$

Similarly we find that if the test particles are protons

$$\tau_{s,p} = \frac{3m_p^{1/2}(kT_p)^{3/2}}{2^{7/2}\pi^{1/2}e^4 n_0 \ln\Lambda} \quad \text{for protons.} \quad (10.5.23)$$

For low-speed test particles, protons play the dominant role in scattering the test particles, since they have the lower velocity, and scattering is greater for particles with smaller velocity differentials.

If we now refer to (10.5.18) and (10.5.19), and represent the energy of the test particles by a temperature, we find that

$$\tau_{s,e} \approx 10^{-2.0} n_0^{-1} T_{t,e}^{3/2} \quad (10.5.24)$$

and

$$\tau_{s,p} \approx 10^{-0.3} n_0^{-1} T_{t,p}^{3/2} \quad (10.5.25)$$

where we have adopted, for definiteness, the typical value $\ln\Lambda = 20$. These equations were derived under the assumption that the speed of the test particle is large compared with that of the mean thermal electrons and mean thermal protons. However, the equations give reasonable estimates of the scattering times even if the test particles refer to a typical group of particles drawn from the Maxwellian distributions of the electrons and protons comprising the plasma.

We now indicate briefly how other relaxation times can be defined, and give the results of calculations for these quantities.

As an example, we may introduce a 'deflection time' τ_D by the definition

$$\frac{dV_\perp^2}{dt} = \frac{V^2}{\tau_D},$$ (10.5.26)

where $V_\perp^2$ denotes the mean-square velocity component normal to the original velocity vector. Very roughly speaking, τ_D is the time it takes for test particles to be deflected through an angle $\pi/2$. Hence τ_D is a reasonable estimate of the characteristic time for an anisotropic distribution of test particles to become isotropized.

Similarly, introducing the following notation for electron and proton kinetic energies,

$$u_e = \tfrac{1}{2} m_e v_e^2, \quad u_p = \tfrac{1}{2} m_p v_p^2,$$ (10.5.27)

we can define the quantity τ_E through the equation

$$\left| \frac{du}{dt} \right| = \frac{u}{\tau_E}.$$ (10.5.28)

This represents a characteristic time for energy exchange between test particles and the plasma.

As an example, yielding more definite expressions for these relaxation rates, we consider a fully ionized plasma in which, initially, the electron and proton distributions are slightly anisotropic, slightly non-Maxwellian, and have slightly different temperatures. We may calculate the time-scale $\tau_{I,e}$ for the electrons to become isotropized by setting $m_t = m_e$, $\alpha_e V = (1.5)^{1/2}$, and $\alpha_p V = (1.5 m_p/m_e)^{1/2}$. Hence we arrive at the expression

$$\tau_{I,e} = \frac{m_e^{1/2}(3kT)^{3/2}}{8\pi e^4 \ln\Lambda [(1.5\pi)^{-1/2} e^{-1.5} + 1 - \tfrac{2}{3}\Phi(\sqrt{1.5})]}.$$ (10.5.29)

We may also calculate the characteristic time $\tau_{E,e}$ that the electron distribution takes to relax to Maxwellian form by calculating τ_E for the same assumptions as we used in deriving (10.5.29). Hence we obtain

$$\tau_{E,e} = \frac{m_e^{1/2}(3kT)^{3/2}}{8\pi e^4 \ln\Lambda (\Phi((1.5)^{1/2}) - 4(1.5/\pi)^{-1/2} e^{-1.5})}.$$ (10.5.30)

We may calculate the characteristic time $\tau_{E,p}$ that it takes for the protons to tend to a Maxwellian distribution by evaluating τ_E for the case $m_t = m_p$, $\alpha_e V = (1.5 m_e/m_p)^{1/2}$ and $\alpha_p V = 1.5$. We then obtain

$$\tau_{E,p} = \frac{m_p^{1/2}(3kT)^{3/2}}{8\pi e^4 \ln\Lambda [\Phi((1.5)^{1/2}) - 4(1.5/\pi)^{-1/2} e^{-1.5}]}.$$ (10.5.31)

Finally, we may calculate the quantity τ_{EQ}, the time it takes electrons and ions to achieve temperature equilibrium, by calculating τ_E for the case $m_t = m_e$, $\alpha_e V = (1.5)^{1/2}$ and $\alpha_p \gg 1$. We now consider that the test particles are electrons and the 'field particles' are only protons. In this way we arrive at the expression

$$\tau_{EQ} = \frac{m_p m_e^{-1/2} (3kT)^{3/2}}{8\pi e^4 \ln\Lambda}. \tag{10.5.32}$$

Once again adopting the typical value $\ln\Lambda = 20$, we find the following approximate numerical expressions for the above relaxation times

$$\tau_{I,e} = 10^{-1.70} n_0^{-1} T^{3/2} \tag{10.5.33}$$

$$\tau_{E,e} = 10^{-1.49} n_0^{-1} T^{3/2} \tag{10.5.34}$$

$$\tau_{E,p} = 10^{0.14} n_0^{-1} T^{3/2} \tag{10.5.35}$$

and

$$\tau_{EQ} = 10^{1.25} n_0^{-1} T^{3/2}. \tag{10.5.36}$$

We see that, approximately, the last three terms are in the following ratio:

$$\tau_{E,e} : \tau_{E,p} : \tau_{EQ} = \left(\frac{m_e}{m_p}\right) : \left(\frac{m_e}{m_p}\right)^{1/2} : 1. \tag{10.5.37}$$

Electrons reach a Maxwellian distribution rapidly because of their high thermal velocity and corresponding high collision frequency. The ions take somewhat longer, owing to their smaller velocities. Electrons take still longer to share energy with the ions, since only a small fraction of their energy is exchanged at each collision due to the large difference in mass.

We now consider briefly the concept of 'electron run-away.' Considering, as a simplifying factor, only one velocity component, we see that the effect of collisions on an electron may be written as

$$\frac{dv}{dt} = -\frac{v}{\tau} = -F(v). \tag{10.5.38}$$

If we now consider that an electric field is present, this equation must be modified to the form

$$\frac{dv}{dt} = -\frac{e}{m_e} E - F(v). \tag{10.5.39}$$

Since $f \propto V$ for small values of the velocity and $F \propto V^{-2}$ for large values, we see that K must have the form shown in Fig. 10.2. If the electric field is not too large, there will be two equilibrium values of the velocity, V_a and V_b, where

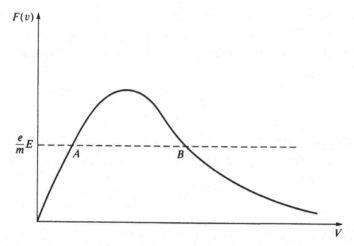

Fig. 10.2. Schematic representation of the two 'points' of balance between the force due to the electric field and the frictional force due to Coulomb collisions.

$$F(V_a) = F(V_b) = -\frac{e}{m}E. \qquad (10.5.40)$$

Let us now consider behavior in the neighborhood of the equilibrium point A in Fig. 10.2. If V increases slightly beyond V_a, K increases and the particle is slowed down, so the velocity tends to return to its original value. Conversely, if V decreases below V_a, K decreases so that the particle tends to speed up and return to its original value. Hence point A on the curve is a point of stable equilibrium.

Let us now consider the situation at point B. If V decreases below value V_b, F increases and the particle slows down and moves to even lower speeds. However, if the velocity increases slightly beyond V_b, F decreases and the particle speeds up so that B is a point of unstable equilibrium. From this point on, the effect of collisions become smaller and smaller so that acceleration will continue indefinitely. Hence any particle that attains a velocity of order V_b is likely to 'run away' (Dreicer, 1959; 1960).

The rate at which electrons run away may be represented by a rate P_r such that $P_r dt$ is the fraction of electrons that will run away in the time interval dt. Spitzer (1962) has shown that, for a plasma composed of ions of charge Ze and electrons, and for small values of the electric field ($E \ll E_D$), P_r may be expressed, to good approximation, as

$$P_r = \frac{24\,eE_D}{(2m_e kT)^{1/2}}\,e^{-(8E_D/E)^{1/2}} \approx 10^{13.36}\,T^{-1/2}E_D e^{-(8E_D/E)^{1/2}},$$

$$(10.5.41).$$

where the 'Dreicer' value of the electric field E_D is given by

$$E_D = \frac{2\pi Z e^3 n_e \ln\Lambda}{kT} \approx 10^{-11.30} Z n_e T^{-1} \ln\Lambda. \qquad (10.5.42)$$

Section 10.5 on relaxation rates owes much to Montgomery and Tidman (1964), to which the reader is referred for further information.

Problems

Problem 10.1.

(a) Evaluate the slowing-down time $\tau_{s,e}$ for a thermal electron in an electron-proton plasma by setting $m_t = m_e$, $Z_t = 1$, $\alpha_e V = 1$ and $\alpha_b V = 1$ and $\alpha_b V = (m_p/m_e)^{1/2}$.

(b) Evaluate the slowing-down time for a thermal proton by setting $m_t = m_p$, $Z_t = 1$, $\alpha_e V = (m_e/m_p)^{1/2}$ and $\alpha_b V = 1$.

(c) Consider the high-velocity and low-velocity approximations for the slowing-down time for electrons. Find the speed at which the two curves intersect and the value of the slowing-down time at that intersection. How does this compare with the value found in (a)?

(d) Repeat part (c) for protons, and compare the result with the value found in (b).

Problem 10.2. One may obtain a measure of the 'deflection time' τ_D of a particle by considering a beam of test particles that begins with the distribution

$$g_t(\mathbf{v}, t) = \delta^3(\mathbf{v} - \mathbf{V}) \quad \text{at } t = 0, \qquad (1)$$

and then evaluating τ_D, defined by

$$\frac{dV_\perp^2}{dt} = \frac{V_\perp^2}{\tau_D}, \qquad (2)$$

(a) Use the Fokker–Planck equation to evaluate τ_D. Note that

$$\int d^3v \, f_e |v - V| = \frac{1}{4\alpha_e^2 V} \{ 4\pi^{-1/2} \alpha_e V \exp(-\alpha_e^2 V^2)$$

$$+ 2(1 + 2\alpha_e^2 V^2)\Phi(\alpha_e V)\}, \qquad (3)$$

and so arrive at the expression

$$\tau_D^{-1} = \frac{4\pi n_0 Z_t^2 e^4 \ln\Lambda}{m_t^2 V^3} \frac{\partial}{\partial V} \left\{ \begin{array}{l} \dfrac{1}{\alpha_e^2 V} [2\pi^{-1/2}\alpha_e V \exp(-\alpha_e^2 V^2) + (1 + 2\alpha_e^2 V^2)\Phi(\alpha_e V)] \\[2ex] + \dfrac{1}{\alpha_p^2 V}[2\pi^{-1/2}\alpha_p V \exp(-\alpha_p^2 V^2) + (1 + 2\alpha_p^2 V^2)\Phi(\alpha_p V)] \end{array} \right\}.$$

$$(4)$$

(b) Hence calculate the deflection time $\tau_{D,e}$ for thermal electrons in an electron-proton plasma by setting $m_t = m_e$, $Z_t = 1$, $\alpha_e V = 1$ and $\alpha_b V = (m_p/m_e)^{1/2}$.

(c) Also calculate the deflection time $\tau_{D,p}$ for thermal protons by setting $m_t = m_p$, $Z_t = 1$, $\alpha_e V = (m_e/m_p)^{1/2}$ and $\alpha_b V = 1$.

(d) Adopting ln $\Lambda = 20$, find numerical expressions for $\tau_{D,e}$ and $\tau_{D,p}$.

11

MHD equations

11.1 The moment equations

The nonrelativistic form of the equation governing a particle distribution function has the form

$$\frac{\partial f}{\partial t} + v_r \frac{\partial f}{\partial x_r} + \frac{q}{m} \left(E_r + \frac{1}{c} \varepsilon_{rst} v_s B_t \right) \frac{\partial f}{\partial v_r} = \left(\frac{\partial f}{\partial t} \right)_c, \qquad (11.1.1)$$

where the term on the right-hand side represents the effects of collisions. The Fokker–Planck form of this term was derived in Chapter 10. We may obtain moment equations from (11.1.1) by multiplying this equation by various functions of velocity $\Psi(\mathbf{v})$ and integrating over velocity space. We shall examine the particular choices $\Psi = 1$, v_r, $v_r v_s$, ..., so obtaining moment equations of progressively higher order. This approach is useful when the plasma is in an approximately thermal state at any place or time, so that its local properties may be represented by a few quantities such as density, mean mass velocity, electric current, etc.

We indicate the average over velocity space of an arbitrary function $\phi(\mathbf{x}, \mathbf{v}, t)$ by the notation

$$\langle \phi(\mathbf{x}, \mathbf{v}, t) \rangle = \frac{1}{n(\mathbf{x}, t)} \int d^3 v \, \phi(\mathbf{x}, \mathbf{v}, t) f(\mathbf{x}, \mathbf{v}, t), \qquad (11.1.2)$$

where the density $n(\mathbf{x}, t)$ is given by

$$n(\mathbf{x}, t) = \int d^3 v \, f(\mathbf{x}, \mathbf{v}, t). \qquad (11.1.3)$$

On multiplying (11.1.1) by $\Psi(\mathbf{v})$ and integrating over velocity space, we find that the first term is given by

$$\int d^3 v \, \Psi \frac{\partial f}{\partial t} = \frac{\partial}{\partial t} \int d^3 v \, \Psi f = \frac{\partial}{\partial t} (n \langle \Psi \rangle). \qquad (11.1.4)$$

The second term is given by

$$\int d^3 v\, \Psi v_r \frac{\partial f}{\partial x_r} = \frac{\partial}{\partial x_r} \int d^3 v\, \Psi v_r f = \frac{\partial}{\partial x_r} (n \langle \Psi v_r \rangle), \qquad (11.1.5)$$

and the third term by

$$\int d^3 v\, \Psi \frac{q}{m} \left(E_r + \frac{1}{c} \varepsilon_{rst} v_s B_t\right) \frac{\partial f}{\partial v_r} = \frac{q}{m} E_r \int d^3 v\, \Psi \frac{\partial f}{\partial v_r} + \frac{q}{mc} \varepsilon_{rst} B_t \int d^3 v\, \Psi v_s \frac{\partial f}{\partial v_r}$$

$$= -\frac{q}{m} E_r \int d^3 v\, \frac{\partial \Psi}{\partial v_r} f - \frac{q}{mc} \varepsilon_{rst} B_t \int d^3 v\, \frac{\partial \Psi}{\partial v_r} v_s f$$

$$= -\frac{q}{m} E_r n \left\langle \frac{\partial \Psi}{\partial v_r} \right\rangle - \frac{q}{mc} \varepsilon_{rst} B_t n \left\langle \frac{\partial \Psi}{\partial v_r} v_s \right\rangle. \qquad (11.1.6)$$

Hence the general moment equation takes the form

$$\frac{\partial}{\partial t} (n \langle \Psi \rangle) + \frac{\partial}{\partial x_r} (n \langle \Psi v_r \rangle) - \frac{qn}{m} E_r \left\langle \frac{\partial \Psi}{\partial v_r} \right\rangle - \frac{qn}{mc} \varepsilon_{rst} B_t \left\langle \frac{\partial \Psi}{\partial v_r} v_s \right\rangle = \left[\frac{\partial}{\partial t} (n \langle \Psi \rangle) \right]_c, \qquad (11.1.7)$$

where

$$\left[\frac{\partial}{\partial t} (n \langle \Psi \rangle) \right]_c = \int d^3 v\, \Psi \left(\frac{\partial f}{\partial t} \right)_c. \qquad (11.1.8)$$

This moment equation is sometimes known also as a 'transfer equation' since, depending on the choice of Ψ, it represents the transfer of mass, momentum, energy, etc.

11.2 Fluid description of an electron–proton plasma

We wish to use the above transfer equation to obtain useful equations describing the evolution of a plasma that is almost charge-neutral. We shall also make the assumption that each species is described approximately by a Maxwellian velocity distribution, but we shall allow for a difference in temperature between the electrons and protons.

Rather than work in terms of the densities n_p and n_e of electrons and protons, it is convenient to introduce the mass density ρ and charge density ζ. These are related to n_e and n_p by

$$\rho = n_p m_p + n_e m_e \qquad (11.2.1)$$

and

$$\zeta = e(n_p - n_e). \qquad (11.2.2)$$

The first moments, formed by taking $\phi = v_r$, lead to the mean velocities $\langle v_{p,r} \rangle$ and $\langle v_{e,r} \rangle$, that we also write as $u_{p,r}$ and $u_{e,r}$:

$$u_{p,r} = \langle v_{p,r} \rangle, \quad u_{e,r} = \langle v_{e,r} \rangle. \qquad (11.2.3)$$

It is now convenient to introduce the mass velocity, defined by

$$U_r = \frac{n_p m_p u_{p,r} + n_e m_e u_{e,r}}{n_p m_p + n_e m_e}, \qquad (11.2.4)$$

and the current density that is given by

$$j_r = \frac{e}{c}(n_p u_{p,r} - n_e u_{e,r}). \qquad (11.2.5)$$

In considering higher moments of the distribution function, it is convenient to refer the velocity of a particle to the mean mass velocity by introducing the notation

$$w_{p,r} = v_{p,r} - U_r, \quad w_{e,r} = v_{e,r} - U_r. \qquad (11.2.6)$$

Clearly, the first moments of these contributions are given

$$\langle w_{p,r} \rangle = u_{p,r} - U_r, \quad \langle w_{e,r} \rangle = u_{e,r} - U_r. \qquad (11.2.7)$$

We now introduce the second moments of w_r to define the pressure tensors:

$$p_{p,rs} = m_p \int \mathrm{d}^3 v f_p w_{p,r} w_{p,s}, \quad p_{e,rs} = m_e \int \mathrm{d}^3 v f_e w_{e,r} w_{e,s}. \qquad (11.2.8)$$

These may be written alternatively as

$$p_{p,rs} = n_p m_p \langle w_{p,r} w_{p,s} \rangle, \quad p_{e,rs} = n_e m_e \langle w_{e,r} w_{e,s} \rangle. \qquad (11.2.9)$$

Clearly each pressure tensor is symmetric.

In order to derive an energy equation, it will be necessary to introduce the following third moment of the random velocity components, that represents the heat-flow vector:

$$Q_r = \tfrac{1}{2} n_p m_p \langle w_p^2 w_{p,r} \rangle + \tfrac{1}{2} n_e m_e \langle w_e^2 w_{e,r} \rangle. \qquad (11.2.10)$$

11.3 The collision term

We are considering a fully-ionized hydrogen plasma, and therefore neglecting ionization and recombination. Hence

$$\left(\frac{\partial n_p}{\partial t}\right)_c = 0, \quad \left(\frac{\partial n_e}{\partial t}\right)_c = 0. \tag{11.3.1}$$

Since collisions between electrons and protons do not change the total momentum density, we may write

$$K_r = m_p \int d^3 v \, v_r \left(\frac{\partial f_p}{\partial t}\right)_c = -m_e \int d^3 v \, v_r \left(\frac{\partial f_e}{\partial t}\right)_c. \tag{11.3.2}$$

Similarly, since collisions do not change the total energy density, we may write

$$H = \frac{1}{2} m_p \int d^3 v \, v^2 \left(\frac{\partial f_p}{\partial t}\right)_c = -\frac{1}{2} m_e \int d^3 v \, v^2 \left(\frac{\partial f_e}{\partial t}\right)_c. \tag{11.3.3}$$

K_r represents the relative 'frictional' force between the electrons and protons, and H represents the rate of transfer of energy from electrons to protons.

11.4 Moment equations for each species

By adopting $\Psi = 1$ and noting (11.3.1), we find that (11.1.7) becomes

$$\frac{\partial n}{\partial t} + \frac{\partial}{\partial x_r} (n u_r) = 0. \tag{11.4.1}$$

This is the conservation equation for each particle species.

By adopting $\Psi = v_r$, and using (11.3.2), we find that (11.1.7) becomes

$$\frac{\partial}{\partial t} (nm u_r) + \frac{\partial}{\partial x_s} (nm \langle v_r v_s \rangle) - nq E_r - \frac{nq}{c} \varepsilon_{rst} u_s B_t = \pm K_r, \tag{11.4.2}$$

where the sign of K_r depends on the particle species.

By adopting $\Psi = \frac{1}{2} m v^2$ and using (11.3.3), we find that (11.1.7) takes the form

$$\frac{\partial}{\partial t} \left(\frac{1}{2} nm \langle v^2 \rangle\right) + \frac{\partial}{\partial x_r} \left(\frac{1}{2} nm \langle v^2 v_r \rangle\right) - nq E_r u_r = \pm H, \tag{11.4.3}$$

where the sign on the right-hand side again depends on the particle species.

Equations (11.4.2) and (11.4.3) are the momentum and energy transfer equations for each species.

11.5 Fluid description

We now wish to express the equations derived in Section 11.4 in terms of the variables introduced in Section 11.2. Before proceeding, we need to consider the second and third order moments.

Using (11.2.6), we see that

$$\langle v_r v_s \rangle = \langle (U_r + w_r)(U_s + w_s) \rangle. \tag{11.5.1}$$

On using (11.2.7) and (11.2.8), we find that this becomes

$$\langle v_r v_s \rangle = \frac{1}{nm} p_{rs} + U_r u_s + U_s u_r - U_r U_s. \tag{11.5.2}$$

Hence, in particular,

$$\langle v^2 \rangle = \frac{1}{nm} p_{rr} + 2U_r u_r - U^2. \tag{11.5.3}$$

By a similar procedure, we find that

$$\langle v^2 v_r \rangle = \langle w^2 w_r \rangle + \frac{1}{nm} p_{ss} U_r + \frac{2}{nm} U_s p_{sr} + 2 U_r U_s u_s + U^2 u_r - 2 U^2 u_r. \tag{11.5.4}$$

On using the above forms of the second and third moments of the velocity, we find that (11.4.2) may be expressed as

$$\frac{\partial}{\partial t}(nmu_r) + \frac{\partial}{\partial x_s} \{p_{rs} + nm(U_r u_s + U_s u_r - U_r U_s)\} - nqE_r - \frac{nq}{c} \varepsilon_{rst} u_s B_t = \pm K_1 \tag{11.5.5}$$

and (11.4.3) as

$$\frac{\partial}{\partial t}\left(\frac{1}{2} p_{ss} + nm U_s u_s - \frac{1}{2} nm U^2\right)$$

$$+ \frac{\partial}{\partial x_r}\left\{\frac{1}{2} nm\langle w^2 w_r \rangle + \frac{1}{2} p_{ss} U_r + p_{rs} U_s + nmu^2 U_r + \frac{1}{2} nm U^2 u_r - nm U^2 U_r\right\}$$

$$- nqE_r u_r = \pm H. \tag{11.5.6}$$

On noting the definitions (11.2.1), (11.2.2), (11.2.4) and (11.2.5), we easily find from the conservation equation (11.4.1), that holds for each species, that

$$\frac{\partial \rho}{\partial t} + \frac{\partial}{\partial x_r}(\rho U_r) = 0 \tag{11.5.7}$$

and

and

$$\frac{1}{c}\frac{\partial \zeta}{\partial t} + \frac{\partial j_r}{\partial x_r} = 0. \qquad (11.5.8)$$

That is, we obtain the equations of mass conservation and charge conservation.

We now consider the electron and proton form of the equation of motion (11.4.2) and add these two equations. Then the interaction force K_r drops out of the resulting equation, that then takes the form

$$\frac{\partial}{\partial t}(\rho U_r) + \frac{\partial}{\partial x_s}(p_{rs} + \rho U_r U_s) - \zeta E_r - \varepsilon_{rst} j_s B_t = 0, \qquad (11.5.9)$$

where we have now introduced the total pressure tensor defined by

$$p_{rs} = p_{p,rs} + p_{e,rs}. \qquad (11.5.10)$$

On using (11.5.7), we find that (11.5.9) may be expressed alternatively as

$$\rho \frac{dU_r}{dt} = -\frac{\partial p_{rs}}{\partial x_s} + \zeta E_r + \varepsilon_{rst} j_s B_t, \qquad (11.5.11)$$

where the term on the left-hand side represents the convective derivative:

$$\frac{dU_r}{dt} = \frac{\partial U_r}{\partial t} + U_s \frac{\partial U_r}{\partial x_s}. \qquad (11.5.12)$$

Equation (11.5.11) has just the form we would expect for the equation of motion of a fluid, taking into account pressure and electric and magnetic forces.

By adding the two energy equations (11.4.3), we obtain the following equation

$$\frac{\partial}{\partial t}\left(\frac{1}{2}p_{ss} + \frac{1}{2}\rho U^2\right) + \frac{\partial}{\partial x_r}\left(Q_r + \frac{1}{2}p_{ss}U_r + p_{rs}U_s + \frac{1}{2}\rho U^2 U_r\right) - cj_r E_r = 0.$$

$$(11.5.13)$$

If we write

$$p_{ss} = 3NkT, \qquad (11.5.14)$$

where N is the total particle density,

$$N = n_e + n_p \approx 2n_e, \qquad (11.5.15)$$

so that T is the mean temperature of the plasma, (11.5.13) may be rewritten as

$$\frac{\partial}{\partial t}\left(\frac{3}{2}NkT + \frac{1}{2}\rho U^2\right) + \frac{\partial}{\partial x_r}\left(\frac{3}{2}NkTU_r + \frac{1}{2}\rho U^2 U_r + p_{rs}U_s + Q_r\right) - cj_r E_r = 0.$$

$$(11.5.16)$$

On using (11.5.9), and using the notation of (11.5.12), we find that this energy equation may be expressed alternatively as

$$\frac{d}{dt}\left(\frac{3}{2}NkT\right)+\frac{3}{2}NkT\frac{\partial U_r}{\partial x_r}+p_{rs}\frac{\partial U_r}{\partial x_s}+\frac{\partial Q_r}{\partial x_r}+(\zeta U_r-cj_r)\left(E_r+\frac{1}{c}\varepsilon_{rst}U_sB_t\right)=0.$$

$$(11.5.17)$$

11.6 Ohm's law

By combining the two equations (11.4.2), we arrived at the equation of motion (11.5.9). However, the current density also is related to the first velocity moments of the two species through (11.2.5). Hence, by taking another combination of the two equations (11.4.2), we should be able to obtain an equation determining the current density. This will represent a generalized form of Ohm's law.

We multiply the proton form of (11.4.2) by ce/m_p, the electron form by $-ce/m_e$, and then add the two equations. The resulting equation does not fall neatly into combinations of ρ, U, etc. If we neglect terms that are quadratic in U_r, $u_{p,r}$ and $u_{e,r}$, we obtain the following equation:

$$c\frac{\partial j_r}{\partial t}=-\frac{e}{m_p}\frac{\partial p_{p,rs}}{\partial x_s}+\frac{e}{m_e}\frac{\partial p_{e,rs}}{\partial x_s}+e^2\left(\frac{n_p}{m_p}+\frac{n_e}{m_e}\right)E_r$$

$$+\frac{e^2}{c}\varepsilon_{rst}\left(\frac{n_pu_{p,s}}{m_p}+\frac{n_eu_{e,s}}{m_e}\right)B_t+e\left(\frac{1}{m_p}+\frac{1}{m_e}\right)K_r. \qquad (11.6.1)$$

On noting that $m_e \ll m_p$, we see that we may neglect the first term on the right-hand side of (11.6.1) and simplify the last term on the right-hand side. We may also note from (11.2.1) and (11.2.2) that

$$n_e\approx n_p\approx m_p^{-1}\rho, \qquad (11.6.2)$$

and from (11.2.4) and (11.2.5) that

$$u_{p,r}\approx U_r, \quad u_{e,r}\approx U_r-\frac{cm_p}{e\rho}j_r. \qquad (11.6.3)$$

Hence we may re-express equation (11.6.1) as

$$c\frac{\partial j_r}{\partial t}=\frac{e}{m_e}\frac{\partial p_{e,rs}}{\partial x_s}+\frac{e^2\rho}{m_pm_e}E_r+\frac{e^2\rho}{cm_pm_e}\varepsilon_{rst}U_sB_t-\frac{e}{m_e}\varepsilon_{rst}j_sB_t+\frac{e}{m_e}K_r.$$

$$(11.6.4)$$

In the approximation that $|\mathbf{u}_p-\mathbf{u}_e|$ is small, we may express K_r in the following form

$$K_r=\nu m_en(u_{e,r}-u_{p,r}), \qquad (11.6.5)$$

where ν is the momentum collision frequency for electrons scattering on protons. We now introduce the resistivity coefficient η and the conductivity coefficient σ defined by

$$\eta^{-1} = \sigma = \frac{ne^2}{m_e c \nu}, \tag{11.6.6}$$

so that

$$K_r = -\frac{\rho e}{m_p} \eta j_r. \tag{11.6.7}$$

On making this substitution in (11.6.4), we find that the equation may be re-expressed as

$$\frac{m_p m_e c}{\rho e^2} \frac{\partial j_r}{\partial t} = E_r + \frac{1}{c} \varepsilon_{\text{rst}} U_s B_t - \frac{m_p}{\rho e} \varepsilon_{\text{rst}} j_s B_t + \frac{m_p}{\rho e} \frac{\partial p_{e,\text{rs}}}{\partial x_s} - \eta j_r. \tag{11.6.8}$$

This is the generalized form of Ohm's law that determines the current density j_r in terms of the electric and magnetic fields, the mass velocity, and the gradient of the electron pressure tensor. As a result of the various approximations we have made, this equation is linear in j_r.

From (10.5.24), that gives an approximate expression for the electron momentum relaxation time for a fully ionized hydrogen plasma, we see that the collision frequency is given by

$$\nu = 10^{2.0} n T^{-3/2}. \tag{11.6.9}$$

Hence we may obtain approximate expressions for the conductivity σ and the resistivity η. However, as Spitzer (1962) has pointed out, to obtain an accurate expression for η, it is necessary to take electron–electron encounters into account. We found, in Chapter 10, that isotropization due to electron–electron encounters (10.5.33) is more rapid than deceleration due to electron–ion encounters (10.5.24). On taking electron–electron encounters into account, the resistivity is increased by a factor of about 2.7, so that we obtain, for a fully ionized hydrogen plasma,

$$\sigma = 10^{-3.7} T^{3/2}, \quad \eta = 10^{3.7} T^{-3/2}. \tag{11.6.10}$$

We should remember that σ and η are expressed in modified Gaussian units. They relate the current density, measured in emu, to the electric field strength, measured in esu (see Appendix A).

It is interesting to note that the plasma density does not appear in expressions (11.6.10). This is because the carrier density is proportional to n, but the collision frequency also is proportional to n so that these two effects cancel out in (11.6.6).

11.7 The ideal MHD equations

We can simplify (11.6.8) by noting that, for a wide range of cases of interest, some of the terms in that equation are negligible. For instance, consider the ratio of the term on the left-hand side to the last term on the right-hand side:

$$\mathscr{R}_1 = \frac{\left| \dfrac{m_p m_e c}{\rho e^2} \dfrac{\partial j_r}{\partial t} \right|}{|\eta j_r|}. \tag{11.7.1}$$

On considering a wave-like disturbance of frequency ω, and on using (11.6.6), we find that

$$\mathscr{R}_1 = \frac{\omega}{\nu}. \tag{11.7.2}$$

Hence if we consider only 'low-frequency' disturbances for which the typical frequency is small compared with the electron–proton collision frequency, so that $\mathscr{R}_1 \ll 1$, we may neglect the term on the left-hand side of (11.6.8) so that the equation becomes

$$j_r + \frac{m_p}{\rho e} \sigma \varepsilon_{rst} j_s B_t = \sigma \left[E_r + \frac{1}{c} \varepsilon_{rst} U_s B_t \right] + \frac{\sigma m_p}{\rho e} \frac{\partial p_{e, rs}}{\partial x_s}. \tag{11.7.3}$$

We now consider the ratio of the third term on the right-hand side of (11.7.3) to the second term. We may estimate the magnitude of this ratio as follows:

$$\mathscr{R}_2 = \frac{\left| \dfrac{\sigma m_p}{\rho e} \dfrac{\partial p_{e, rs}}{\partial x_s} \right|}{\sigma c^{-1} |\varepsilon_{rst} U_s B_t|}. \tag{11.7.4}$$

If we write

$$\left| \frac{\partial p_{e, rs}}{\partial x_s} \right| \approx L^{-1} p_e, \tag{11.7.5}$$

where L is the length-scale determining the pressure gradient,

$$p_e = n_e k T_e = \tfrac{1}{3} n_e m_e v_{th, e}^2, \tag{11.7.6}$$

then we find that

$$\mathfrak{R}_2 \approx \frac{v_{\text{th,e}}}{U}\frac{r_{\text{g,e}}}{L}. \qquad (11.7.7)$$

Hence if $L/r_{\text{g,e}}$ is large compared with $v_{\text{th,e}}/U$, then $\mathfrak{R}_2 \ll 1$, and we may neglect the third term on the right-hand side of (11.7.3), so that the equation reduces to

$$j_r + \frac{m_p}{\rho e}\sigma\varepsilon_{\text{rst}}j_s B_t = \sigma\left[E_r + \frac{1}{c}\varepsilon_{\text{rst}}U_s B_t\right]. \qquad (11.7.8)$$

Let us now consider the ratio of the second term on the left-hand side of (11.7.8) to the first term:

$$\mathfrak{R}_3 = \frac{\left|\dfrac{m_p}{\rho e}\sigma\varepsilon_{\text{rst}}j_s B_t\right|}{|j_r|}. \qquad (11.7.9)$$

On using (11.6.6), we find that

$$\mathfrak{R}_3 \approx \frac{\omega_{\text{g,e}}}{\nu}. \qquad (11.7.10)$$

In many cases, the electron gyrofrequency will not be small compared with the collision frequency, so that the second term in (11.7.8) cannot be neglected. However, for a sufficiently dense plasma, ν will be sufficiently high that $\mathfrak{R}_3 \ll 1$. Equation (11.7.8) then becomes

$$j_r = \sigma\left[E_r + \frac{1}{c}\varepsilon_{\text{rst}}U_s B_t\right]. \qquad (11.7.11)$$

This equation is the form of Ohm's law embodied in the assumption of 'ideal MHD' systems.

Let us now consider the charge continuity equation (11.5.8). On using Poisson's equation, this may be rewritten as

$$\frac{\partial}{\partial x_r}\left(\frac{1}{4\pi c}\frac{\partial E_r}{\partial t} + j_r\right) = 0. \qquad (11.7.12)$$

The ratio of the terms inside the parentheses may be written as

$$\mathfrak{R}_4 = \frac{\left|\dfrac{1}{4\pi c}\dfrac{\partial \mathbf{E}}{\partial t}\right|}{|\mathbf{j}|}. \qquad (11.7.13)$$

We see from (11.7.11) that, in the high-conductivity limit, we may replace E by $c^{-1}UB$. If we consider a wave-like disturbance, of frequency ω and wave-number k, (11.7.13) becomes

$$\mathcal{R}_4 = \frac{\omega U}{c^2 k}.$$

(11.7.14)

Hence for the usual case that the fluid motions and the phase velocities are both sub-relativistic, $\mathcal{R}_4 \ll 1$, and the charge continuity equation reduces to

$$\frac{\partial j_r}{\partial x_r} = 0.$$

(11.7.15)

This argument also has the consequence that we may neglect the displacement current $(1/c)(\partial E/\partial t)$ compared with the plasma current $\mathbf{j}$ in (2.1.4).

We now consider the equation of motion in the form (11.5.11) and consider the ratio of the second term on the right-hand side to the third term:

$$\mathcal{R}_5 = \frac{|\zeta \mathbf{E}|}{|\mathbf{j} \times \mathbf{B}|}.$$

(11.7.16)

From Poisson's equation, the magnitude of the charge density ζ may be taken to be $(4\pi)^{-1}kE$. From our previous estimate for the magnitude of E, we obtain

$$\mathcal{R}_5 = \frac{1}{4\pi}\left(\frac{U}{c}\right)^2.$$

(11.7.17)

Hence if the motions are sub-relativistic, $\mathcal{R}_5 \ll 1$ and the equation of motion (11.5.11) may be simplified to

$$\rho \frac{dU_r}{dt} = -\frac{\partial p_{rs}}{\partial x_s} + \varepsilon_{rst} j_s B_t.$$

(11.7.18)

Finally, we consider the energy equation (11.5.17). In the limiting high-conductivity approximation, we see from (11.7.11) that the last term in (11.5.17) may be neglected. If, in addition, we consider a sufficiently dense plasma that the heat-flux term may be neglected, and if we further consider that the pressure tensor is isotropic, (11.5.17) reduces to

$$\frac{d}{dt}\left(\frac{3}{2}p\right) + \frac{5}{2}p\frac{\partial U_r}{\partial x_r} = 0.$$

(11.7.19)

Since the continuity equation (11.5.17) may be written as

$$\frac{d\rho}{dt} = -\rho\frac{\partial U_r}{\partial x_r},$$

(11.7.20)

(11.7.19) may be expressed as

$$\frac{1}{p}\frac{dp}{dt} = \frac{5}{3}\frac{1}{\rho}\frac{d\rho}{dt}.$$

(11.7.21)

Hence, for the 'ideal' case, we obtain the adiabatic equation of state:

$$p \propto \rho^{5/3}. \tag{11.7.22}$$

We now gather together equations describing the behavior of a plasma that incorporate the above approximations, for which the conductivity is taken to be infinite, and the pressure tensor is taken to be isotropic:

$$\frac{\partial \rho}{\partial t} + \nabla \cdot (\rho \mathbf{U}) = 0 \tag{11.7.23}$$

$$\rho \frac{d\mathbf{U}}{dt} = -\nabla p + \mathbf{j} \times \mathbf{B} \tag{11.7.24}$$

$$p\rho^{-5/3} = \text{const.} \tag{11.7.25}$$

$$\frac{\partial \mathbf{B}}{\partial t} = \nabla \times (\mathbf{U} \times \mathbf{B}) \tag{11.7.26}$$

and

$$\nabla \times \mathbf{B} = 4\pi \mathbf{j}. \tag{11.7.27}$$

(11.7.26) and (11.7.27) are two of Maxwell's equations, in the former of which we have eliminated the electric field by using the infinite-conductivity condition.

11.8 The conductivity tensor

In a wide range of plasmas, we will find that Ohm's law reduces to the form (11.7.8), but no further simplification is possible. It is interesting to examine the implications of this equation by examining the relationship between the current density $\mathbf{j}$ and the electric field $\mathbf{E}$. For simplicity of notation, we neglect the second term on the right-hand side of (11.7.8), but this term can always be reinstated if required. Hence we consider the equation

$$\mathbf{j} + \frac{\sigma}{ne}\mathbf{j} \times \mathbf{B} = \sigma \mathbf{E}. \tag{11.8.1}$$

We see from (11.6.6) that this may be re-expressed alternatively as

$$\mathbf{j} + \frac{e}{m_e c \nu}\mathbf{j} \times \mathbf{B} = \sigma \mathbf{E}. \tag{11.8.2}$$

On introducing the unit vector defined by

$$\mathbf{b} = \frac{\mathbf{B}}{B}, \tag{11.8.3}$$

(11.8.2) may be expressed as

$$\mathbf{j}+\omega_{g,e}\tau\mathbf{j}\times\mathbf{b}=\sigma\mathbf{E}, \tag{11.8.4}$$

where $\omega_{g,e}$ is the electron gyrofrequency, and τ is the mean collision time given by

$$\nu\tau=1. \tag{11.8.5}$$

We may seek a solution of this equation of the form

$$\mathbf{j}=\alpha\mathbf{E}+\beta\mathbf{b}\times\mathbf{E}+\gamma(\mathbf{b}\cdot\mathbf{E})\mathbf{b}. \tag{11.8.6}$$

On substituting this expression in (11.8.4), we obtain

$$\alpha\mathbf{E}+\beta\mathbf{b}\times\mathbf{E}+\gamma(\mathbf{b}\cdot\mathbf{E})\mathbf{b}+(\omega_{g,e}\tau)[-\alpha\mathbf{b}\times\mathbf{E}+\beta\mathbf{E}-\beta(\mathbf{b}\cdot\mathbf{E})\mathbf{b}]=\sigma\mathbf{E}. \tag{11.8.7}$$

On examining separately the coefficients of $\mathbf{E}$, $\mathbf{b}\times\mathbf{E}$, and $\mathbf{b}$, we obtain three equations that lead to the following values of α, β, and γ:

$$\alpha=\frac{\sigma}{1+(\omega_{g,e}\tau)^2}, \quad \beta=\frac{(\omega_{g,e}\tau)\sigma}{1+(\omega_{g,e}\tau)^2}, \quad \gamma=\frac{(\omega_{g,e}\tau)^2\sigma}{1+(\omega_{g,e}\tau)^2}. \tag{11.8.8}$$

An equivalent representation of the relationship between current density and electric field is

$$\mathbf{j}=\sigma_\perp\mathbf{E}+(\sigma_\parallel-\sigma_\perp)(\mathbf{b}\cdot\mathbf{E})\mathbf{b}+\sigma_H\mathbf{b}\times\mathbf{E}, \tag{11.8.9}$$

where the parallel ($\sigma_\parallel$), perpendicular ($\sigma_\perp$) and Hall (σ_H) components of the conductivity tensor are given by

$$\sigma_\parallel=\sigma, \quad \sigma_\perp=\frac{\sigma}{1+(\omega_{g,e}\tau)^2}, \quad \sigma_H=\frac{(\omega_{g,e}\tau)\sigma}{1+(\omega_{g,e}\tau)^2}. \tag{11.8.10}$$

If the electric field is parallel to the magnetic field, then $\mathbf{j}=\sigma\mathbf{E}$. If $\mathbf{E}$ is perpendicular to $\mathbf{b}$, there are two components to the current. The component parallel to $\mathbf{E}$ is given by $\sigma_\perp\mathbf{E}$ but, in addition, there is a component perpendicular to both $\mathbf{b}$ and $\mathbf{E}$ given by $\sigma_H\mathbf{b}\times\mathbf{E}$.

If we choose axes such that

$$\mathbf{b}=(0,0,1), \tag{11.8.11}$$

then (11.8.9) may be expressed in matrix form as

$$\begin{pmatrix}j_1\\j_2\\j_3\end{pmatrix}=\begin{pmatrix}\sigma_\perp&-\sigma_H&0\\\sigma_H&\sigma_\perp&0\\0&0&\sigma_\parallel\end{pmatrix}\begin{pmatrix}E_1\\E_2\\E_3\end{pmatrix}. \tag{11.8.12}$$

We see that if the collision frequency is very high, so that $\omega_{g,e}\tau \to 0$, $\sigma_{\parallel} \to \sigma$, $\sigma_{\perp} \to \sigma$, and $\sigma_H \to 0$. Hence the conductivity tensor becomes isotropic, as one would expect. On the other hand, if the collision frequency is very low so that $\omega_{g,e}\tau \to \infty$, then $\sigma_{\perp} \to 0$, and $\sigma_H \to 0$. In this case, an electron current is driven only in the $\mathbf{b}$ direction, by the component of electric field parallel to the magnetic field.

Problems

Problem 11.1. Consider the steady-state motion of electrons in a plasma, assuming that the force due to an applied electric field $\mathbf{E}$ is balanced by a frictional force

$$\mathbf{F} = -m_e \nu_c (\mathbf{v}_e - \mathbf{v}_i),$$

where ν_c is the momentum-exchange collision frequency for electrons with velocity $\mathbf{v}_e$ being scattered by ions with velocity $\mathbf{v}_i$. Hence, assuming that

$$|\mathbf{v}_e| \gg |\mathbf{v}_i|, \tag{1}$$

obtain the following expression for the conductivity

$$\sigma = \frac{ne^2}{m_e c \nu_c} \tag{2}$$

where n_e is the electron density.

Problem 11.2. Examine the derivation of the generalized form of Ohm's law and show that, if the phase velocity is comparable with $|\mathbf{U}|$, the conditions for the neglect of terms that are quadratic in $\mathbf{U}$ and in $\mathbf{j}$ may be expressed as

$$(\omega/\omega_P)^2 \ll (U/c)^2 \tag{1}$$

and

$$(\omega/\omega_P)^2 \ll 1. \tag{2}$$

Problem 11.3. Consider the following form of the generalized Ohm's law

$$\mathbf{E} + \frac{1}{c}\mathbf{U} \times \mathbf{B} - \frac{m_p}{\rho e c}(\mathbf{j} \times \mathbf{B}) - \eta \mathbf{j} = 0.$$

Find expressions for the components of $\mathbf{j}$ parallel to $\mathbf{B}$ and transverse to $\mathbf{B}$.

Problem 11.4. Determine the simplest form of the MHD equations that are appropriate for each of the following situations:

(a) A small-amplitude wave in the solar chromosphere for which $n_e = n_p = 10^{12}$ cm^{-3}, $T = 10^4$ K and $B = 10^2$ G, the wave frequency is 10^{-2} Hz, and the phase velocity varies from the sound speed to the Alfvén speed.

(b) A small-amplitude wave in the solar corona for which $n_e = n_p = 10^8$ cm^{-3}, $T = 10^6$ K and $B = 10$ G, the wave frequency is 10^{-3} Hz, and the phase velocity varies from the sound speed to the Alfvén speed.

(c) A small-amplitude wave in the Earth's magnetosphere for which $n_e = n_p = 10^4$ cm^{-3}, $T = 10^3$ K and $B = 10^{-1}$ G, the wave frequency is 10 Hz, and the phase velocity varies from the sound speed to the Alfvén speed.

12
Magnetohydrodynamics

12.1 Evolution of the magnetic field

In laboratory electromagnetic experiments, one normally regards the electric field as the 'applied' or 'given' quantity, and the current as a quantity 'produced' by the electric field. This is the normal interpretation of Ohm's law. However, in astrophysics and many other situations of plasma physics, it is convenient to interpret Ohm's law rather differently. This is exemplified by (11.7.26) where, for the infinite-conductivity approximation, we find that Ohm's law leads to an equation for the time evolution of the magnetic field in terms of the magnetic field and the velocity field.

We now explore in a little more detail the consequences of Ohm's law, considering the form (11.7.11) rather than the infinite-conductivity form. On rewriting (11.7.11) as

$$\mathbf{E} = \eta \mathbf{j} - \frac{1}{c} \mathbf{v} \times \mathbf{B}, \tag{12.1.1}$$

where we now use $\mathbf{v}$ for the fluid velocity, we obtain from the induction equation (2.1.1) the following equation for the evolution of the magnetic field:

$$\frac{\partial \mathbf{B}}{\partial t} = -\nabla \times (c \eta \mathbf{j}) + \nabla \times (\mathbf{v} \times \mathbf{B}). \tag{12.1.2}$$

On using (11.7.27), this may be re-expressed as

$$\frac{\partial \mathbf{B}}{\partial t} = \nabla \times (\mathbf{v} \times \mathbf{B}) - \frac{c}{4\pi} \nabla \times (\eta \nabla \times \mathbf{B}). \tag{12.1.3}$$

Given the initial magnetic-field distribution, and given the velocity field and resistivity as functions of space and time, this equation would determine the evolution of the magnetic field.

On using the fact that the magnetic field is divergence-free, this equation may be expressed alternatively (in rectangular coordinates) as

$$\frac{\partial \mathbf{B}}{\partial t} = \nabla \times (\mathbf{v} \times \mathbf{B}) - \frac{c}{4\pi} \nabla \eta \times (\nabla \times \mathbf{B}) + \frac{c\eta}{4\pi} \nabla^2 \mathbf{B}. \qquad (12.1.4)$$

The second term on the right-hand side represents the effect of the spatial variation of resistivity on the currents generated by the magnetic field. For simplicity, we will now assume that the resistivity is uniform so that this term disappears.

For reasons that will become clear, we now define a *diffusion coefficient* D by

$$D = \frac{c\eta}{4\pi}. \qquad (12.1.5)$$

Equation (12.1.4) now becomes

$$\frac{\partial \mathbf{B}}{\partial t} = \nabla \times (\mathbf{v} \times \mathbf{B}) + D\nabla^2 \mathbf{B}. \qquad (12.1.6)$$

The first term on the right-hand side is called the *convective* term and will be seen to represent the tendency of magnetic field lines to be 'frozen' in the fluid. The second term is called the *diffusive* term and represents the resistive leakage of magnetic field lines across the conducting fluid.

Either of these two competing effects can become dominant, depending on the relevant time- and length-scales. It is therefore useful to introduce the following dimensionless parameter called the *magnetic Reynolds number*, in analogy with the hydrodynamic Reynolds number:

$$R_{\mathrm{M}} = \frac{|\text{convective term}|}{|\text{diffusive term}|} = \frac{L^{-1}vB}{DL^{-2}B}, \qquad (12.1.7)$$

so that

$$R_{\mathrm{M}} = \frac{Lv}{D}. \qquad (12.1.8)$$

In this equation we are, of course, referring only to the orders of magnitude of the relevant quantities. L is a length characteristic of the spatial variation of the magnetic field. It is clear that either diffusion or convection becomes dominant depending on whether $R_{\mathrm{M}} \ll 1$ or $R_{\mathrm{M}} \gg 1$. For laboratory experiments involving liquids such as mercury or sodium, $R_{\mathrm{M}} \ll 1$ for typical laboratory velocities. However, in geophysics and astrophysics, R_{M} is normally a very large number.

For the special case of a fully ionized hydrogen plasma, we may use the expressions of (11.6.10). We then find that the diffusion coefficient is expressible as

$$D = 10^{13.1} T^{-3/2} \qquad\qquad (12.1.9)$$

so that (12.1.8) becomes

$$R_M = 10^{-13.1} T^{3/2} L v. \qquad\qquad (12.1.10)$$

Consider, as an example, the situation in the Earth's magnetosphere. A typical temperature is $T = 10^4$ K, and we may take as a typical length-scale the Earth's radius so that $L \sim 10^9$ cm. If we take, for the velocity, the speed of the jet stream, about 10^4 cm s^{-1}, we find that $R_M \sim 10^6$.

As another example, consider the evolution of the magnetic field in an active region of the solar corona. In this situation, $T \sim 10^6$ K, $L \sim 10^9$ cm and typical motions of the photosphere lead to $v \sim 10^5$ cm s^{-1}. We find that this leads to $R_M \sim 10^{10}$.

We see that, in both the geophysical and solar-physics examples, the effect of the diffusive term is completely negligible provided the relevant length-scales are those of the macroscopic system. However, we must be cautious in making this assumption. For instance, if we were to consider a situation in which one pair of sunspots moves towards another pair of sunspots, the fact that the plasma is highly conducting inhibits the field from adopting a potential configuration. Instead, a region with very high magnetic-field gradient, termed a 'current sheet,' tends to develop.

12.2 Frozen magnetic field lines

Let us now consider that the conductivity may be assumed to be infinite or, equivalently, the magnetic Reynolds number may be taken to be infinite. Then (12.1.6) reduces to

$$\frac{\partial \mathbf{B}}{\partial t} = \nabla \times (\mathbf{v} \times \mathbf{B}). \qquad\qquad (12.2.1)$$

Let Γ be a closed contour bounding a surface S in a time varying, non-uniform magnetic field $\mathbf{B}(\mathbf{x}, t)$. Our goal is to evaluate the rate of change of the flux passing through this surface, assuming that the contour moves with the fluid that has been assumed to be perfectly conducting. In a small time interval Δt, each point on the closed curve Γ moves a small distance $\mathbf{v}\Delta t$, to become the new contour Γ'. Each point on the surface S has now been displaced and maps out a new surface S' that is bounded by the contour Γ'.

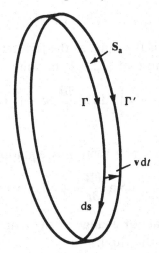

Fig. 12.1. The contour Γ moves with the plasma to become the contour Γ' after time dt. The area of the annular region between the two contours is S_a.

(See Fig. 12.1.) The thin annular surface connecting the contours Γ and Γ' is denoted by S_a.

We now wish to compare the magnetic flux embraced by the contour Γ at time t and that embraced by the contour Γ' at time $t+\Delta t$. We write

$$\Phi(S, t) = \int_S dS\,\mathbf{n}\cdot\mathbf{B}(\mathbf{x}, t) \qquad (12.2.2)$$

and

$$\Phi(S', t+\Delta t) = \int_{S'} dS\,\mathbf{n}\cdot\mathbf{B}(\mathbf{x}, t+\Delta t), \qquad (12.2.3)$$

where n is the unit vector normal to the surface. By expanding $\mathbf{B}(\mathbf{x}, t+\Delta t)$ in (12.2.3), we see that, to first order in Δt,

$$\Phi(S', t+\Delta t) = \int_{S'} dS\,\mathbf{n}\cdot\mathbf{B}(\mathbf{x}, t) + \int_S dS\,\mathbf{n}\cdot\frac{\partial\mathbf{B}}{\partial t}\Delta t. \qquad (12.2.4)$$

The change introduced by integrating the second term over the area S rather than the area S' is of second order and so may be neglected.

The magnetic flux that passes through the surface S_a is given, to first order in Δt, by

$$\Phi(S_a, t) = \oint_\Gamma (\mathbf{ds}\times\mathbf{v}\Delta t)\cdot\mathbf{B}(\mathbf{x}, t). \qquad (12.2.5)$$

Since the magnetic field is divergence-free, we see that

$$\Phi(S, t) = \Phi(S', t) + \Phi(S_a, t). \qquad (12.2.6)$$

The first term in (12.2.6) is the same as the first term on the right-hand side of (12.2.4). Hence we find that (12.2.4) becomes

$$\Phi(S', t+\Delta t) - \Phi(S, t) = \int_S dS\, \mathbf{n} \cdot \frac{\partial \mathbf{B}}{\partial t} \Delta t - \oint_\Gamma (d\mathbf{s} \times \mathbf{v}\Delta t) \cdot \mathbf{B}(\mathbf{x}, t).$$

$$(12.2.7)$$

In the limit that $\Delta t \to 0$, we may write this equation as

$$\frac{d\Phi}{dt} = \int_S dS\, \mathbf{n} \cdot \frac{\partial \mathbf{B}}{\partial t} - \oint_\Gamma d\mathbf{s} \cdot (\mathbf{v} \times \mathbf{B}), \qquad (12.2.8)$$

where $d\Phi/dt$ means the rate of change of the magnetic flux embraced by a contour as it moves with the fluid.

On using Stokes' theorem, (12.2.8) becomes

$$\frac{d\Phi}{dt} = \iint_S dS\, \mathbf{n} \cdot \left[\frac{\partial \mathbf{B}}{\partial t} - \nabla \times (\mathbf{v} \times \mathbf{B}) \right] = 0 \qquad (12.2.9)$$

where we have used (12.2.1).

We have now shown that the magnetic flux threading any surface S is unchanged as the surface moves with the fluid. Although we cannot deduce directly from this statement that 'magnetic-field lines' are 'frozen' in the plasma, we see that this interpretation is perfectly compatible with (12.2.9).

Another approach to this problem is to expand the right-hand side of (12.2.1) to become (on using $\nabla \cdot \mathbf{B} = 0$),

$$\frac{\partial \mathbf{B}}{\partial t} = (\mathbf{B} \cdot \nabla)\mathbf{v} - (\mathbf{v} \cdot \nabla)\mathbf{B} - (\nabla \cdot \mathbf{v})\mathbf{B}. \qquad (12.2.10)$$

This leads to the equation

$$\frac{d\mathbf{B}}{dt} = (\mathbf{B} \cdot \nabla)\mathbf{v} - (\nabla \cdot \mathbf{v})\mathbf{B} \qquad (12.2.11)$$

for the convective derivative of the magnetic field vector. Since, from the equation of continuity

$$\frac{d\rho}{dt} = -(\nabla \cdot \mathbf{v})\rho, \qquad (12.2.12)$$

we see that

$$\frac{d\mathbf{B}}{dt} = (\mathbf{B} \cdot \nabla)\mathbf{v} + \frac{\mathbf{B}}{\rho} \frac{d\rho}{dt}. \qquad (12.2.13)$$

This may be re-expressed as

$$\frac{\mathrm{d}}{\mathrm{d}t}\left(\frac{\mathbf{B}}{\rho}\right) = \left(\left(\frac{\mathbf{B}}{\rho}\right)\cdot\nabla\right)\mathbf{v}. \qquad (12.2.14)$$

Let us now consider the evolution of a small vector $\delta\mathbf{x}$ connecting two neighboring points in the fluid, as the fluid moves with the prescribed velocity field. The point initially at position $\mathbf{x}$ at time t will be displaced to the position $\mathbf{x} + \mathbf{v}(\mathbf{x})\Delta t$ at time $t + \Delta t$. The point initially at $\mathbf{x} + \delta\mathbf{x}$ at time t will be displaced to the position $\mathbf{x} + \delta\mathbf{x} + \mathbf{v}(\mathbf{x} + \delta\mathbf{x})\Delta t$ at time Δt. Hence we find that

$$\frac{\mathrm{d}}{\mathrm{d}t}(\delta\mathbf{x}) = (\delta\mathbf{x}\cdot\nabla)\mathbf{v}. \qquad (12.2.15)$$

We now see from (12.2.14) and (12.2.15) that $\rho^{-1}\mathbf{B}$ and $\delta\mathbf{x}$ satisfy the same differential equation (for all time). Hence, if initially $\delta\mathbf{x} = \varepsilon\rho^{-1}\mathbf{B}$, this same relation will hold for all time. Hence, if two neighboring particles of the plasma are initially on a field line, one will find subsequently that they are again on a field line. Note that we cannot assert that they remain on the 'same' magnetic-field line unless we have a way of identifying a field line, but we see that we are making no error in asserting that a given field line moves with the plasma.

Another viewpoint is to *define* the 'velocity' of a magnetic field as the velocity of the frame in which the electric field is zero. The velocity $\mathbf{v}_M$ of this frame must satisfy

$$\mathbf{E} + \frac{1}{c}\mathbf{v}_M\times\mathbf{B} = 0. \qquad (12.2.16)$$

Clearly, this is possible only if $\mathbf{E}\cdot\mathbf{B} = 0$, and only the components of $\mathbf{v}_M$ transverse to $\mathbf{B}$ are defined in this way. Hence, if we stipulate that

$$\mathbf{B}\cdot\mathbf{v}_M = 0 \qquad (12.2.17)$$

(noting that a component of $\mathbf{v}$ parallel to $\mathbf{B}$ does not 'move' field lines), $\mathbf{v}_M$ is given by

$$\mathbf{v}_M = c\frac{\mathbf{E}\times\mathbf{B}}{B^2}. \qquad (12.2.18)$$

However, in the infinite-conductivity approximation, $\mathbf{E}$ is given by

$$\mathbf{E} = -\frac{1}{c}\mathbf{v}\times\mathbf{B} \qquad (12.2.19)$$

so that (12.2.18) gives

$$\mathbf{v}_M = \mathbf{v}_\perp \qquad (12.2.20)$$

where the $\mathbf{v}_\perp$ is the component of $\mathbf{v}$ normal to $\mathbf{B}$. Hence, defining the velocity of magnetic field by (12.2.16), it follows immediately that the magnetic field moves with the plasma.

It is sometimes advantageous to describe the magnetic field by parameters that 'label' the field lines (Sturrock and Woodbury, 1967). Since $\nabla \cdot \mathbf{B} = 0$, we can represent $\mathbf{B}$ as follows:

$$\mathbf{B} = \nabla\alpha \times \nabla\beta. \tag{12.2.21}$$

The scalar quantities α and β are known as Clebsch variables. Clearly,

$$\mathbf{B} \cdot \nabla\alpha = 0, \quad \mathbf{B} \cdot \nabla\beta = 0, \tag{12.2.22}$$

so that α and β are constant on each field line. That is, each field line can be labeled by the two scalar quantities α and β.

Since we have found it possible to conceive of field lines as moving with the fluid, we may suspect that it should be possible to describe an evolving magnetic field by scalar quantities α and β such that α and β take constant values for a point that moves with the fluid. We therefore suppose that

$$\left.\begin{aligned} \frac{d\alpha}{dt} &\equiv \frac{\partial\alpha}{\partial t} + \mathbf{v} \cdot \nabla\alpha = 0, \\ \frac{d\beta}{dt} &\equiv \frac{\partial\beta}{\partial t} + \mathbf{v} \cdot \nabla\beta = 0, \end{aligned}\right\} \tag{12.2.23}$$

and we shall seek to see if these equations lead to (12.2.1).

Note that we cannot expect that these equations will be satisfied for all permissible sets of functions $\alpha(\mathbf{x}, t)$, $\beta(\mathbf{x}, t)$ that describe the magnetic field through (12.2.21). For instance, the magnetic field defined by (12.2.21) is unchanged if α and β are changed to α' and β' where

$$\left.\begin{aligned} \alpha'(\mathbf{x}, t) &= \alpha(\mathbf{x}, t) + f(t) + F(\beta), \\ \beta'(\mathbf{x}, t) &= \beta(\mathbf{x}, t) + g(t) + G(\alpha), \end{aligned}\right\} \tag{12.2.24}$$

in which $f(t)$ and $g(t)$ are arbitrary functions of t, $F(\beta)$ is an arbitrary function of β, and $G(\alpha)$ is an arbitrary function of α. Even if equations (12.2.23) are satisfied by α and β, they will not be satisfied by α' and β' for an arbitrary choice of $f(t)$ and $g(t)$.

We calculate the left-hand side of (12.2.1), now using $\dot{\alpha}$ for $\partial\alpha/\partial t$, etc. Then we obtain equation

$$\frac{\partial \mathbf{B}}{\partial t} = \nabla\dot{\alpha} \times \nabla\beta + \nabla\alpha \times \nabla\dot{\beta} = \nabla \times (\dot{\alpha}\nabla\beta - \dot{\beta}\nabla\alpha). \tag{12.2.25}$$

We also find that the right-hand side of (12.2.1) may be expressed as

$$\nabla \times (\mathbf{v} \times \mathbf{B}) = \nabla \times [\mathbf{v} \times (\nabla \alpha \times \nabla \beta)] \qquad (12.2.26)$$

$$= \nabla \times [(\mathbf{v} \cdot \nabla \beta) \nabla \alpha - (\mathbf{v} \cdot \nabla \alpha) \nabla \beta].$$

Hence equation (12.2.1) becomes

$$\nabla \times [(\dot{\alpha} + v \cdot \nabla \alpha) \nabla \beta - (\dot{\beta} + v \cdot \nabla \beta) \nabla \alpha] = 0. \qquad (12.2.27)$$

This equation is certainly satisfied if equations (12.2.23) are satisfied. Hence assuming that α and β retain constant values at each point in the fluid is equivalent to assuming that (12.2.1) is satisfied. However, to repeat a point made earlier, (12.2.1) does not *require* that equations (12.2.23) should be satisfied.

12.3 Diffusion of magnetic field lines

In cases for which $R_M \ll 1$, that is, in cases for which the convective term in (12.1.6) may be neglected, the evolution of the magnetic field (for the case of uniform conductivity) is given by

$$\frac{\partial \mathbf{B}}{\partial t} = D \nabla^2 \mathbf{B}. \qquad (12.3.1)$$

This is the well known diffusion equation, the solution of which is given in most textbooks on partial differential equations.

To obtain some insight into the behavior of the solution, we consider an initial distribution of the magnetic field in the form

$$B_1 = 0, \quad B_2 = 0, \quad B_3(\mathbf{x}, t) = \delta(x_1) \quad \text{at } t = 0. \qquad (12.3.2)$$

That is, we are considering a thin sheet of dense magnetic field oriented in the direction of the x_3 axis and initially confined in the plane $x_1 = 0$. We find that the solution of (12.3.1), for $t > 0$, has the form

$$B_1 = 0, \quad B_2 = 0, \quad B_3(\mathbf{x}, t) = \left(\frac{D}{4\pi t}\right)^{1/2} \exp\left[-\frac{x_1^2}{4Dt}\right]. \qquad (12.3.3)$$

That is, magnetic flux progressively diffuses away from the plane $x_1 = 0$.

The above function is plotted in Fig. 12.2. We see that the width of the flux distribution, that is given approximately by $4(Dt)^{1/2}$, increases as $t^{1/2}$, as is typical for diffusion problems.

It is often helpful to make an order-of-magnitude estimate of the diffusion time of a magnetic-field configuration. If the diffusion time is denoted by τ_D, we see from (12.3.1) that

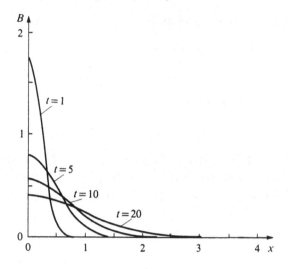

Fig. 12.2. Example of the diffusion of magnetic field, as given by (12.3.3).

$$\frac{B}{\tau_D} \approx \frac{DB}{L^2}, \tag{12.3.4}$$

where L is the characteristic length for the spatially varying magnetic-field distribution. Hence we see that

$$\tau_D \approx D^{-1}L^2. \tag{12.3.5}$$

On using (12.1.5), we see that this may be expressed alternatively as

$$\tau_D \approx 10^{-9.4}\sigma L^2. \tag{12.3.6}$$

For the special case of a fully ionized hydrogen plasma, we see from (12.1.9) that (12.3.5) may be expressed as

$$\tau_D \approx 10^{-13.1}T^{3/2}L^2. \tag{12.3.7}$$

On inserting into (12.3.6) the magnitude $10^{7.3}$ of σ for copper, we find that τ_D is of order 1 s for a copper sphere of radius 10 cm. However, on noting that $T \sim 10^7$ K in the interior of the Sun, and on noting that the radius of the Sun is $10^{10.8}$ cm, we find that the decay time for a dipole field contained in the Sun would be of order $10^{19.0}$ s, i.e. $10^{11.5}$ y, much longer than the age of the Sun. This suggests that any large-scale magnetic field in the Sun is likely to be a 'fossil' magnetic field, that was part of the solar nebula when it formed from the interstellar medium. However, such a conclusion is valid only if our assumptions are valid, that the convective term is negligible and that the characteristic length of variation of the magnetic field is comparable with the solar radius.

In fact, the outer layer of the Sun, from $0.8 R_{\odot}$ out, is in a state of convection. There may also be comparatively slow velocity fields in the internal 'radiative' zone. Hence diffusion theory, of the type we have just considered, is completely inappropriate for discussion of the evolution of the magnetic field in the convection zone. Indeed, the prevalent theory for the origin of the magnetic field that is seen at the Sun's surface is that it is produced by 'dynamo' action. The gist of this theory is that small-scale vortical motion and differential rotation have the combined effect of amplifying a small initial poloidal component of magnetic field.

12.4 The virial theorem

If a plasma cylinder carries a current flowing along its length, this current will generate a magnetic field outside the cylinder, and the pressure of this external magnetic field can balance the gas pressure of the plasma. This is known as 'magnetic confinement,' which we shall discuss further in Chapter 15. This raises the important question of whether it is possible for a finite plasma configuration (sometimes called a 'plasmoid') to be self-confined. In general, one could ask whether the plasma could develop electric and magnetic fields that lead to confinement. At this time, we consider only the more limited question of whether (within the context of MHD theory) the plasma can develop a magnetic field configuration that leads to self-confinement. We now show, with the aid of an important theorem called the 'virial theorem,' that this is not possible.

We consider the equation of motion in the form (11.7.24), that we derived for the ideal MHD approximation. Then the force density on the plasma may be expressed as

$$\mathbf{F} = -\nabla p + \mathbf{j} \times \mathbf{B}. \tag{12.4.1}$$

On using (11.7.27), we find that the force is expressible as the divergence of a stress tensor,

$$F_r = \frac{\partial T_{rs}}{\partial x_s}, \tag{12.4.2}$$

where

$$T_{rs} = -p\delta_{rs} + \frac{1}{4\pi} B_r B_s - \frac{1}{8\pi} B^2 \delta_{rs}. \tag{12.4.3}$$

We wish to examine the possibility that a configuration can be found for which $\mathbf{F} = 0$ everywhere. In order to examine this possibility, we form the quantity

$$V = \int d^3x \, x_r F_r.$$ (12.4.4)

On using (12.4.2), and carrying out an integration by parts, we find that

$$V = \int dS \, n_r x_r T_{rs} - \int d^3x \, T_{rr},$$ (12.4.5)

in which the surface integral is to be taken over a sphere of infinite radius. We see that, since the magnetic field of a dipole falls off as r^{-3}, and since we are assuming that the plasma is confined to a finite volume, the surface integral vanishes. Hence, using (12.4.3), we see that

$$V = \int d^3x \left[3p + \frac{1}{8\pi} B^2 \right].$$ (12.4.6)

If it were possible to find a static configuration for which $\mathbf{F}=0$ everywhere, then clearly $V=0$. However, we see from (12.4.6) that V is necessarily a positive quantity. This contradiction shows that it is not possible to construct a self-confined magnetohydrostatic system.

12.5 Extension of the virial theorem

By retracing the steps that led from (11.5.9) to (11.5.11), we find that the equation of motion (11.7.24) is expressible as

$$\frac{\partial}{\partial t}(\rho v_r) = \frac{\partial T_{rs}}{\partial x_s},$$ (12.5.1)

where we now denote the plasma fluid velocity by v_r, and

$$T_{rs} = -\rho v_r v_s - p\delta_{rs} + \frac{1}{4\pi} B_r B_s - \frac{1}{8\pi} B^2 \delta_{rs}.$$ (12.5.2)

We now consider the symmetric tensor defined by

$$V_{rs} = -\int d^3x \left[x_l \left\{ \frac{\partial}{\partial t}(\rho v_s) - \frac{\partial T_{st}}{\partial x_t} \right\} + x_s \left\{ \frac{\partial}{\partial t}(\rho v_r) - \frac{\partial T_{rt}}{\partial x_t} \right\} \right].$$ (12.5.3)

Clearly, in view of the equation of motion (12.5.1), $V_{rs} \equiv 0$.

We now consider the moment of inertia tensor defined by

$$I_{rs} = \int d^3x \, \rho x_r x_s.$$ (12.5.4)

The time derivative is given by

$$\frac{\mathrm{d}}{\mathrm{d}t} I_{rs} = \int \mathrm{d}^3x \frac{\partial \rho}{\partial t} x_r x_s. \qquad (12.5.5)$$

On using the conservation equation (11.7.23), this becomes

$$\frac{\mathrm{d}}{\mathrm{d}t} I_{rs} = -\int \mathrm{d}^3x \frac{\partial}{\partial x_t} (\rho v_t) x_r x_s. \qquad (12.5.6)$$

On integrating by parts and neglecting the integral over the surface at infinity, we obtain

$$\frac{\mathrm{d}}{\mathrm{d}t} I_{rs} = \int \mathrm{d}^3x [\rho x_r v_s + \rho x_s v_r]. \qquad (12.5.7)$$

Hence

$$\frac{\mathrm{d}^2}{\mathrm{d}t^2} I_{rs} = \int \mathrm{d}^3x \left[x_r \frac{\partial}{\partial t} (\rho v_s) + x_s \frac{\partial}{\partial t} (\rho v_r) \right] \qquad (12.5.8)$$

so that, from (12.5.1),

$$\frac{\mathrm{d}^2}{\mathrm{d}t^2} I_{rs} = \int \mathrm{d}^3x \left[x_r \frac{\partial T_{st}}{\partial x_t} + x_s \frac{\partial T_{rt}}{\partial x_t} \right]. \qquad (12.5.9)$$

On integrating by parts (and again neglecting the term at infinity), we obtain

$$\frac{1}{2} \frac{\mathrm{d}^2}{\mathrm{d}t^2} I_{rs} = -\int \mathrm{d}^3x \, T_{rs}, \qquad (12.5.10)$$

since T_{rs} is symmetric.

If we now contract the tensors in (12.5.10), using the notation

$$I \equiv I_{rr} = \int \mathrm{d}^3x \, \rho x^2, \qquad (12.5.11)$$

and use (12.5.2), we obtain

$$\frac{1}{2} \frac{\mathrm{d}^2 I}{\mathrm{d}t^2} = \int \mathrm{d}^3x \left[\rho v^2 + 3p + \frac{1}{8\pi} B^2 \right]. \qquad (12.5.12)$$

We see that the integrand in (12.5.12) is positive definite, so this equation shows that $\mathrm{d}^2I/\mathrm{d}t^2$ is necessarily positive. This implies that a plasma configuration will necessarily eventually expand.

If the plasma behaves as a perfect gas, the total thermal energy is given by

$$\Theta = \int \mathrm{d}^3x \, \tfrac{3}{2} NkT = \tfrac{3}{2} \int \mathrm{d}^3x \, p. \qquad (12.5.13)$$

Hence, denoting the kinetic energy and magnetic energy as follows,

$$K = \int d^3x \, \tfrac{1}{2} \rho v^2 \tag{12.5.14}$$

and

$$M = \int d^3x \frac{1}{8\pi} B^2, \tag{12.5.15}$$

we see that (12.5.12) may be re-expressed as

$$\frac{1}{2} \frac{d^2 I}{dt^2} = 2K + 2\Theta + M. \tag{12.5.16}$$

In systems that involve gravitation, one must add to the force in (12.4.1) the gravitational force density given by

$$F_{g,r} = -\rho \frac{\partial \Psi}{\partial x_r}. \tag{12.5.17}$$

In this equation, Ψ is the scalar gravitational potential that satisfies the equation

$$\nabla^2 \Psi = 4\pi G \rho. \tag{12.5.18}$$

Since the gravitational potential is expressible as

$$\Psi(x) = -G \int \frac{d^3x' \rho(x')}{|x - x'|}, \tag{12.5.19}$$

the virial expression (12.4.4) now includes the term

$$V_g = -\int d^3x \, x_r \rho \frac{\partial \Psi}{\partial x_r}, \tag{12.5.20}$$

that is

$$V_g = -G \int d^3x \, \rho(x) x_r \int \frac{d^3x' \, (x_r - x_r') \rho(x')}{|x - x'|^3}. \tag{12.5.21}$$

This is expressible as

$$V_g = -\frac{1}{2} G \int \int d^3x \, d^3x' \frac{\rho(x)\rho(x')}{|x - x'|^3} [x_r(x_r - x_r') + x_r'(x_r' - x_r)] \tag{12.5.22}$$

so that V_g may be expressed as

$$V_g = -\frac{1}{2} G \int \int d^3x \, d^3x' \frac{\rho(x)\rho(x')}{|x - x'|}. \tag{12.5.23}$$

Since the gravitational potential energy Ω is defined by

$$\Omega = \tfrac{1}{2} \int d^3x \, \rho(\mathbf{x}) \, \Psi(\mathbf{x}), \qquad (12.5.24)$$

that is

$$\Omega = \tfrac{1}{2} \int d^3x \, d^3x' \, \frac{\rho(\mathbf{x}) \, \rho(\mathbf{x}')}{|\mathbf{x} - \mathbf{x}'|}, \qquad (12.5.25)$$

we see that

$$V_{\mathrm{g}} = -\Omega. \qquad (12.5.26)$$

If this term is included in (12.5.16), the more general form of the virial theorem (including gravitational effects) becomes

$$\frac{1}{2} \frac{d^2 I}{dt^2} = 2K + 2\Theta + M - \Omega. \qquad (12.5.27)$$

We see from (12.5.25) that Ω is positive, so that it is possible to form a static system that involves a gravitational field. Furthermore, this system can be stable. Stars are examples of such a system.

12.6 Stability analysis using the virial theorem

The virial theorem may be used for simple stability analyses, as we shall now show.

Suppose that the system considered in Section 12.5 is in equilibrium. Then, from (12.5.27),

$$2K + 2\Theta + M - \Omega = 0. \qquad (12.6.1)$$

Now suppose that the system is changed in such a way that it expands uniformly by the factor $1 + \varepsilon$, where ε is allowed to be a function of time. Then

$$\mathbf{x} \to \mathbf{x}^* = (1 + \varepsilon)\mathbf{x}. \qquad (12.6.2)$$

Since the mass is unchanged by the perturbation,

$$\rho \to \rho^* = (1 + \varepsilon)^{-3} \rho. \qquad (12.6.3)$$

If we now assume that

$$p \propto \rho^\gamma, \qquad (12.6.4)$$

we find from the expression

$$\Theta = \frac{1}{\gamma-1}\int d^3x\, p, \tag{12.6.5}$$

that

$$\Theta \to \Theta^* = (1+\varepsilon)^{-3(\gamma-1)}\Theta. \tag{12.6.6}$$

If we allow for rotational fluid motions, such as oscillations or rotation, an adiabatic change in scale has the result that

$$v \to v^* = (1+\varepsilon)^{-1}v, \tag{12.6.7}$$

so that

$$K \to K^* = (1+\varepsilon)^{-2}K. \tag{12.6.8}$$

If magnetic flux is conserved,

$$B \to B^* = (1+\varepsilon)^{-2}B, \tag{12.6.9}$$

so that

$$M \to M^* = (1+\varepsilon)^{-1}M. \tag{12.6.10}$$

We see from (12.5.25) that

$$\Omega \to \Omega^* = (1+\varepsilon)^{-1}\Omega. \tag{12.6.11}$$

Finally, we note from (12.5.4) and (12.5.11) that

$$I \to I^* = (1+\varepsilon)^2 I. \tag{12.6.12}$$

On substituting the above expressions in (12.5.27), we see that

$$I\ddot{\varepsilon} = 2(1-2\varepsilon)K + 2[1-3(\gamma-1)\varepsilon]\Theta + (1-\varepsilon)M - (1-\varepsilon)\Omega, \tag{12.6.13}$$

where we have retained only terms of zero order and first order in ε. Terms of zero order cancel out, by virtue of (12.6.1). Terms of first order yield

$$I\ddot{\varepsilon} = -[4K + 6(\gamma-1)\Theta + M - \Omega]\varepsilon. \tag{12.6.14}$$

On using (12.6.1), this reduces to

$$\frac{d^2\varepsilon}{dt^2} + \left[I^{-1}K + 6\left(\gamma - \frac{4}{3}\right)I^{-1}\Theta\right]\varepsilon = 0. \tag{12.6.15}$$

We see from this equation that if in a star the gravitational force is balanced by a combination of gas and magnetic pressures, the star will be unstable if $\gamma \leqslant 4/3$. If $\gamma > 4/3$, we expect that the star can exhibit oscillatory modes of compression and dilation, the period of oscillation being determined by the time it takes a sound wave to traverse the star. Moreover, rotational motions can help to stabilize the star.

Problems

Problem 12.1. Consider the evolution of the magnetic field in a static plasma, using equation (12.1.3).

(a) Show that the right-hand side of the equation may be re-expressed as a diffusion term plus an additional term:

$$\frac{\partial B_r}{\partial t} = \frac{\partial}{\partial x_s}\left(D\frac{\partial B_r}{\partial x_s}\right) - \frac{\partial D}{\partial x_s}\frac{\partial B_s}{\partial x_r}. \tag{1}$$

(b) Examine the significance of the second term on the right-hand side by supposing that $\eta = \eta(x_3)$ and assuming that, at $t=0$, **B** is uniform in the x_3 direction.

(c) Now suppose that $\eta = \eta(x_1)$ and $B = (0, 0, B_3\,(x_1, t)\,)$. Show that (1) reduces to the diffusion equation

$$\frac{\partial B_3}{\partial t} = \frac{\partial}{\partial x_1}\left(D\frac{\partial B_3}{\partial x_1}\right), \tag{2}$$

and show that, in the limit that $t \to \infty$ for which B_3 becomes stationary, the magnetic field becomes homogeneous, provided that $D \neq 0$ everywhere.

Problem 12.2. Consider the equation that describes the diffusion of magnetic field in a plasma.

(a) Suppose that the field has only one component B_y that is a function only of x and t. Show that if D, the diffusion coefficient, is constant and the field configuration at $t=0$ is given by

$$B_y(x, 0) = f(x), \tag{1}$$

the evolution of the field for $t > 0$ may be obtained from

$$B(x, t) = (4\pi Dt)^{-1/2} \int_{-\infty}^{\infty} f(s) \exp\left[-\frac{(s-x)^2}{4Dt}\right] ds. \tag{2}$$

(b) Now consider the situation that the plasma is confined to the region $x > 0$, and that the region $x < 0$ is a vacuum. Suppose that at $t=0$ a uniform magnetic field permeates the vacuum region but does not extend into the plasma. Obtain an expression for the evolution of the magnetic field.

(c) Can you offer any physical interpretation of the results of parts (a) and (b)?

Problem 12.3. Consider an inviscid fluid of uniform conductivity that is in a state of steady (but possibly non-uniform) rotation about the z-axis in the presence of a magnetic field that has no azimuthal component: $\mathbf{B} = (B_z, B_r; 0)$.

(a) Show that

$$\nabla \times \mathbf{j} = \frac{\sigma}{c} \nabla \times (\mathbf{v} \times \mathbf{B}).$$

(b) Noting that the current is wholly azimuthal and that $\nabla \cdot \mathbf{B} = 0$, show that the fluid has constant uniform angular velocity along a field line.

Problem 12.4. Extend the argument of Section 12.4 to show that it is not possible to construct a self-confined plasma system, even when one takes into account a possible electric field.

13

Force-free magnetic-field configurations

13.1 Introduction

In many astrophysical situations, such as stellar atmospheres and stellar interiors, we need to include in the equation of motion for a plasma the gravitational field. If we introduce such a term, then the ideal MHD equation of motion (11.7.24) becomes

$$\rho \frac{d\mathbf{v}}{dt} = \mathbf{j} \times \mathbf{B} - \nabla p + \rho \mathbf{g} \qquad (13.1.1)$$

where $\mathbf{g}$ is the gravitational acceleration vector that may be written in terms of a gravitational scalar potential Ψ as

$$\mathbf{g} = -\nabla \Psi. \qquad (13.1.2)$$

We now restrict ourselves to the consideration of stationary states, that is, to magnetostatic configurations. The condition for such equilibria is that

$$\mathbf{j} \times \mathbf{B} - \nabla p + \rho \mathbf{g} = 0. \qquad (13.1.3)$$

In a stellar interior, it is likely that the dominant terms in this equation will be the ∇p and $\rho \mathbf{g}$ terms. However, in stellar atmospheres such as the solar corona, it is frequently the case that the magnetic field plays the dominant role. If this is the case, then the equation for magnetostatic equilibrium reduces to the equation

$$\mathbf{j} \times \mathbf{B} = 0. \qquad (13.1.4)$$

A magnetic field that satisfies this equation is said to be 'force free,' meaning that the 'self force,' i.e. the Lorentz force, is zero.

Equation (13.1.4) is deceptively simple. It is not an easy matter to calculate force-free magnetic-field configurations. At the root of this difficulty is the fact that the above equation, that may be rewritten as

$$(\nabla \times \mathbf{B}) \times \mathbf{B} = 0, \tag{13.1.5}$$

is nonlinear. If $\mathbf{B}^{(1)}$ and $\mathbf{B}^{(2)}$ are solutions of (13.1.5), it does *not* follow that $\mathbf{B}^{(1)} + \mathbf{B}^{(2)}$ is a solution.

In a real situation, equations (13.1.4) will clearly be true only approximately. In general, even if p and ρ are small, $\mathbf{j}$ and $\mathbf{B}$ will not be exactly parallel. If χ is the angle between $\mathbf{j}$ and $\mathbf{B}$, so that

$$\sin \chi = \frac{|\mathbf{j} \times \mathbf{B}|}{jB}, \tag{13.1.6}$$

it is clear that we may estimate the magnitude of χ from the equation

$$|\sin \chi| \approx \frac{|\nabla p|}{|jB|} + \frac{|\rho g|}{|jB|}. \tag{13.1.7}$$

If we first ignore the gravitational field and suppose that the spatial variations of the magnetic field and gas pressure may be characterized by the lengths L_{m} and L_{p}, respectively, then the order of magnitude of $\sin \chi$ may be estimated from

$$\sin \chi \approx \frac{4\pi p L_{\mathrm{m}}}{B^2 L_{\mathrm{p}}}. \tag{13.1.8}$$

Hence

$$\sin \chi \approx \tfrac{1}{2} \beta \frac{L_{\mathrm{m}}}{L_{\mathrm{p}}}, \tag{13.1.9}$$

where β is the ratio of plasma pressure to magnetic pressure, p_{m}, defined by

$$\beta = \frac{p}{p_{\mathrm{m}}} = \frac{8\pi p}{B^2}. \tag{13.1.10}$$

Hence we see that, unless L_{p} is much less than L_{m}, the force-free assumption will be a good approximation in a 'low-beta' plasma.

We next consider the role of a gravitational field. The length-scale for the variation of the plasma density and pressure with height is determined by the 'scale height' H, defined by

$$H = \frac{p}{\rho g} = \frac{kT}{m_{\mathrm{av}} g}. \tag{13.1.11}$$

Hence we find that the influence of the gravitational term on the angle χ may be estimated from

$$\sin x \approx 2\beta \frac{L_{\mathrm{m}}}{H}. \tag{13.1.12}$$

Hence we see that, unless H is much less than L_m, the force-free assumption will be a good approximation in a 'low-beta' plasma.

It is found that, in many astrophysical situations of interest, the force-free approximation is not unrealistic.

Force-free magnetic fields cannot be simple in structure. There are limitations on the possible topological structures of such configurations. For instance, it is not possible that a force-free field contain a set of nested field lines, that all lie in a common surface, as depicted in Fig. 13.1. We can see this by supposing that such a set of field lines does exist in a surface S.

If one such field line runs along the closed contour Γ in the surface S, then

$$\oint_\Gamma d\mathbf{s} \cdot \mathbf{B} = \int dS\, \mathbf{n} \cdot (\nabla \times \mathbf{B}). \tag{13.1.13}$$

This may be rewritten as

$$\oint_\Gamma d\mathbf{s} \cdot \mathbf{B} = 4\pi \int dS\, \mathbf{n} \cdot \mathbf{j}. \tag{13.1.14}$$

However, since the field is force-free, $\mathbf{j}$ is parallel to $\mathbf{B}$. Since $\mathbf{B}$ lies in the surface S, $\mathbf{n} \cdot \mathbf{B} = 0$ so that $\mathbf{n} \cdot \mathbf{j} = 0$. Hence the right-hand side of (13.1.14) is zero. However, we began by assuming that the configuration is such that the left-hand side of the equation is nonzero. Hence we have a contradiction, showing that our basic assumption is incorrect. It is not possible for a force-free field to contain a set of nested field lines that all lie in a common surface.

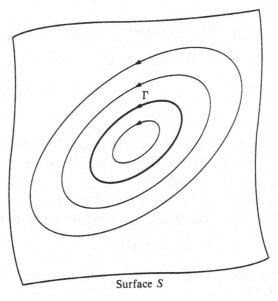

Fig. 13.1. Schematic representation of a set of nested field lines lying in a surface S.

13.2 Linear force-free fields

We have pointed out that (13.1.5) defines a nonlinear problem. However, there is an important subclass of force-free fields that are the solution of a *linear* equation.

It is clear from (13.1.5) that

$$\nabla \times \mathbf{B} = \lambda(\mathbf{x})\mathbf{B} \qquad (13.2.1)$$

where we have emphasized that, in general, λ is a function of the spatial coordinates. On taking the divergence of both sides of the above equation and using the fact that the magnetic field is divergence-free, we see that

$$\mathbf{B} \cdot \nabla \lambda(\mathbf{x}) = 0, \qquad (13.2.2)$$

showing that, quite generally, the function λ that relates $\nabla \times \mathbf{B}$ to $\mathbf{B}$ must be constant on any field line. Let us now consider two particular solutions $\mathbf{B}^{(1)}$ and $\mathbf{B}^{(2)}$, with corresponding forms $\lambda^{(1)}$ and $\lambda^{(2)}$ of the coefficient $\lambda(\mathbf{x})$. Then

$$\left.\begin{array}{l} \nabla \times \mathbf{B}^{(1)} = \lambda^{(1)}(\mathbf{x})\,\mathbf{B}^{(1)}, \\ \nabla \times \mathbf{B}^{(2)} = \lambda^{(2)}(\mathbf{x})\,\mathbf{B}^{(2)}. \end{array}\right\} \qquad (13.2.3)$$

On using these equations, we find that

$$[\nabla \times (\mathbf{B}^{(1)} + \mathbf{B}^{(2)})] \times (\mathbf{B}^{(1)} + \mathbf{B}^{(2)}) = [\lambda^{(1)}(\mathbf{x}) - \lambda^{(2)}(\mathbf{x})](\mathbf{B}^{(1)} \times \mathbf{B}^{(2)}).$$
$$(13.2.4)$$

Hence we see that $\mathbf{B}^{(1)} + \mathbf{B}^{(2)}$ also represents a force-free field if $\lambda^{(1)}(\mathbf{x}) = \lambda^{(2)}(\mathbf{x})$ everywhere.

If we now consider the class of force-free fields for which

$$\lambda(\mathbf{x}) = \text{constant}, \qquad (13.2.5)$$

so that

$$\nabla \times \mathbf{B} = \lambda \mathbf{B}, \qquad (13.2.6)$$

where λ is now constant, we see that the magnetic-field vector is now defined by a *linear* partial differential equation. Hence any linear combination of solutions of (13.2.6) is itself a solution.

On taking the curl of (13.2.6), we see that a necessary (but not sufficient) condition for a linear force-free field is that it satisfy the Helmholtz equation

$$\nabla^2 \mathbf{B} + \lambda^2 \mathbf{B} = 0. \qquad (13.2.7)$$

In order to determine a magnetic-field configuration, it is necessary to specify not only the equation that the field satisfies, but also the boundary conditions on the field. In considering linear force-free fields, for a specified value of the coefficient λ, it is sufficient to specify the values of the components of **B** on a closed surface, or the derivatives of the components of **B** on that surface. However, for the more general and more realistic case of nonlinear force-free fields, the choice of boundary conditions is not so simple. This matter will be discussed further in Section 13.5.

Before leaving the subject of linear force-free fields, it is worth pointing out that certain types of model, that one might wish to construct, cannot be found within the context of *linear* force-free-field theory. For instance, consider the case of a sunspot of dimensions D and assume that the spot is unipolar, is surrounded by an annular region of similar dimensions, and that the flux through the annulus is equal and opposite to that through the spot (see Fig. 13.2). If the field is current-free, then the field will be similar to that of a dipole with moment of order $B_0 D^3$ at the origin, so that the magnitude of the field at a large distance r will be given approximately by

$$B(r) \approx B_0 D^3 r^{-3}. \tag{13.2.8}$$

Now suppose that the spot is rotated with respect to the annulus. Then the field will no longer be current-free, but we can assume that it remains force-free. We expect that, at a large distance r, the field will still fall off with radius, with a scale distance that is of order r. The important point is that such a behavior cannot be modeled by a linear force-free field. We see, from (13.2.6), that if the field falls off as r^{-n}, then, approximately,

$$r^{-1} \approx \lambda. \tag{13.2.9}$$

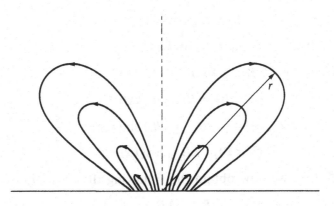

Fig. 13.2. Schematic representation of a cylindrically symmetric force-free field produced by the differential rotation of an internal region and an external flux region of opposite polarity.

However, λ is assumed to be a constant, whereas r is a variable. Hence the above relationship cannot be true for all values of r. Hence the field cannot be modeled, over the entire space, as a linear force-free field.

The fact is that any linear force-free field cannot become 'smooth' at large distances. The magnetic field gradients tend to have the same length-scale over all space. This is obviously a severe handicap in many cases, such as the one we have just discussed.

13.3 Examples of linear force-free fields

In order to understand some properties of force-free fields in general, and some of the limitations of linear force-free fields in particular, it is helpful to consider some special examples. The first that we consider can be constructed in rectangular coordinates.

Consider a field configuration that is uniform in the y dimension, and for which each component is sinusoidal in the x-direction with wave number k. On examining (13.2.7), it appears that the field can vary exponentially in z, for example as e^{-lz}, if λ satisfies the condition $\lambda^2 < k^2$. On the other hand, if $\lambda^2 > k^2$, we could construct a solution by allowing for sinusoidal variation in the z-direction also, with wave number $(\lambda^2 - k^2)^{1/2}$.

If we are attempting to model the magnetic field of a solar active region, for instance, we will seek field configurations that extend throughout the half space $z > 0$, and decrease in strength with z. For this purpose, we must restrict our attention to values of λ that are, in magnitude, less than the magnitude of k. With this assumption, we may seek solutions of the form

$$\left.\begin{aligned}
B_x &= B_{x,0}\sin(kx)e^{-lz}, \\
B_y &= B_{y,0}\sin(kx)e^{-lz}, \\
B_z &= B_0\cos(kx)e^{-lz}.
\end{aligned}\right\} \tag{13.3.1}$$

The components of $\nabla \times \mathbf{B}$ are found to be

$$\left.\begin{aligned}
(\nabla \times \mathbf{B})_x &= lB_{y,0}\sin(kx)e^{-lz}, \\
(\nabla \times \mathbf{B})_y &= (-lB_{x,0} + kB_0)\sin(kx)e^{-lz}, \\
(\nabla \times \mathbf{B})_z &= kB_{y,0}\cos(kx)e^{-lz}.
\end{aligned}\right\} \tag{13.3.2}$$

On using (13.2.6), we now obtain the following three equations

$$\left.\begin{aligned}
lB_{y,0} &= \lambda B_{x,0}, \\
-lB_{x,0} + kB_0 &= \lambda B_{y,0}, \\
kB_{y,0} &= \lambda B_0.
\end{aligned}\right\} \tag{13.3.3}$$

Hence the field (13.3.1) may be expressed as

$$
\left.\begin{array}{l}
B_x = k^{-1}lB_0 \sin(kx)e^{-lz}, \\[4pt]
B_y = k^{-1}\lambda B_0 \sin(kx)e^{-lz}, \\[4pt]
B_z = B_0 \cos(kx)e^{-lz},
\end{array}\right\}
\tag{13.3.4}
$$

where k, l and λ must be related by

$$
l^2 = k^2 - \lambda^2.
\tag{13.3.5}
$$

This is the relationship that we would expect on the basis of the Helmholtz equation (13.2.7).

We see from (13.3.4) that the projection of magnetic field lines on the x-y plane are all parallel straight lines defined by

$$
B_y = \frac{\lambda}{(k^2 - \lambda^2)^{1/2}} B_x.
\tag{13.3.6}
$$

The projections of magnetic field lines on the x-z plane are as shown in Fig. 13.3(a). A physical interpretation of this model is that we begin with a potential field of the form shown in (13.3.4) with $\lambda = 0$. We then shear the x-y plane in such a way that strips parallel to the y-axis are displaced in the y-direction. This shear is uniform, as shown in Fig. 13.3(b). If θ is the angle of inclination of field lines to the x-axis, we see from (13.3.6) that

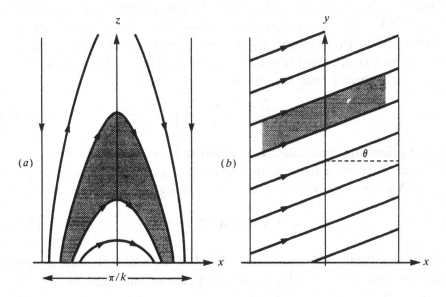

Fig. 13.3. Panel (a) shows the projection in the x-z plane of the force-free field defined by (13.3.4). Panel (b) shows the projection in the x-y plane. (Figure reproduced with kind permission from Priest 1982.)

$$\sin \theta = \lambda / k. \qquad (13.3.7)$$

Hence increasing the shear (increasing θ) amounts to increasing λ. However this equation also shows that there is a limiting value λ: it cannot increase beyond the value $\lambda = k$. At this point, the field lines have the form

$$\left.\begin{aligned} B_x &= 0, \\ B_y &= B_0 \sin(kx), \\ B_z &= B_0 \cos(kx). \end{aligned}\right\} \qquad (13.3.8)$$

This describes a magnetic-field configuration of constant strength, for which the direction of the magnetic-field vector rotates steadily with increasing x.

In this model, field lines extend to greater and greater height as the shear increases. The limiting case $\theta = \pi/2$ corresponds to infinite shear. In this case, all field lines extend to infinite height. On the other hand, there is no change in the horizontal magnetic-field configuration as shear increases. This is due to the fact that we have assumed an infinite array of line dipoles. We shall see, in a later section, that if we consider only a single line dipole, increasing shear causes field lines to expand horizontally as well as vertically.

13.4 The generating-function method

We again consider a magnetic-field configuration that has translational symmetry. (A similar analysis could be derived for a configuration of rotational (i.e. cylindrical) symmetry.) In terms of rectangular coordinates x, y, z, we assume that the field is uniform in the z-direction. Since $\nabla \cdot \mathbf{B} = 0$, we see that the magnetic field may be expressed as

$$\mathbf{B} = \left(\frac{\partial A}{\partial y}, \ -\frac{\partial A}{\partial x}, \ B_z \right). \qquad (13.4.1)$$

In the following calculation, it is instructive to allow for a possible gas pressure in order to see the relationship between magnetic stress and plasma pressure. We therefore consider that the following force equation is satisfied

$$\frac{1}{4\pi} (\nabla \times \mathbf{B}) \times \mathbf{B} - \nabla p = 0. \qquad (13.4.2)$$

On examining the three components of (13.4.2), we find that, since no quantity depends upon the variable z, the third component becomes

$$\frac{\partial B_z}{\partial x} \frac{\partial A}{\partial y} - \frac{\partial B_z}{\partial y} \frac{\partial A}{\partial x} = 0. \qquad (13.4.3)$$

This shows that the vectors $\nabla_\perp B_z$ and $\nabla_\perp A$ are parallel, where $\nabla_\perp$ refers to the gradient of a quantity in the x–y plane. Hence B_z is expressible as a function of A,

$$B_z(x, y) = G(A(x, y)). \tag{13.4.4}$$

Using (13.4.4), we find that the x and y equations of (13.4.2) are expressible as

$$\left.\begin{aligned}
\frac{1}{4\pi}(GG' + \nabla_\perp^2 A)\frac{\partial A}{\partial x} + \frac{\partial p}{\partial x} &= 0, \\[2mm]
\frac{1}{4\pi}(GG' + \nabla_\perp^2 A)\frac{\partial A}{\partial y} + \frac{\partial p}{\partial y} &= 0.
\end{aligned}\right\} \tag{13.4.5}$$

Hence $\nabla_\perp p$ is parallel to $\nabla_\perp A$, so that the pressure is expressible as

$$p(x, y) = P(A(x, y)). \tag{13.4.6}$$

We now see that both equations (13.4.5) are satisfied if

$$\frac{1}{4\pi}\nabla_\perp^2 A + \frac{1}{4\pi}GG' + P' = 0. \tag{13.4.7}$$

Note that this can be expressed as

$$\frac{1}{4\pi}\nabla_\perp^2 A + P_t' = 0 \tag{13.4.8}$$

where P_t, that is expressible as $P_t(A)$, is the *total* pressure:

$$P_t = \frac{1}{8\pi}B_z^2 + p. \tag{13.4.9}$$

Equation (13.4.8) is known as the 'Grad-Shafranov equation.'

We see from (13.4.9) that the z-component of magnetic field plays the same role in determining the magnetic-field configuration as does plasma pressure. We expect that increasing gas pressure will tend to 'inflate' a magnetic-field configuration. Hence we should also expect that shearing the footpoints of a magnetic-field configuration will also lead to inflation. We found this to be the case in Section 13.3, and we shall find, from later studies, that this conjecture is generally correct.

From this point on, we neglect the role of plasma pressure, and so return to the consideration of force-free magnetic-field configurations. The 'generating-function' method of computing force-free magnetic-field configurations is the procedure of defining a *family* of solutions of the Grad-Shafranov equation, which now takes the form

$$\nabla_\perp^2 A + GG' = 0. \tag{13.4.10}$$

By assuming that

$$B_z \equiv G(A) = \lambda F(A), \tag{13.4.11}$$

(13.4.10) takes the form

$$\nabla_\perp^2 A + \lambda^2 f(A) = 0, \tag{13.4.12}$$

where

$$f(A) = F(A)F'(A). \tag{13.4.13}$$

We may refer to $f(A)$ as a 'generating function,' since (13.4.12) 'generates' a family of magnetic-field configurations for a range of values of the parameter λ^2.

A specific example of a generating-function model is one discussed originally by Low (1977) and later by Birn, Goldstein and Schindler (1978) and by Priest and Milne (1980). In our notation, the generating function is given by

$$f(A) = -k^2 \exp(-2A), \tag{13.4.14}$$

corresponding to the following constraint on the B_z component of the magnetic field:

$$B_z = \lambda F(A) \equiv \lambda k \exp(-A). \tag{13.4.15}$$

In order to determine a specific magnetic-field configuration, it is of course necessary to specify the boundary conditions. The following condition is adopted on the plane $y = 0$:

$$A(x, 0) = \ln(1 + k^2 x^2). \tag{13.4.16}$$

Above the plane, at large distances from the origin, it is assumed that

$$|\nabla_\perp A| \to 0 \quad \text{as} \quad x^2 + y^2 \to \infty. \tag{13.4.17}$$

The following form for $A(x, y)$ is found to satisfy both (13.4.12) and the boundary conditions (13.4.16) and (13.4.17):

$$A(x, y) = \ln\left[1 + k^2 x^2 + 2\left(\frac{1 - \mu^2}{1 + \mu^2} \right) ky + k^2 y^2 \right], \tag{13.4.18}$$

where λ and μ are related by

$$\lambda = \frac{4\mu}{1 + \mu^2}. \tag{13.4.19}$$

An important point concerning magnetic-field configurations produced by the generating-function method is that the topology of the field can change as the multiplying parameter is changed. If we allow the parameter μ to increase from zero to infinity, we find that λ increases from zero to a maximum of 2 (at $\mu = 1$) and then decreases back down to zero. For $0 \leqslant \mu \leqslant 1$, the magnetic-field configuration is that of a simple arcade in which the distribution of footpoints at the surface $y = 0$ is given by

$$kz = \frac{2\mu}{1 + \mu^2} \arcsin \left\{ \left[\left(\frac{1 - \mu^2}{1 + \mu^2} \right)^2 + k^2 x^2 \right]^{-1/2} kx \right\}. \qquad (13.4.20)$$

However, for $\mu > 1$, one finds that the magnetic-field configuration is no longer that of a simple arcade. It contains a flux tube that runs above and parallel to the surface $y = 0$. This flux tube may be referred to as 'floating flux.' The distribution of footpoints is now given by

$$kz = \frac{2\mu}{1 + \mu^2} \left(\pi \frac{x}{|x|} - \arcsin \left\{ \left[\left(\frac{1 - \mu^2}{1 + \mu^2} \right)^2 + k^2 x^2 \right]^{-1/2} kx \right\} \right). \qquad (13.4.21)$$

Fig. 13.4 shows the distribution of footpoint z-displacements for various values of μ, as defined by (13.4.20) and (13.4.21).

There has been a great deal of discussion concerning the possible physical significance of magnetic-field sequences produced by the generating-function method. Consider, for instance, the possible relevance of the above sequence of configurations to the situation in the solar corona, where the distribution of magnetic flux at the photosphere (the plane $y = 0$) is as specified by

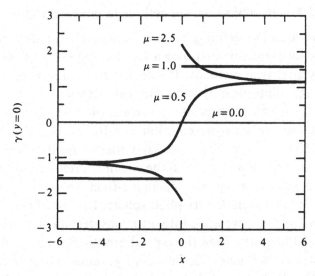

Fig. 13.4. The distribution of footpoint z-displacements, for four different values of μ, as determined by (13.4.20) and (13.4.21). (Figure reproduced with kind permission from Low 1977.)

(13.4.16). We may now ask how the magnetic field will evolve if the foot-points are sheared as specified by the sequence (13.4.20) for $0 \leqslant \mu \leqslant 1$, and then by the continuing function (13.4.21) for $\mu > 1$. Over the range $0 \leqslant \mu \leqslant 1$, there is no apparent problem, and one can infer that the magnetic field would be that described by the vector potential form (13.4.18). On the other hand, when μ increases beyond unity, the topology of the field must change to incorporate 'floating flux.' Since there is no way that a simple shearing motion at the photosphere could introduce such flux, it has been proposed by Low (1977) and others that there is a 'loss of equilibrium,' and that this loss of equilibrium may be relevant to solar 'eruptive events,' such as 'coronal mass ejections.'

This problem has been discussed by Klimchuk and Sturrock (1989), who conclude that the proposed 'loss of equilibrium' has no physical basis. The magnetic-field sequence terminates not because some instability necessarily sets in, but because unphysical requirements are being made on the magnetic-field configuration.

In support of the above claim, Klimchuk and Sturrock have shown that it is possible to calculate a well behaved sequence of magnetic-field configurations (without floating flux) by specifying the magnetic-field configuration at the photosphere through (13.4.16), and by specifying the movement of footpoints through the two sequences (13.4.20) and (13.4.21).

13.5 Calculation of magnetic-field configurations

Since the basic equation defining force-free magnetic fields is intrinsically nonlinear, it is not surprising that only a very limited range of solutions may be obtained by analytical methods. Hence we can obtain more general, and more interesting, solutions only by numerical methods.

In order to devise an appropriate numerical method, we need to decide upon the most appropriate representation of a force-free field. This should be decided upon the basis of the physical systems that need to be investigated.

A typical problem is that related to the example discussed in the previous section. In the solar atmosphere, magnetic-field lines are rooted in the highly conducting and high-density photosphere. In view of the 'frozen-flux' theorem of Section 12.2, we are entitled to consider magnetic field lines as having a physical identity. Then the typical problem is that of determining the evolution of the total magnetic-field configuration that is brought about by motion of the footpoints of the magnetic field lines. For such problems, it is appropriate to describe the magnetic field by Clebsch variables, that were introduced in Section 12.2, since these variables 'label' magnetic field lines. We therefore consider once more the description

$$\mathbf{B} = \nabla\alpha \times \nabla\beta. \tag{13.5.1}$$

We may determine a set of nonlinear partial differential equations governing the variables α and β by substituting (13.5.1) in (13.1.5). These equations are of fourth order in α and β, and involve derivatives of up to second order.

The equations can be simplified, and the simpler set of equations may be derived by noting that, since $\mathbf{j}$ and $\mathbf{B}$ are parallel, equations (12.2.22) lead to

$$\mathbf{j}\cdot\nabla\alpha = 0, \quad \mathbf{j}\cdot\nabla\beta = 0. \tag{13.5.2}$$

On expressing $\mathbf{j}$ in terms of the Clebsch variables, these equations become

$$[\nabla \times (\nabla\alpha \times \nabla\beta)]\cdot\nabla\alpha = 0, \quad [\nabla \times (\nabla\alpha \times \nabla\beta)]\cdot\nabla\beta = 0. \tag{13.5.3}$$

These equations are of only third order in α and β.

When written out explicitly, these equations look rather formidable. Nevertheless, they may be solved by a relaxation procedure introduced by Sturrock and Woodbury (1967) for a problem in rectangular Cartesian coordinates.

Another procedure is to consider the variational equation

$$\delta\left[\frac{1}{8\pi}\int d^3x (\nabla\alpha \times \nabla\beta)^2\right] = 0. \tag{13.5.4}$$

In this equation, it is assumed that α and β retain fixed values on a bounding surface.

The above equation may be re-expressed as

$$\int d^3x\, \mathbf{B}\cdot(\nabla\delta\alpha \times \nabla\beta + \nabla\alpha \times \nabla\delta\beta) = 0. \tag{13.5.5}$$

On carrying out integrations by parts, this equation may be expressed as

$$\int d^3x\{[\nabla\beta\cdot(\nabla\times\mathbf{B})]\delta\alpha - [\nabla\alpha\cdot(\nabla\times\mathbf{B})]\delta\beta\} = 0. \tag{13.5.6}$$

Hence requiring that the integral in (13.5.4) should be stationary for arbitrary variations $\delta\alpha$, $\delta\beta$ (subject to the boundary conditions) leads back to (13.5.3). Hence the variational equation (13.5.4) leads to force-free magnetic-field configurations.

Sakurai (1979) has used the above variational equation as a basis for the calculation of force-free fields. One may use the Rayleigh–Ritz method (Courant and Hilbert, 1953) of representing the field in terms of a set of base functions and solving for the coefficients arising in this representation.

Another procedure has been introduced more recently that is closely related to a method developed by Chodura and Schluter (1981) for application to MHD problems. In the present context, the procedure may be termed the 'magneto-frictional' method.

Suppose that the magnetic field is imbedded in a highly conducting medium and that the parameters are such that the 'frozen-flux' condition is satisfied. However, suppose also that the system contains a medium that is fixed in space such that the plasma experiences a frictional force when it moves with respect to that medium. Then the equation of motion of the plasma is

$$\rho \frac{d\mathbf{v}}{dt} = -\nabla p + \rho \mathbf{g} + \mathbf{F} - \nu \mathbf{v}. \qquad (13.5.7)$$

In this equation, $\mathbf{F}$ represents the Lorentz force,

$$\mathbf{F} = \mathbf{j} \times \mathbf{B}, \qquad (13.5.8)$$

and ν is the coefficient of friction. In situations that lead to force-free fields, ρ and p are negligible so that (13.5.7) leads to

$$\mathbf{v} = \nu^{-1} \mathbf{F}. \qquad (13.5.9)$$

On combining this equation with equations (12.2.23), we obtain the following equations for α and β:

$$\frac{\partial \alpha}{\partial t} = -\nu^{-1} \mathbf{F} \cdot \nabla \alpha, \quad \frac{\partial \beta}{\partial t} = -\nu^{-1} \mathbf{F} \cdot \nabla \beta. \qquad (13.5.10)$$

We are free to choose the parameter ν as an arbitrary function of space and time. In practice, it has proved convenient to choose ν in such a way that we obtain the following increments for α and β for each time step in the numerical calculation:

$$\delta \alpha = -\mu \frac{\mathbf{F} \cdot \nabla \alpha}{B^2}, \quad \delta \beta = -\mu \frac{\mathbf{F} \cdot \nabla \beta}{B^2}. \qquad (13.5.11)$$

In these equations, μ is an adjustable parameter. More details concerning this procedure, and its application to simple examples, may be found in Yang, Sturrock and Antiochos (1986).

13.6 Linear force-free fields of cylindrical symmetry

We saw that force-free fields of translational symmetry must have a field component in the direction of translational symmetry. A similar statement is true for force-free fields of cylindrical symmetry. Suppose that this were not true. Suppose that a force-free field of cylindrical symmetry were expressible as

$$\mathbf{B} = (B_z, B_r, 0).\qquad(13.6.1)$$

Then we find that the current density has only one nonzero component, namely

$$j_\phi = \frac{1}{4\pi}\left(\frac{\partial B_r}{\partial z} - \frac{\partial B_z}{\partial r}\right).\qquad(13.6.2)$$

Hence the Lorentz force is given by

$$\mathbf{F} \equiv \mathbf{j} \times \mathbf{B} = (-j_\phi B_r, j_\phi B_z, 0).\qquad(13.6.3)$$

It follows that the field can be force-free only if the current density vanishes identically.

Let us now seek a solution of the linear force-free equation (13.2.6). Assuming that the field is of cylindrical symmetry, we find that the components of (13.2.6) become

$$\left.\begin{aligned}\frac{1}{r}\frac{\partial}{\partial r}(rB_\phi) &= \lambda B_z,\\[6pt]-\frac{\partial B_\phi}{\partial z} &= \lambda B_r,\\[6pt]\frac{\partial B_r}{\partial z} - \frac{\partial B_z}{\partial r} &= \lambda B_\phi.\end{aligned}\right\}\qquad(13.6.4)$$

Since the magnetic field is divergence-free, the z and r components may be expressed as

$$B_z = \frac{1}{r}\frac{\partial}{\partial r}(rA), \quad B_r = -\frac{\partial A}{\partial z}.\qquad(13.6.5)$$

On substituting these expressions into (13.6.4), we find that equations (13.6.4) are satisfied if

$$B_\phi = \lambda A\qquad(13.6.6)$$

and if A satisfies the equation

$$\left(\frac{\partial^2}{\partial z^2} + \frac{\partial^2}{\partial r^2} + \frac{1}{r}\frac{\partial}{\partial r} - \frac{1}{r^2} + \lambda^2\right)A = 0.\qquad(13.6.7)$$

We may find Bessel-function solutions of the above equation, of which the simplest has the form

$$B_z = B_0 J_0(kr) e^{-lz},$$

$$B_r = lk^{-1} B_0 J_1(kr) e^{-lz},$$

$$B_\phi = (1 - l^2 k^{-2})^{1/2} B_0 J_1(kr) e^{-lz}, \tag{13.6.8}$$

where k and l must be related by

$$k^2 - l^2 = \lambda^2. \tag{13.6.9}$$

A solution of the type shown in (13.6.8), for which the field components decrease exponentially with distance from the plane $z = 0$, exists only if $k^2 > \lambda^2$. The behavior of cylindrical linear force-free fields is clearly analogous to that of linear force-free fields of translational symmetry.

13.7 Uniformly twisted cylindrical force-free field

In solar physics and in other areas of astrophysics, one often deals with the concept of 'thin flux tubes.' In general, these tubes will be twisted, due for instance to vortical motion of the surface or surfaces in which the flux tube terminates. This problem was considered some time ago by Gold and Hoyle (1958) in connection with a model of the magnetic-field configurations relevant to solar flares.

We consider a magnetic-field configuration that is uniform in z and in ϕ, so that all quantities are functions of r only. We see from the divergence equation that, if B_r is to be well behaved, it must be zero. Hence

$$\mathbf{B} = (B_z, 0, B_\phi). \tag{13.7.1}$$

We now assume that the magnetic field has uniform twist in the sense that each field line rotates the same number of radians per unit length along the axis. Hence B_ϕ and B_z will be related as follows:

$$\frac{B_\phi}{B_z} = r \frac{d\phi}{dz} = br, \tag{13.7.2}$$

where b is taken to be a constant.

The current density is expressible as

$$j_z = -\frac{1}{4\pi} \frac{1}{r} \frac{d}{dr}(rB_\phi),$$

$$j_r = 0, \tag{13.7.3}$$

$$j_\phi = -\frac{1}{4\pi} \frac{dB_z}{dr}.$$

On requiring the Lorentz force to be zero, we now obtain only one equation (for the r-direction), namely,

$$B_z \frac{dB_z}{dr} + B_\phi \frac{1}{r} \frac{d}{dr} (rB_\phi) = 0. \qquad (13.7.4)$$

On substituting for B_ϕ from (13.7.2), we now obtain the equation

$$\frac{d}{dr} [(1 + b^2 r^2) B_z] = 0. \qquad (13.7.5)$$

In this way, we find that this model for the magnetic field is expressible as

$$\left.\begin{array}{l} B_z = \dfrac{B_0}{1 + b^2 r^2}, \\[2ex] B_r = 0, \\[2ex] B_\phi = \dfrac{B_0 br}{1 + b^2 r^2}. \end{array}\right\} \qquad (13.7.6)$$

The above model does not describe a finite flux tube. It has no outer boundary, and the total magnetic flux is infinite. Let us therefore consider a model in which the flux tube has a boundary at $r = R$. We suppose that, outside this surface, there is gas of given pressure P. Then we find, from (13.7.6), that

$$\frac{1}{8\pi} \frac{B_0^2}{(1 + b^2 R^2)} = P. \qquad (13.7.7)$$

The total magnetic flux, that we denote by Φ, is found to be

$$\Phi = \frac{\pi B_0}{b^2} \ln(1 + b^2 R^2). \qquad (13.7.8)$$

It is now interesting to see what happens as we vary the twist parameter b under the condition that the total flux Φ and the outer radius R are both kept constant. We see from (13.7.8) that the field strength at $r = 0$ is given by

$$B_0 = \frac{\Phi b^2}{\pi \ln(1 + b^2 R^2)}, \qquad (13.7.9)$$

so that the field strength at the center of the tube increases with increasing twist. However, the pressure at the boundary is found to be given by

$$P = \frac{1}{8\pi} \frac{\Phi^2 b^4}{(1 + b^2 R^2) [\ln(1 + b^2 R^2)]^2}. \qquad (13.7.10)$$

This quantity is found to increase fairly slowly as a function of b.

Perhaps a more interesting discussion is to inquire into the behavior of the finite flux tube when the magnetic flux Φ and the external gas pressure P are both held constant. On multiplying (13.7.10) by R^4, we find that

$$\frac{R(b)}{R(0)} = \frac{\beta^{1/2}}{(1+\beta)^{1/4}[\ln(1+\beta)]^{1/2}}, \qquad (13.7.11)$$

where

$$\beta = b^2 R^2. \qquad (13.7.12)$$

We see from (13.7.11) that the twist parameter b is expressible in terms of β as

$$R_0 b = (1+\beta)^{1/4}[\ln(1+\beta)]^{1/2}. \qquad (13.7.13)$$

We also find, from (13.7.9), that the field strength at the center of the flux tube is expressible as

$$\frac{B_0(b)}{B_0(0)} = (1+\beta)^{1/2}. \qquad (13.7.14)$$

We find from the above equations that, as β increases progressively, the twist parameter b increases, the field strength at $r=0$ increases, and the outer radius R increases, although only slowly. For instance, as β increases from 0 to 100, bR_0 increases from 0 to 6.81, and B_0 increases by a factor of 10.05. On the other hand, R increases only by a factor 1.47. The behavior of this model is shown in more detail in Table 13.1 and in Figs. 13.5 and 13.6.

As we shall see in Chapter 15, it is possible to confine a plasma cylinder

Table 13.1. *Evolution of the Gold–Hoyle model of a uniformly twisted cylindrical force-free magnetic field configuration*

β	$R_0 b$	R/R_0	$B_0(b)/B_0(0)$
0	0	1	1
0.1	0.3162	1.0002	1.0488
0.2	0.4469	1.0007	1.0954
0.5	0.7047	1.0034	1.2247
1	0.9901	1.01	1.4142
2	1.3794	1.0252	1.7321
5	2.095	1.0674	2.4495
10	2.8201	1.1213	3.3166
20	3.7352	1.1973	4.5826
50	5.2989	1.3344	7.1414
100	6.8104	1.4683	10.05
200	8.6711	1.631	14.177
500	11.796	1.8956	22.383
1000	14.785	2.1389	31.639

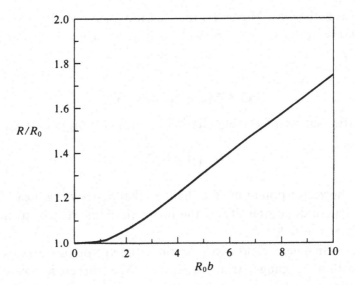

Fig. 13.5. The variation of the outer radius of the flux tube as a function of the amount of twist, for the model described in Section 13.7.

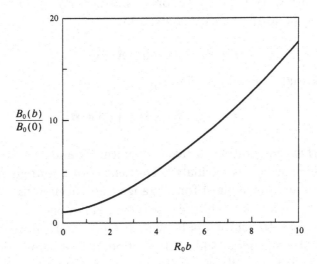

Fig. 13.6. The variation of the magnetic field strength on the axis of the flux tube as a function of the amount of twist, for the model described in Section 13.7.

by means of the external azimuthal magnetic field produced by surface current parallel to the cylinder axis. This is known as the 'pinch effect.' Knowledge of the pinch effect might suggest that a flux tube, of the type we are considering, would tend to contract as the twist is increased. We see that the central field strength does indeed increase, implying that field lines are moved towards the axis. However, the flux tube, as a whole, does not contract. As we see from Fig. 13.5, the outer radius increases with twist, although quite slowly.

13.8 Magnetic helicity

The magnetic helicity of a magnetic-field configuration is defined as

$$H = \int d^3x \, \mathbf{B} \cdot \mathbf{A}, \tag{13.8.1}$$

where **A** is the vector potential. We shall see that helicity is a measure of the degree of structural complexity of the magnetic field, related to the 'interconnection' of the field.

For a given magnetic-field configuration, the vector potential is defined only to within a gauge transformation. We therefore consider the transformation

$$\mathbf{A} \rightarrow \mathbf{A}' = \mathbf{A} + \nabla\chi, \tag{13.8.2}$$

where χ is an arbitrary scalar function of position. We find that the corresponding change in the helicity is given by

$$H \rightarrow H' = H + \int d^3x \, \mathbf{B} \cdot \nabla\chi. \tag{13.8.3}$$

Using the fact that $\nabla \cdot B = 0$, we find that

$$H' - H = \int d^3x \, \nabla \cdot (\chi\mathbf{B}) = \int dS \, \mathbf{n} \cdot \mathbf{B}\chi, \tag{13.8.4}$$

where the surface integration is taken over the boundary containing the region over which helicity is calculated. The helicity of a configuration is well defined if its value is unchanged for all gauge transformations of the vector potential. We see that this will be the case if the field extends over all of space and if B decreases sufficiently rapidly with distance (and χ does not increase too rapidly). For a magnetic-field configuration of finite dimensions, we see that magnetic helicity is well defined if and only if $\mathbf{n} \cdot \mathbf{B} = 0$ on the bounding surface.

We next show that the helicity of a magnetic-field configuration is conserved if the field is confined within a closed surface S, if $\mathbf{n} \cdot \mathbf{B} = 0$ at that surface, and if the field permeates a perfectly conducting medium that moves in such a way that $\mathbf{n} \cdot \mathbf{v} = 0$ on S. From (12.2.1), i.e.

$$\frac{\partial \mathbf{B}}{\partial t} = \nabla \times (\mathbf{v} \times \mathbf{B}), \tag{13.8.5}$$

we see that the evolution of the vector potential is defined, to within a gauge transformation, by

$$\frac{\partial \mathbf{A}}{\partial t} = \mathbf{v} \times \mathbf{B}. \tag{13.8.6}$$

(We have just shown that a gauge transformation has no effect on the estimated helicity.) From

$$\frac{\mathrm{d}H}{\mathrm{d}t} = \int \mathrm{d}^3 x \left(\frac{\partial \mathbf{A}}{\partial t} \cdot \mathbf{B} + \mathbf{A} \cdot \frac{\partial \mathbf{B}}{\partial t} \right), \tag{13.8.7}$$

we see that

$$\frac{\mathrm{d}H}{\mathrm{d}t} = \int \mathrm{d}^3 x \left(\frac{\partial \mathbf{A}}{\partial t} \cdot (\nabla \times \mathbf{A}) + \mathbf{A} \cdot \left(\nabla \times \frac{\partial \mathbf{A}}{\partial t} \right) \right). \tag{13.8.8}$$

From (13.8.6), we see that $\partial \mathbf{A}/\partial t$ is normal to $\mathbf{B}$, so that the first term in the integral is zero. We may therefore change the sign of that term without changing the value of the integral. We then find that the integral may be expressed as follows:

$$\frac{\mathrm{d}H}{\mathrm{d}t} = \int \mathrm{d}^3 x \, \nabla \cdot \left(\frac{\partial \mathbf{A}}{\partial t} \times \mathbf{A} \right) = \int \mathrm{d}S \, \mathbf{n} \cdot \left(\frac{\partial \mathbf{A}}{\partial t} \times \mathbf{A} \right). \tag{13.8.9}$$

Since we are assuming that both $\mathbf{B}$ and $\mathbf{v}$ are normal to $\mathbf{n}$ on the surface, it follows from (13.8.6) that $\partial \mathbf{A}/\partial t$ is parallel to $\mathbf{n}$ at the surface. Hence the surface integral in (13.8.9) is zero, showing that for the conditions we have specified, H is a constant of the motion.

In order to get a feeling for the significance of the quantity we are referring to as helicity, consider the simple example, shown in Fig. 13.7, of two thin flux tubes that are linked together. Clearly, the total magnetic helicity is the sum of the contribution from each tube taken separately:

$$H = H_1 + H_2. \tag{13.8.10}$$

For thin flux tubes, H_1 may be written approximately as

$$H_1 = \oint \mathrm{d}\mathbf{s} \cdot \mathbf{A} \int \mathrm{d}S \, \mathbf{n} \cdot \nabla \times \mathbf{A}. \tag{13.8.11}$$

In writing the contribution in this way, we have used the fact that $\nabla \times \mathbf{A}$ is approximately normal to S, that is a cross section of flux tube 1. Hence the surface integral is Φ_1, the flux of tube 1, and the first integral is Φ_2, the flux of tube 2:

$$H_1 = \Phi_1 \Phi_2. \tag{13.8.12}$$

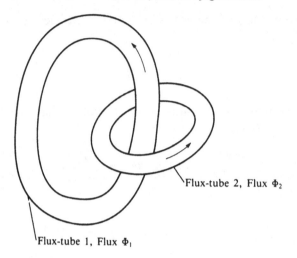

Fig. 13.7. Schematic representation of two interlinking flux tubes.

Clearly, H_2 has the same value so that

$$H = 2\Phi_1\Phi_2. \qquad (13.8.13)$$

If two flux tubes wind around each other N times, then

$$H = \pm 2N\Phi_1\Phi_2, \qquad (13.8.14)$$

where the sign depends upon the relative orientation of the magnetic field in the two flux tubes.

In Section 12.2, and earlier in this chapter, we represented magnetic-field configurations in terms of Clebsch variables:

$$\mathbf{B} = \nabla\alpha \times \nabla\beta. \qquad (13.8.15)$$

This choice of $\mathbf{B}$ clearly corresponds to the following choice for the vector potential $\mathbf{A}$:

$$\mathbf{A} = \alpha\nabla\beta + \nabla\chi. \qquad (13.8.16)$$

We have learned already that the term $\nabla\chi$, corresponding to a gauge transformation, does not change the total magnetic helicity. If we ignore this term, it is clear that $\mathbf{A} \cdot \mathbf{B} = 0$, so that the magnetic helicity of a field that can be described in terms of Clebsch potentials is necessarily zero.

Conversely, any magnetic-field configuration that is multiply connected, so that the helicity is nonzero, cannot be described by Clebsch variables. On the other hand, it might be possible to divide the field into two or more regions, each of which is simply connected, so that each region may be described by its own set of Clebsch variables.

13.9 Woltjer's theorem

Woltjer (1958) considered the following problem. *What is the magnetic-field configuration when the plasma relaxes to a state that minimizes the magnetic energy, subject to the constraint that the magnetic helicity is constant?*

We may approach this problem by using the method of Lagrange multipliers. The magnetic field will satisfy the variational expression

$$\delta\left[\int d^3x\, B^2 - \mu \int d^3x\, \mathbf{B}\cdot\mathbf{A}\right] = 0 \qquad (13.9.1)$$

where μ is the Lagrange multiplier. This equation leads to

$$\int d^3x(2\mathbf{B}\cdot\delta\mathbf{B} - \mu\mathbf{A}\cdot\delta\mathbf{B} - \mu\mathbf{B}\cdot\delta\mathbf{A}) = 0, \qquad (13.9.2)$$

that may also be expressed as

$$\int d^3x[2\nabla\cdot(\delta\mathbf{A}\times\mathbf{B}) + 2(\nabla\times\mathbf{B})\cdot\delta\mathbf{A} - 2\mu\mathbf{B}\cdot\delta\mathbf{A} - \mu\nabla\cdot(\delta\mathbf{A}\times\mathbf{A})] = 0. \qquad (13.9.3)$$

This may be expressed as

$$\int dS\, \mathbf{n}\cdot[2\delta\mathbf{A}\times\mathbf{B} - \mu\delta\mathbf{A}\times\mathbf{A}] + 2\int d^3x[\nabla\times\mathbf{B} - \mu\mathbf{B}]\cdot\delta\mathbf{A} = 0, \qquad (13.9.4)$$

where the surface integral is over the surface bounding the volume of integration.

If we now assume that the plasma is perfectly conducting, so that the magnetic field is 'frozen' into the plasma, we may infer from (13.8.6) that the perturbation in the vector potential may be expressed as

$$\delta\mathbf{A} = \delta\boldsymbol{\xi}\times\mathbf{B}, \qquad (13.9.5)$$

where $\delta\boldsymbol{\xi}$ represents the displacement of the plasma during the perturbation.

If we assume that the boundary S is fixed, and if we require that the normal components of $\mathbf{B}$ are zero over the surface, then it is clear that $\delta\mathbf{A}$ is parallel to $\mathbf{n}$, so that the surface integral in (13.9.4) is zero.

We now require that the remaining integral in (13.9.4) should be zero, for arbitrary choice of the vector $\delta\mathbf{A}$. This leads to the result that

$$\nabla\times\mathbf{B} = \mu\mathbf{B}. \qquad (13.9.6)$$

Since μ is a constant parameter, we have shown that the magnetic-field configuration that has stationary energy for given helicity is a linear force-free field.

For an imperfect plasma, with small but nonzero resistivity, Taylor (1974) suggests that the helicity is approximately invariant so that, by Woltjer's theorem, the minimum energy configuration is a linear force-free field.

13.10 Useful relations for semi-infinite force-free magnetic-field configurations

In solar physics, we are interested in the total magnetic energy that can be stored above the photosphere, since this is believed to be the source of energy for activities such as solar flares. Observational data, made with magnetographs, gives us information about the line-of-sight magnetic-field component. If the active region is near disk center, this corresponds to knowledge of the normal magnetic-field component at the photosphere.

Adopting rectangular Cartesian coordinates, let us consider the magnetic-field configuration in the half space $x_3 > 0$, and assume that we have knowledge of B_3 at the plane $x_3 = 0$. We donote this also by B_n, since it is the component normal to the plane. The lower limit to the energy of any magnetic field in the space $x_3 > 0$ is the energy of the current-free field with the same boundary condition on B_n. By using the virial theorem, we may also obtain an upper limit on the energy of a force-free field with the same distribution of B_n.

We see from Section 12.4 that we may express the force-free condition in the form

$$\frac{\partial T_{rs}}{\partial x_s} = 0 \tag{13.10.1}$$

where T_{rs} is the Maxwell tensor for the magnetic field:

$$T_{rs} = \frac{1}{4\pi} B_r B_s - \frac{1}{8\pi} B^2 \delta_{rs}. \tag{13.10.2}$$

The volume integral of (13.10.1) is given by

$$\int d^3x \frac{\partial T_{rs}}{\partial x_s} = -\int dS \, n_s T_{rs} \tag{13.10.3}$$

if, in this section, we denote by $\mathbf{n}$ the *inward* normal on the bounding surface, so that $\mathbf{n} = (0, 0, 1)$. Here and elsewhere, we assume that the magnetic field decreases sufficiently rapidly with distance that the integral over the sphere at infinity may be neglected.

On using (13.10.1) and (13.10.2), (13.10.3) yields

$$\int dS\,[2B_r n_s B_s - B^2 n_r] = 0. \tag{13.10.4}$$

On taking the scalar product of this equation with **n**, we obtain

$$\int dS\,[2(\mathbf{n}\cdot\mathbf{B})^2 - B^2] = 0. \tag{13.10.5}$$

On writing $\mathbf{B}_\perp$ for the part of the vector **B** that is transverse to **n**,

$$\mathbf{B}_\perp \doteq \mathbf{B} - (\mathbf{n}\cdot\mathbf{B})\mathbf{n}, \tag{13.10.6}$$

we see from (13.10.5) that

$$\int dS\,B_\perp{}^2 = \int dS\,B_n{}^2. \tag{13.10.7}$$

We next note that

$$\int d^3x\,T_{rs} = \int d^3x\,T_{rt}\frac{\partial x_s}{\partial x_t}. \tag{13.10.8}$$

Since, by (13.10.1),

$$\int d^3x\,T_{rt}\frac{\partial x_s}{\partial x_t} = \int d^3x\,\frac{\partial}{\partial x_t}(T_{rt}x_s), \tag{13.10.9}$$

we see that

$$\int d^3x\,T_{rs} = -\int dS\,n_t T_{rt}x_s. \tag{13.10.10}$$

On contracting the tensors in this equation, we see that

$$U_M = \frac{1}{4\pi}\int dS\,(\mathbf{n}\cdot\mathbf{B})(\mathbf{x}\cdot\mathbf{B}) - \frac{1}{8\pi}\int dS\,B^2\mathbf{n}\cdot\mathbf{x}, \tag{13.10.11}$$

where U_M is the magnetic energy:

$$U_M = \frac{1}{8\pi}\int d^3x\,B^2. \tag{13.10.12}$$

Since $\mathbf{n}\cdot\mathbf{x}=0$ on the plane $x_3=0$, (13.10.11) becomes

$$U_M = \frac{1}{4\pi}\int dS\,B_n\mathbf{x}\cdot\mathbf{B}_\perp. \tag{13.10.13}$$

If we now replace the rectangular coordinates x_1, x_2, with cylindrical coordinates R, ϕ, in the plane $x_3 = 0$, we see that the above equation may be written as

$$U_M = \frac{1}{4\pi} \int dS\, R B_n B_R. \tag{13.10.14}$$

We may now use Schwartz's inequality to infer that

$$U_M \leqslant \frac{1}{4\pi} \left[\int dS\, R^2 B_n^2 \int dS\, B_R^2 \right]^{1/2}. \tag{13.10.15}$$

Clearly

$$\int dS\, B_R^2 < \int dS\, B_\perp^2. \tag{13.10.16}$$

On using this relation and (13.10.7), we see that

$$U_M \leqslant \frac{1}{4\pi} \left[\int dS\, R^2 B_n^2 \int dS\, B_n^2 \right]^{1/2}. \tag{13.10.17}$$

The above relations were derived by Aly (1984).

Clearly, the limit given by (13.10.17) depends upon the choice of the origin of the coordinate system in the plane $x_3 = 0$. A displacement of the origin to $(-X_1, -X_2, 0)$ is equivalent to replacing $\mathbf{x}$ by $\mathbf{x} + \mathbf{X}$. On introducing the suffix i, $i = 1, 2$, (13.10.17) now becomes

$$U_M \leqslant \frac{1}{4\pi} \left[\int dS\, (x_i + X_i)^2 B_n^2 \int dS\, B_n^2 \right]^{1/2}. \tag{13.10.18}$$

The first integral in this expression may be expanded as

$$\int dS\, (x_i + X_i)^2 B_n^2 = \int dS\, x_i^2 B_n^2 + 2X_i \int dS\, x_i B_n^2 + X^2 \int dS\, B_n^2. \tag{13.10.19}$$

For each component X_i, this expression clearly is a parabolic expression in X_i that becomes infinite for infinite values of X_i, and therefore has a minimum value where the expression is stationary with respect to X_i. The appropriate values of X_i are found to be

$$X_i = \frac{\int dS\, B_n^2 x_i}{\int dS\, B_n^2} \tag{13.10.20}$$

On combining (13.10.18), (13.10.19) and (13.10.20) we finally obtain the following 'best upper bound' for the magnetic energy:

$$U_{\rm M} \leqslant \frac{1}{4\pi} \left[\left(\int {\rm d}S \, B_{\rm n}^2 x_i^2 \right) \left(\int {\rm d}S \, B_{\rm n}^2 \right) - \left(\int {\rm d}S \, B_{\rm n}^2 x_i \right)^2 \right]^{1/2} . \quad (13.10.21)$$

This expression is now independent of the choice of coordinate system.

It is clear from the above relation that the magnetic energy of a magnetic-field configuration that is limited in extent, in the x_1–x_2 plane, is finite. We should therefore expect that, for a given distribution $B_{\rm n}(x_1, x_2, 0)$, there is one particular magnetic-field configuration that has maximum energy. We saw in Section 13.3 that the effect of shearing of a magnetic-field configuration is to cause the field to inflate. This drives the field towards the open configuration in which all field lines extend out to infinity. Aly has advanced the conjecture that the state of maximum energy is the corresponding open-field configuration (Aly, 1984). We now examine this conjecture.

We wish to consider all field configurations that have the same distribution of $B_{\rm n}$ in the plane $x_3 = 0$. These configurations may be generated by displacing the footpoints of field lines subject to the requirement that the displacements do not change $B_{\rm n}$.

If the displacement vector is taken to be $\xi_i(\mathbf{x})$, where ξ_i is taken to be small, we see from the Lagrange expansion (p. 145) that

$$\delta B_{\rm n} = -\frac{\partial}{\partial x_i} (B_{\rm n} \xi_i), \quad (13.10.22)$$

to first order in the displacement. Hence we must consider only displacements that satisfy the condition

$$\frac{\partial}{\partial x_i} (B_{\rm n} \xi_i) = 0. \quad (13.10.23)$$

The magnetic field above the surface $x_3 = 0$ exerts a force per unit area on the surface given by

$$F_i = \frac{1}{4\pi} B_{\rm n} B_i. \quad (13.10.24)$$

Hence the change in energy due to the displacement is given by

$$\delta W = -\frac{1}{4\pi} \int {\rm d}S \, B_{\rm n} B_i \xi_i. \quad (13.10.25)$$

Our problem is to find a state of the magnetic field for which $\delta W = 0$ for any distribution of displacements $\xi_i(\mathbf{x})$ that satisfy the constraint (13.10.23). We may examine this problem by using the method of Lagrange multipliers. We therefore examine the equation

$$\int dS \left[\frac{1}{4\pi} B_n B_i \xi_i + \mu(\mathbf{x}) \frac{\partial}{\partial x_i} (B_n \xi_i) \right] = 0. \qquad (13.10.26)$$

On performing an integration by parts, and requiring that $|B_n(x)| \to 0$ sufficiently rapidly as $|\mathbf{x}| \to \infty$, we see that this equation is equivalent to

$$\int dS \left[\frac{1}{4\pi} B_n B_i - \frac{\partial \mu}{\partial x_i} B_n \right] \xi_i = 0. \qquad (13.10.27)$$

Hence the magnetic-field configuration that has stationary energy, subject to the constraint (13.10.23), satisfies the equation

$$B_n \left[B_i - 4\pi \frac{\partial \mu}{\partial x_i} \right] = 0 \qquad (13.10.28)$$

at every point in the plane $x_3 = 0$. It follows that, at each point, either $B_n = 0$ or B_i is expressible as

$$B_i = -\frac{\partial \Psi}{\partial x_i}. \qquad (13.10.29)$$

However, if (13.10.29) is satisfied, then $j_3 = 0$. Hence the required limiting form of the magnetic-field configuration satisfies the following condition

$$B_n = 0 \quad \text{or} \quad j_n = 0 \quad \text{at each point on } S. \qquad (13.10.30)$$

If $B_n \neq 0$, then $j_n = 0$. However, since we are considering force-free magnetic-field configurations, $\mathbf{j} = \lambda \mathbf{B}$. Hence $\lambda = 0$ and $\mathbf{j} = 0$. It follows that regions of the magnetic field that connect to the surface $x_3 = 0$ in regions where $B_n \neq 0$ are current-free regions. Hence the only regions of nonzero current density must be related to field lines that connect to the surface where $B_n = 0$. That is, the only regions of nonzero current are sheets that extend above the field-reversal lines.

The open-field configuration is a state in which the field is current free except at the surfaces between flux systems of opposite polarity, where there can be current sheets. Hence the open flux system is compatible with our requirements for the field of maximum energy to be one that has the open-field configuration at that line. Hence it appears that the maximum-energy state, for a given distribution of B_n, is indeed the open-field configuration.

A more general demonstration of this result is to be found in Sturrock (1991).

Problems

Problem 13.1. Show that the Helmholtz equation

$$\nabla^2 \mathbf{B} + \lambda^2 \mathbf{B} = 0 \tag{1}$$

is satisfied by **B** if **B** satisfies the equation

$$\nabla \times \mathbf{B} = \lambda \mathbf{B} + \mathbf{F}, \tag{2}$$

provided that the vector $\mathbf{F(x)}$ satisfies the equation

$$\nabla \times \mathbf{F} = -\lambda \mathbf{F}. \tag{3}$$

Problem 13.2. Consider a force-free magnetic-field configuration that is uniform in the z-direction, so that it can be described as

$$\mathbf{B} = \nabla \alpha \times \nabla \beta, \tag{1}$$

where

$$\alpha = \alpha(x, y) \quad \text{and} \quad \beta = \gamma(x, y) - z. \tag{2}$$

(a) Show that

$$(\mathbf{B} \cdot \nabla) B_z = 0. \tag{3}$$

(b) Can you find a similar result for configurations of cylindrical symmetry?

Problem 13.3. Consider the following model of a linear force-free field:

$$\left.\begin{array}{l} B_x = a_x \sin(kx) e^{-lz} \\ B_y = a_y \sin(kx) e^{-lz} \\ B_z = B_0 \cos(kx) e^{-lz} \end{array}\right\} \tag{1}$$

defined for the region $0 < z < \infty$.

(a) Verify that

$$\nabla \times \mathbf{B} = \lambda \mathbf{B}. \tag{2}$$

(b) Consider one section of the field between the planes $x = -\pi/2k$ and $x = \pi/2k$. Regarding k as given and l as determined by k and λ, evaluate the energy of the magnetic field per unit length in the y-direction in the section indicated. Verify that the energy is a minimum for $\lambda = 0$.

(c) Now suppose that the field below the plane $z = 0$ is of the form

$$\left.\begin{array}{l} B_x = a_x \sin(kx)e^{lz} \\[4pt] B_y = a_y \sin(kx)e^{lz} \\[4pt] B_z = B_0 \cos(kx)e^{lz} \end{array}\right\} \qquad (3)$$

for which

$$\nabla \times \mathbf{B} = -\lambda \mathbf{B}. \qquad (4)$$

We can now regard the field specified for $z>0$ and $z<0$ as being produced by currents flowing in a sheet located at $x=0$. Calculate the surface current densities J_x and J_y (current per unit length, rather than current per unit area) flowing in that sheet by considering the abrupt changes in B_x and B_y at the plane.

(d) The condition

$$\nabla \cdot \mathbf{j} = 0 \qquad (5)$$

leads to a relation between j_z just above and just below the plane $z=0$ and J_x and J_y in the plane. Verify that this relation is satisfied.

(e) Verify that the force-free field may be regarded as being produced from an initially potential field with $\lambda=0$ by a shear displacement in the y-direction, the displacement being proportional to x and a function of λ.

(f) Calculate the Lorentz force F_y per unit area acting on the sheet located at $z=0$. Hence calculate the work done in starting with an unsheared field and increasing the shear to the amount determined by λ. How does this estimate compare with the result of part (b)?

Problem 13.4. Consider a compact force-free magnetic-field configuration above the plane $z=0$ of a rectangular coordinate system (x, y, z). Suppose that there is a similar field configuration below the sheet defined by $z=0$ such that

$$B_x(x, y, -z) = -B_x(x, y, z),$$
$$B_y(x, y, -z) = -B_y(x, y, z),$$

and

$$B_z(x, y, -z) = B_z(x, y, z).$$

(a) Find an expression for the Lorentz force acting on each point of the sheet $z=0$.

(b) Now suppose that the entire configuration is allowed to expand uniformly by supposing that each point $\mathbf{x}$ in the sheet is moved to $\mathbf{x}^* = \Gamma \mathbf{x}$. Assume that field lines are 'frozen' into the sheet, so that the field configuration on the sheet expands uniformly. Show that one may construct the new force-free field by supposing that each field line expands uniformly by the same rule.

(c) By considering the Lorentz force that the field exerts on the sheet, calculate the work done by the field as the sheet expands, with Γ changing to $\Gamma + d\Gamma$.

(d) Integrate this expression to find the total energy expended by the field as it expands to infinity. Hence obtain an expression for the energy of the magnetic field in each half-space, expressed as an integral of field components over the plane $z = 0$.

Problem 13.5. Adopting cylindrical coordinates z, r, ϕ, consider a thin straight cylindrically symmetric flux tube of length L and radius $R(z)$, such that $R \ll L$ for all z. Assume further that the length-scale that characterizes the variation of the field configuration along its length is of order L. Simplify the model by considering only the lowest order significant terms in expansions of the field components in powers of r, so that

$$
\left.
\begin{aligned}
B_z(z, r) &= B_0(z) + \frac{r^2}{2} \left.\frac{\partial^2 B_z}{\partial r^2}\right|_{r=0} + O(r^4), \\[2mm]
B_r(z, r) &= r\left.\frac{\partial B_r}{\partial r}\right|_{r=0} + O(r^3),
\end{aligned}
\right\} \tag{1}
$$

and

$$
B_\phi(z, r) = r\left.\frac{\partial B_\phi}{\partial r}\right|_{r=0} + O(r^3).
$$

(a) Why must B_z be an even function of r and B_r, B_ϕ be odd functions of r?

(b) Find the form for $\mathbf{j}$ that results from the above form for $\mathbf{B}$. Assume that the field is force-free so that $\mathbf{j}$ and $\mathbf{B}$ may be related by

$$
\mathbf{j} = \lambda(z, r)\mathbf{B}(z, r). \tag{2}
$$

Assume that the flux tube is twisted and define a parameter b by

$$
B_\phi/B_0 = r\,d\phi/dz = br. \tag{3}
$$

Expand in r and consider that λ is constant near the axis. Show that $b = \lambda/2$ and thus find expressions for the components of $\mathbf{B}$, valid for small r, in terms of b, B_0' and B_0'', where a prime denotes differentiation with respect to z. Note that each field line is rotated about the axis by an angle $\Delta\chi = bL$.

(c) Find the total flux through the tube

$$
\Phi = \int_0^R 2\pi r B_z\,dr \tag{4}
$$

and the magnetic pressure p at the surface of the tube.

(d) Consider a flux tube that is initially untwisted so that $b = 0$, and immersed in a plasma so that the magnetic pressure on the surface of the tube is balanced by the (uniform) gas pressure P outside. Now allow the flux tube to become twisted (b increases from zero), but require that $p(R)$ remain constant ($=P$). Note also that Φ remains constant. Find ΔB_0 and ΔR, the changes in B_0 and R, to order b^2.

(e) The magnetic energy in the tube is given by

$$W = \int_0^L dz \int_0^R 2\pi r \frac{B^2}{8\pi} dr. \tag{5}$$

Calculate ΔW, the increase in the energy W due to the twist imposed on the flux tube. Show that this quantity may be expressed as

$$\Delta W = \frac{\Phi^2 (\Delta \chi)^2}{16\pi^2 L}.$$

(f) Show that this increase is equal to the work done in rotating the ends of the flux tube.

(g) Do you expect that this equality will be true to higher powers in the quantity b? Justify your opinion.

14

Waves in MHD systems

14.1 MHD waves in a uniform plasma

We derived the equations that govern an ideal MHD system in Section 11.7. Since the current density is of no direct interest, we express these equations in the following form:

$$\frac{\partial \rho}{\partial t} + \nabla \cdot (\rho \mathbf{v}) = 0, \tag{14.1.1}$$

$$\rho \left(\frac{\partial}{\partial t} + \mathbf{v} \cdot \nabla \right) \mathbf{v} = -\nabla p + \frac{1}{4\pi} (\nabla \times \mathbf{B}) \times \mathbf{B}, \tag{14.1.2}$$

$$\left(\frac{\partial}{\partial t} + \mathbf{v} \cdot \nabla \right) (p \rho^{-\gamma}) = 0, \tag{14.1.3}$$

$$\frac{\partial \mathbf{B}}{\partial t} = \nabla \times (\mathbf{v} \times \mathbf{B}), \tag{14.1.4}$$

and

$$\nabla \cdot \mathbf{B} = 0. \tag{14.1.5}$$

In these equations, we have neglected the gravitational force density $\rho \mathbf{g}$ in the equation of motion, since the gravitational field necessarily introduces inhomogeneity into the plasma. We shall consider the role of a gravitational field in Section 14.2.

Our goal in this section is to explore time-dependent disturbances of a static, homogeneous magnetoplasma, as described by the above set of equations. These equations are nonlinear, but we will restrict our attention to small-amplitude disturbances, and we therefore expect to find a linear set of differential equations that describe small-amplitude perturbations. We shall then seek wave-like solutions of these equations.

If the unperturbed velocity is nonzero but uniform, we can always transform the plasma to a frame in which it is at rest. We therefore assume that the initial velocity is zero and we also assume that the plasma is in equilibrium. If we disturb this static equilibrium by introducing a small perturbation in the velocity of the plasma, this perturbation will give rise to small perturbations of the magnetic field, the fluid pressure, and the mass density. We therefore write

$$\left. \begin{aligned} \mathbf{v} &\to \delta\mathbf{v}, \\ \mathbf{B} &\to \mathbf{B} + \delta\mathbf{B}, \\ p &\to p + \delta p, \\ \rho &\to \rho + \delta\rho. \end{aligned} \right\} \qquad (14.1.6)$$

On substituting these expressions into equations (14.1.1) through (14.1.4), we obtain the following equations

$$\frac{\partial}{\partial t}\delta\rho + \rho\nabla\cdot(\delta\mathbf{v}) = 0, \qquad (14.1.7)$$

$$\rho\frac{\partial}{\partial t}\delta\mathbf{v} = -\nabla(\delta p) + \frac{1}{4\pi}(\nabla\times\delta\mathbf{B})\times\mathbf{B}, \qquad (14.1.8)$$

$$\frac{\partial}{\partial t}\delta\mathbf{B} = \nabla\times(\delta\mathbf{v}\times\mathbf{B}), \qquad (14.1.9)$$

and

$$\frac{\delta p}{\delta\rho} = v_s^2, \qquad (14.1.10)$$

where v_s is the sound speed defined by

$$v_s^2 = \frac{\gamma p}{\rho}. \qquad (14.1.11)$$

Since the unperturbed system is uniform and static, and since the above equations are linear, it is convenient to use Fourier transformation. For present purposes, we consider only one component, and we therefore introduce the following change of notation

$$\delta\mathbf{v}(\mathbf{x}, t) = \delta\mathbf{v}\, e^{i(\mathbf{k}\cdot\mathbf{x} - \omega t)}, \text{ etc.} \qquad (14.1.12)$$

Using this notation, equations (14.1.7) through (14.1.9) become

$$-i\omega\,\delta\rho + i\rho\mathbf{k}\cdot\delta\mathbf{v} = 0, \qquad (14.1.13)$$

$$-i\omega\rho\,\delta\mathbf{v} = -i\mathbf{k}\,\delta p + \frac{1}{4\pi}i(\mathbf{k}\times\delta\mathbf{B})\times\mathbf{B}, \qquad (14.1.14)$$

and

$$-i\omega\,\delta\mathbf{B} = i\mathbf{k}\times(\delta\mathbf{v}\times\mathbf{B}).\qquad(14.1.15)$$

We see from (14.1.10) and (14.1.13) that δp is expressible as

$$\delta p = \omega^{-1}v_s^2\rho\mathbf{k}\cdot\delta\mathbf{v}.\qquad(14.1.16)$$

On using this expression in (14.1.14), we obtain

$$\omega^2\delta\mathbf{v} = v_s^2(\mathbf{k}\cdot\delta\mathbf{v})\mathbf{k} - \frac{\omega}{4\pi\rho}(\mathbf{k}\times\delta\mathbf{B})\times\mathbf{B}.\qquad(14.1.17)$$

On using (14.1.15), this equation becomes

$$\omega^2\delta\mathbf{v} = v_s^2(\mathbf{k}\cdot\delta\mathbf{v})\mathbf{k} + \frac{1}{4\pi\rho}\{\mathbf{k}\times[\mathbf{k}\times(\delta\mathbf{v}\times\mathbf{B})]\}\times\mathbf{B}.\qquad(14.1.18)$$

On taking the scalar product of (14.1.17) with $\mathbf{k}\times\delta\mathbf{B}$, we find that

$$\delta\mathbf{v}\cdot(\mathbf{k}\times\delta\mathbf{B}) = 0.\qquad(14.1.19)$$

On taking the scalar product of equation (14.1.15) with $\delta\mathbf{v}\times\mathbf{B}$, we obtain

$$\delta\mathbf{B}\cdot(\delta\mathbf{v}\times\mathbf{B}) = 0.\qquad(14.1.20)$$

These two relations show that $\delta\mathbf{v}$, $\delta\mathbf{B}$, $\mathbf{k}$ and $\mathbf{B}$ are all coplanar, unless $\delta\mathbf{v}$ is parallel to $\delta\mathbf{B}$, in which case there is no constraint on $\mathbf{k}$ and $\mathbf{B}$. We shall find that the former case describes magnetosonic waves, while the latter case corresponds to Alfvén waves.

We now choose a coordinate system in which the unperturbed magnetic field is parallel to the x_3 axis,

$$\mathbf{B} = (0, 0, B),\qquad(14.1.21)$$

and the wave vector lies in the 1–3 plane:

$$\mathbf{k} = (k\sin\theta,\ 0,\ k\cos\theta).\qquad(14.1.22)$$

We also introduce the phase velocity v_ϕ,

$$v_\phi = \frac{\omega}{k},\qquad(14.1.23)$$

and the Alfvén speed v_A:

$$v_A^2 = \frac{B^2}{4\pi\rho}.\qquad(14.1.24)$$

We find, incidentally, that (14.1.15) now takes the form

$$\begin{pmatrix} \delta B_1 \\ \delta B_2 \\ \delta B_3 \end{pmatrix} = \begin{pmatrix} -Bv_\phi^{-1}\cos\theta & 0 & 0 \\ 0 & -Bv_\phi^{-1}\cos\theta & 0 \\ Bv_\phi^{-1}\sin\theta & 0 & 0 \end{pmatrix} \begin{pmatrix} \delta v_1 \\ \delta v_2 \\ \delta v_3 \end{pmatrix}. \qquad (14.1.25)$$

This shows explicitly that the velocity component along the unperturbed magnetic field has no effect in perturbing the magnetic field.

On using the above notation in (14.1.18), we find that this equation reduces to

$$\begin{bmatrix} v_\phi^2 - v_A^2 - v_s^2\sin^2\theta & 0 & -v_s^2\sin\theta\cos\theta \\ 0 & v_\phi^2 - v_A^2\cos^2\theta & 0 \\ -v_s^2\sin\theta\cos\theta & 0 & v_\phi^2 - v_s^2\cos^2\theta \end{bmatrix} \begin{pmatrix} \delta v_1 \\ \delta v_2 \\ \delta v_3 \end{pmatrix} = 0.$$

$$(14.1.26)$$

Since the equation involving δv_2 is not coupled to the equations involving δv_1 and δv_3, the above matrix equation may be reduced to the pair of equations

$$(v_\phi^2 - v_A^2\cos^2\theta)\delta v_2 = 0 \qquad (14.1.27)$$

and

$$\begin{bmatrix} v_\phi^2 - v_A^2 - v_s^2\sin^2\theta & -v_s^2\sin\theta\cos\theta \\ -v_s^2\sin\theta\cos\theta & v_\phi^2 - v_s^2\cos^2\theta \end{bmatrix} \begin{pmatrix} \delta v_1 \\ \delta v_2 \end{pmatrix} = 0. \qquad (14.1.28)$$

We examine these equations separately.

We see that (14.1.27) is equivalent to the dispersion relation

$$\omega^2 = v_A^2 k_3^2. \qquad (14.1.29)$$

For this mode, the group velocity has the value $(0, 0, \pm v_A)$. That is, the disturbance moves parallel to the magnetic field with the Alfvén speed, no matter what direction the wave vector may have. The eigenvector of this mode may be expressed as

$$\delta \mathbf{v} = \begin{pmatrix} 0 \\ 1 \\ 0 \end{pmatrix}, \quad \delta \mathbf{B} = -Bv_\phi^{-1} \begin{pmatrix} 0 \\ 1 \\ 0 \end{pmatrix}. \qquad (14.1.30)$$

Hence $\delta \mathbf{v}$ and $\delta \mathbf{B}$ are transverse to the magnetic field vector and to the wave vector (and are parallel to each other).

The two remaining modes are governed by (14.1.28). The dispersion relation is found by setting the determinant equal to zero:

$$v_\phi{}^2 - (v_s{}^2 + v_A{}^2) v_\phi{}^2 + v_s{}^2 v_A{}^2 \cos^2\theta = 0. \qquad (14.1.31)$$

The two solutions for $v_\phi{}^2$ are

$$v_\phi{}^2 = \tfrac{1}{2}(v_s{}^2 + v_A{}^2) \pm \tfrac{1}{2}[v_s{}^4 + v_A{}^4 - 2v_s{}^2 v_A{}^2 \cos 2\theta]^{1/2}. \qquad (14.1.32)$$

The eigenvector may be written in the form

$$\delta\mathbf{v} = \begin{pmatrix} v_\phi{}^2 - v_s{}^2 \cos^2\theta \\ 0 \\ v_s{}^2 \sin\theta \cos\theta \end{pmatrix}, \quad \delta\mathbf{B} = B v_\phi{}^{-1}\left[v_\phi{}^2 - v_s{}^2\cos^2\theta\right]\begin{pmatrix} -\cos\theta \\ 0 \\ \sin\theta \end{pmatrix}. \qquad (14.1.33)$$

The perturbation in the magnetic field is normal to the wave vector, as we expect from (14.1.5).

If there is no magnetic field, so that $v_A = 0$, the non-trivial solution of (14.1.31) is $v_\phi = \pm v_s$. As far as (14.1.32) is concerned, we may now regard the magnetic field as leading to a modification of the acoustic modes. We therefore refer to the modes as the *fast and slow magneto-acoustic waves*, according to the sign adopted in (14.1.32).

For the special case of propagation parallel to or orthogonal to the magnetic field, we find that

$$(v_{\phi,\,\text{fast}})^2 = \begin{array}{ll} \max(v_s{}^2, v_A{}^2) & \text{for } \theta = 0, \\ v_s{}^2 + v_A{}^2 & \text{for } \theta = \pm\pi/2; \end{array} \qquad (14.1.34)$$

and

$$(v_{\phi,\,\text{slow}})^2 = \begin{array}{ll} \min(v_s{}^2, v_A{}^2) & \text{for } \theta = 0, \\ 0 & \text{for } \theta = \pm\pi/2. \end{array} \qquad (14.1.35)$$

The solution of (14.1.32) is shown in Fig. 14.1. The same information is shown in polar-diagram form in Fig. 14.2.

We now examine the fast and slow modes in the limits where $\theta \to 0$ or $\pi/2$.

Case 1. Propagation along the magnetic field: $\theta = 0$, $\mathbf{k} = (0, 0, k)$. The matrix equation (14.1.26) becomes

$$\begin{pmatrix} v_\phi{}^2 - v_A{}^2 & 0 & 0 \\ 0 & v_\phi{}^2 - v_A{}^2 & 0 \\ 0 & 0 & v_\phi{}^2 - v_s{}^2 \end{pmatrix} \begin{pmatrix} \delta v_1 \\ \delta v_2 \\ \delta v_3 \end{pmatrix} = 0 \qquad (14.1.36)$$

and the relation (14.1.25) becomes

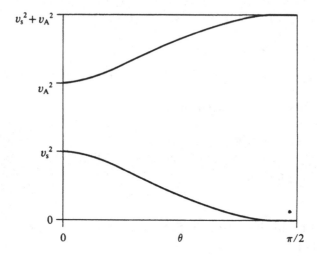

Fig. 14.1. The square of the phase velocity of the fast and slow magnetoacoustic modes as a function of angle between the propagation vector and the magnetic field vector. (Shown for the case that $v_A > v_s$.)

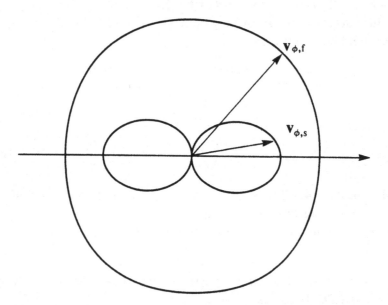

Fig. 14.2. Phase-velocity vector for fast mode and slow mode, as in Fig. 14.1 but shown in the form of a polar diagram.

$$\begin{pmatrix} \delta B_1 \\ \delta B_2 \\ \delta B_3 \end{pmatrix} = \begin{pmatrix} -Bv_\phi^{-1} & 0 & 0 \\ 0 & -Bv_\phi^{-1} & 0 \\ 0 & 0 & 0 \end{pmatrix} \begin{pmatrix} \delta v_1 \\ \delta v_2 \\ \delta v_3 \end{pmatrix}. \tag{14.1.37}$$

We see that there are two transverse modes with $v_\phi^2 = v_A^2$, and one longitudinal mode for which $v_\phi^2 = v_s^2$. If $v_s > v_A$, then the sound wave is the fast wave, and the other two modes correspond to the slow magnetosonic mode and the Alfvén wave. If $v_A > v_s$, then the waves traveling at the Alfvén speed are the Alfvén wave and the fast magnetosonic mode. The wave with phase velocity v_s is then the slow magnetosonic mode.

Case 2. Propagation transverse to the magnetic field: $\theta = \pi/2$, $\mathbf{k} = (k, 0, 0)$. The matrix equation (14.1.26) now becomes

$$\begin{pmatrix} v_\phi^2 - v_s^2 - v_A^2 & 0 & 0 \\ 0 & v_\phi^2 & 0 \\ 0 & 0 & v_\phi^2 \end{pmatrix} \begin{pmatrix} \delta v_1 \\ \delta v_2 \\ \delta v_3 \end{pmatrix} = 0 \tag{14.1.38}$$

and the relation (14.1.25) becomes

$$\begin{pmatrix} \delta B_1 \\ \delta B_2 \\ \delta B_3 \end{pmatrix} = \begin{pmatrix} 0 & 0 & 0 \\ 0 & 0 & 0 \\ Bv_\phi^{-1} & 0 & 0 \end{pmatrix} \begin{pmatrix} \delta v_1 \\ \delta v_2 \\ \delta v_3 \end{pmatrix}. \tag{14.1.39}$$

The mode with phase velocity v_ϕ given by

$$v_\phi^2 = v_s^2 + v_A^2 \tag{14.1.40}$$

is the fast mode. The velocity vector is parallel to $\mathbf{k}$. The magnetic-field component B_3 varies in phase with v_1. This mode is just like a sound wave, with the effect of gas pressure supplemented with that of magnetic pressure.

There are two modes for which $v_\phi^2 = 0$. The one with $\delta v_2 \neq 0$ does not give rise to fluctuations of the magnetic field, nor does it give rise to fluctuations in pressure. It is the limiting case of the Alfvén wave when $k_3 \to 0$. The other case, with $\delta v_3 \neq 0$, is the limiting case of the slow magneto-acoustic mode. For this mode also, there is no change in magnetic field or in gas density or pressure.

14.2 Waves in a barometric medium

In discussing wave motion in the atmosphere of a star, it may be necessary to take into account the effect of gravity. If the gravitational field is taken into account, it is of course impossible to consider the medium as being spatially uniform, since the density will decrease with height above the surface of the star.

If the density scale height is large compared with the wavelength of the MHD waves under consideration, then it is possible to study the properties of the waves by means of the WKB approximation. (See, for instance, Weinberg (1962).) If the scale height is comparable with the wavelengths of the waves, this approximation is invalid. In this case, one must resort to numerical calculation to study the properties of waves in such a system.

It turns out that, if the system does not involve a magnetic field, or if the magnetic field is of such a low intensity that its effects may be neglected, and if the gas has uniform temperature, it is possible to study the wave properties of the system by analytical techniques.

The equations to be solved in addressing the current problem are the continuity equation,

$$\frac{\partial \rho}{\partial t} + \nabla \cdot (\rho \mathbf{v}) = 0, \tag{14.2.1}$$

the equation of motion,

$$\rho \left(\frac{\partial}{\partial t} + \mathbf{v} \cdot \nabla \right) \mathbf{v} = -\nabla p + \rho \mathbf{g}, \tag{14.2.2}$$

and the equation of state,

$$\left(\frac{\partial}{\partial t} + \mathbf{v} \cdot \nabla \right) (p \rho^{-\gamma}) = 0. \tag{14.2.3}$$

The pressure and density may be expressed as

$$p = nkT \tag{14.2.4}$$

and

$$\rho = n m_{\mathrm{av}}, \tag{14.2.5}$$

where m_{av} is the mean particle mass.

We now adopt rectangular coordinates x, y, z, with z in the vertically upward direction, so that $\mathbf{g} = (0, 0, -g)$. On using (14.2.2), (14.2.4) and (14.2.5), and noting that we are assuming the temperature to be constant, we obtain the following equation determining the equilibrium configuration:

$$-kT \frac{\mathrm{d}n}{\mathrm{d}z} - n m_{\mathrm{av}} g = 0. \tag{14.2.6}$$

The solution is expressible as

$$n = n_0 e^{-z/H}, \tag{14.2.7}$$

where H is the *scale height*, given by

$$H = \frac{kT}{m_{av} g}. \tag{14.2.8}$$

For fully ionized hydrogen, this is expressible, in numerical form, as

$$H = 10^{8.22} g^{-1} T. \tag{14.2.9}$$

For instance, for the case of the solar corona, $g = 10^{4.44}$ cm s^{-2} and $T \sim 10^6$ K so that $H \sim 10^{9.8}$ cm. For the case of a neutron star, on the other hand, $g \sim 10^{14}$ cm s^{-2} so that, even if $T \sim 10^6$ K, the scale height is of order only 1 cm.

Following (14.2.7), we may write

$$p = p_0 e^{-z/H} \tag{14.2.10}$$

and

$$\rho = \rho_0 e^{-z/H}, \tag{14.2.11}$$

where H may now be expressed as

$$H = \frac{p_0}{\rho_0 g}. \tag{14.2.12}$$

We now suppose that the system is perturbed so that

$$\left. \begin{aligned} p &\to p + \delta p, \\ \rho &\to \rho + \delta \rho, \\ \mathbf{v} &\to \delta \mathbf{v}. \end{aligned} \right\} \tag{14.2.13}$$

On substituting these expressions into (14.2.1), (14.2.2) and (14.2.3), and retaining only terms of first order in the perturbation, we arrive at the following equations:

$$\frac{\partial}{\partial t} \delta \rho + \nabla \cdot (\rho \delta \mathbf{v}) = 0, \tag{14.2.14}$$

$$\rho \frac{\partial}{\partial t} \delta \mathbf{v} = -\nabla(\delta p) + \delta \rho \, \mathbf{g}, \tag{14.2.15}$$

and

$$\delta \mathbf{v} \cdot \nabla(p \rho^{-\gamma}) + \frac{\partial}{\partial t} (\delta p \, \rho^{-\gamma} - \gamma p \rho^{-\gamma - 1} \delta \rho) = 0. \tag{14.2.16}$$

We note that an extra term must now be included in the perturbed form of the equation of state, since the unperturbed system is no longer spatially homogeneous.

Since the unperturbed system is stationary and uniform in the x- and y-directions, we may certainly Fourier transform in t, x and y. In considering a particular Fourier component, we may without loss of generality assume that the x and y axes have been so chosen that $k_y = 0$. Hence we consider all perturbed quantities to be of the form

$$\delta\rho \to \delta\rho \, e^{i(k_x x - \omega t)}, \text{ etc.} \tag{14.2.17}$$

Equations (14.2.14), (14.2.15) and (14.2.16) now take the form

$$-i\omega \, \delta\rho + \frac{\partial}{\partial z}[\rho_0 e^{-z/H} \delta v_z] - ik_x \rho_0 e^{-z/H} \delta v_x = 0, \tag{14.2.18}$$

$$-i\omega \rho_0 e^{-z/H} \delta v_x = -ik_x \delta p, \tag{14.2.19}$$

$$-i\omega \rho_0 e^{-z/H} \delta v_z = -\frac{\partial}{\partial z} \delta p - g \, \delta\rho, \tag{14.2.20}$$

and

$$\delta v_z \frac{\partial}{\partial z}[p_0 \rho_0^{-\gamma} e^{(\gamma-1)z/H}] - i\omega[\delta p \, \rho_0^{-\gamma} e^{\gamma z/H} - \gamma p_0 \rho_0^{-\gamma-1} e^{\gamma z/H} \delta\rho] = 0, \tag{14.2.21}$$

where we have used (14.2.10) and (14.2.11).

Since the above set of four differential equations have some coefficients that are functions of z, one cannot expect to be able to carry out Fourier transformation in the z-coordinate. However, if we were to replace δv by the variable δw defined by

$$\delta \mathbf{w} = \delta \mathbf{v} \, e^{-z/H}, \tag{14.2.22}$$

we find that the resulting differential equations all have constant coefficients. It would therefore be possible to Fourier transform in the z-coordinate only. The only drawback of this procedure is that we find that the resulting values of the wave vector component k_z are complex. The imaginary part of k_z is found to be $\frac{1}{2}iH^{-1}$.

With this information, we see that it is more appropriate to introduce the following expressions for the perturbations:

$$\delta p = \delta p_0 \exp\left[-\frac{z}{2H} + i(k_x x + k_z z - \omega t)\right], \tag{14.2.23}$$

$$\delta\rho = \delta\rho_0 \exp\left[-\frac{z}{2H} + i(k_x x + k_z z - \omega t)\right], \tag{14.2.24}$$

$$\delta v_x = \delta v_{x,0} \exp\left[\frac{z}{2H} + i(k_x x + k_z z - \omega t)\right], \qquad (14.2.25)$$

and

$$\delta v_z = \delta v_{z,0} \exp\left[\frac{z}{2H} + i(k_x x + k_z z - \omega t)\right]. \qquad (14.2.26)$$

We see that the amplitudes of δp and $\delta \rho$ will decrease exponentially with height, whereas the amplitudes of δv_x and δv_z will increase exponentially with height.

On substituting the above expressions into (14.2.18), (14.2.19), (14.2.20) and (14.2.21), we obtain the following matrix relation for the amplitudes:

$$\begin{pmatrix} ik_x & 0 & -i\omega\rho_0 & 0 \\ 0 & -i\omega & ik_x\rho_0 & \rho_0\left(ik_z - \frac{1}{2H}\right) \\ ik_z - \frac{1}{2H} & g & 0 & -i\omega\rho_0 \\ -i\omega & -i\omega v_s^2 & 0 & \frac{\gamma-1}{H}p_0 \end{pmatrix} \begin{pmatrix} \delta\rho_0 \\ \delta p_0 \\ \delta v_{x,0} \\ \delta v_{z,0} \end{pmatrix} = 0, \qquad (14.2.27)$$

where v_s is the speed of sound:

$$v_s^2 = \frac{\gamma p_0}{\rho_0}. \qquad (14.2.28)$$

By setting the determinant of the matrix in (14.2.27) equal to zero, we obtain the following dispersion relation:

$$\omega^4 - v_s^2(k_x^2 + k_z^2)\omega^2 - \frac{\gamma^2 g^2}{4v_s^2}\omega^2 + (\gamma-1)g^2 k_x^2 = 0. \qquad (14.2.29)$$

The quantities δp_0, etc., are then found to be expressible as

$$\delta\rho_0 = A[\omega^2 k_z - i(\gamma-1)gk_x^2 + \tfrac{1}{2}i\gamma g v_s^{-2}\omega^2]\rho_0, \qquad (14.2.30)$$

$$\delta p_0 = A\gamma\omega^2[k_z + i(1 - \tfrac{1}{2}\gamma)g v_s^{-2}]p_0, \qquad (14.2.31)$$

$$\delta v_{x,0} = A\omega k_x v_s^2[k_z + i(1 - \tfrac{1}{2}\gamma)g v_s^{-2}], \qquad (14.2.32)$$

and

$$\delta v_{z,0} = A\omega(\omega^2 - v_s^2 k_x^2). \qquad (14.2.33)$$

The dispersion relation (14.2.29) may be expressed in a more convenient form by introducing the acoustic resonant frequency ω_a, defined by

$$\omega_a = \frac{v_s}{2H},$$

(14.2.34)

and the 'gravity' resonant frequency ω_g defined by

$$\omega_g = \frac{(\gamma-1)^{1/2} v_s}{\gamma H}.$$

(14.2.35)

The latter frequency is usually referred to as the Brunt-Väisälä frequency. On using these definitions, and on noting that

$$v_s^2 = \gamma g H,$$

(14.2.36)

we find that (14.2.29) may be expressed as

$$\omega^4 - [\omega_a^2 + v_s^2(k_x^2 + k_z^2)]\omega^2 + \omega_g^2 v_s^2 k_x^2 = 0.$$

(14.2.37)

If we write

$$\mathbf{k} = (k\sin\theta, 0, k\cos\theta),$$

(14.2.38)

this equation may be expressed alternatively as

$$\omega^2(\omega^2 - \omega_a^2) = v_s^2 k^2(\omega^2 - \omega_g^2 \sin^2\theta).$$

(14.2.39)

We should note that

$$\frac{\omega_a^2}{\omega_g^2} = 1 + \frac{(\gamma-2)^2}{4(\gamma-1)}$$

(14.2.40)

so that

$$\omega_a \geqslant \omega_g.$$

(14.2.41)

On expressing (14.2.39) as

$$\sin^2\theta = F(\omega^2) \equiv \frac{\omega^2(\omega_a^2 + v_s^2 k^2 - \omega^2)}{\omega_g^2 v_s^2 k^2},$$

(14.2.42)

we see that the mode structure depends on whether or not the maximum value of $F(\omega^2)$ exceeds unity. If we denote by G the maximum value of $F(\omega^2)$, so that

$$G = [F(\omega^2)]_{\max}$$

(14.2.43)

we find that

$$G = \frac{1}{4} \frac{(\omega_a^2 + v_s^2 k^2)^2}{\omega_g^2 v_s^2 k^2}.$$

(14.2.44)

However, we find that

$$G \geqslant G_{\min} = \frac{\omega_a^2}{\omega_g^2}, \tag{14.2.45}$$

so that $G_{\min} > 1$ unless $\gamma = 2$. Hence, ignoring the case $\gamma = 2$, solutions of (14.2.42) always fall into two distinct frequency ranges that we term 'gravity waves' and 'acoustic waves.'

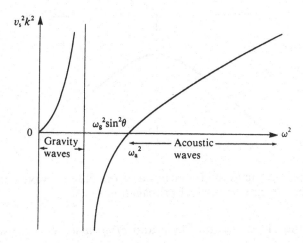

Fig. 14.3. This diagram shows how the square of the wave number varies as a function of the square of the frequency, for a given direction of the wave vector.

An alternative view is to regard (14.2.39) as an equation for $v_s^2 k^2$ as a function of ω^2. It may then be represented as in Fig. 14.3. Once again, we see that there are two branches: one branch with $\omega^2 > \omega_a^2$, the 'acoustic' branch; the other, with $\omega^2 < \omega_g^2 \sin^2 \theta$, the 'gravitational' branch.

In the case of vertical propagation, with $\theta = 0$, (14.2.39) becomes

$$\omega^2 (\omega^2 - \omega_a^2 - v_s^2 k^2) = 0. \tag{14.2.46}$$

The root $\omega^2 = 0$ is a special case of the gravity mode. The more interesting branch is

$$\omega^2 = \omega_a^2 + v_s^2 k^2, \tag{14.2.47}$$

that clearly tends to the dispersion relation for acoustic waves in a uniform medium as $k \to \infty$. The dispersion relation is shown in Fig. 14.4, and is clearly similar in form to the dispersion relation for Langmuir waves in a warm plasma (cf. (9.3.6)). Clearly, there is a cutoff at the acoustic frequency ω_a. Waves with frequency $\omega < \omega_a$ are evanescent, although we find from (14.2.26) that the velocity fluctuation increases with height, and we find

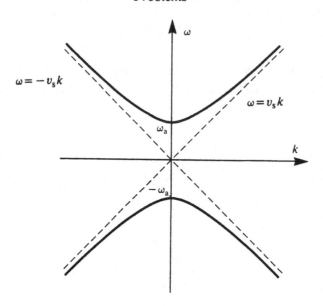

Fig. 14.4. This diagram shows the frequency as a function of wave number for the acoustic mode, for the case of vertical propagation ($\theta = 0$).

from (14.2.23) and (14.2.24) that $\delta\rho/\rho$ and $\delta p/p$ also increase with height. The energy flux for a single evanescent wave will be zero, although two (opppsitely directed) evanescent waves may combine to give a nonzero energy flux.

For the case of horizontal propagation, $\theta = \pi/2$, (14.2.39) may be written as

$$v_s^2 k^2 = \frac{\omega^2(\omega^2 - \omega_a^2)}{\omega^2 - \omega_g^2}. \qquad (14.2.48)$$

The graphical form of this relationship is given by Fig. 14.3, where we now note that $\theta = \pi/2$. We find from (14.2.30) etc., that as $\omega \to \infty$, $|\delta v_{x,0}|/|\delta v_{z,0}| \to 0$, as we would expect, since the wave becomes an acoustic wave. As $\omega \to 0$, $|\delta v_{z,0}|/|\delta v_{x,0}| \to 0$ once more, as is characteristic of gravity waves.

Problems

Problem 14.1. Consider Alfvén waves propagating in a uniform plasma in a uniform field, with wave vector **k** parallel to the applied magnetic field **B**.

(a) Find the resulting dispersion relation when you include the effect of the displacement current in Maxwell's equations.

(b) Obtain expressions that show how energy is partitioned between the magnetic field, the electric field, and the kinetic energy of the plasma.

(c) How is the wave energy partitioned in the limit that $v_A \ll c$ and when v_A approaches c?

Problem 14.2. Consider once more Alfvén waves propagating with wave vector **k** parallel to the applied magnetic field **B** in a uniform plasma in a uniform field.

(a) Find the dispersion relation, assuming that the displacement current may be neglected but it is necessary to take account of finite (but high) conductivity of the plasma.

(b) Consider a wave of real frequency and find the real and imaginary parts of the wave vector.

(c) Express the imaginary part of the wave vector in terms of the electron mean free path.

Problem 14.3. Consider a system comprising a fluid of density ρ_2 supported by a fluid of density ρ_1. In the initial unperturbed state, the fluid of density ρ_1 occupies the space $z < 0$ and the fluid of density ρ_2 occupies the space $z > 0$, where z is directed in the vertical direction so that the gravitational field has components $(0, 0, -g)$.

(a) Show that the dispersion relation for horizontally propagating gravity waves is

$$\omega^2 = \frac{(\rho_1 - \rho_2) g k}{\rho_1 + \rho_2}.$$

(b) Note that the system is unstable if $\rho_2 > \rho_1$. This instability is known as the Rayleigh–Taylor instability. The growth rate of the instability becomes infinite as $k \to \infty$. What physical processes are likely to set a limit on the growth rate?

15

Magnetohydrodynamic stability

15.1 The linear pinch

In this chapter, we shall discuss by means of an example a classical procedure for studying the stability of plasma configurations within the context of MHD theory. The example we consider is one of the more important plasma configurations called a 'linear pinch,' in which the internal gas pressure of a cylindrical column is balanced by the external magnetic pressure due to a current being carried by the plasma column. This procedure is discussed at greater length in books by Bateman (1980), Boyd and Sanderson (1969), and Thompson (1962), among others. The treatment in this chapter follows most closely that of Boyd and Sanderson.

Before studying the stability of the configuration, it is of course necessary to study its equilibrium configuration. We adopt cylindrical coordinates z, r, ϕ, and assume that the equilibrium magnetic field has the form

$$\mathbf{B} = (0, 0, B_\phi). \tag{15.1.1}$$

From the force balance equation

$$-\nabla p + \mathbf{j} \times \mathbf{B} = 0, \tag{15.1.2}$$

using

$$j_z = \frac{1}{4\pi} \frac{1}{r} \frac{\mathrm{d}}{\mathrm{d}r} (rB_\phi), \tag{15.1.3}$$

we arrive at the equation

$$\frac{\mathrm{d}p}{\mathrm{d}r} = -\frac{1}{4\pi} B_\phi \frac{1}{r} \frac{\mathrm{d}}{\mathrm{d}r} (rB_\phi). \tag{15.1.4}$$

On multiplying the terms of (15.1.4) by r^2 and integrating over r, we obtain

248

$$\int_0^R \mathrm{d}r\, r^2 \frac{\mathrm{d}p}{\mathrm{d}r} = -\frac{1}{4\pi} \int_0^R \mathrm{d}r\, rB_\phi \frac{\mathrm{d}}{\mathrm{d}r}(rB_\phi), \tag{15.1.5}$$

where the outer boundary of the system is taken to be $r=R$. If we assume that the gas pressure drops to zero at $r=R$, (15.1.5) becomes

$$2\int_0^R \mathrm{d}r\, rp = \frac{1}{8\pi} R^2 [B_\phi(R)]^2. \tag{15.1.6}$$

This equation may be put into a particularly convenient form for the simple case of an electron–proton plasma. We assume that

$$\begin{aligned} n_e &= n_p = n, \\ T_e &= T_p = T, \end{aligned} \tag{15.1.7}$$

so that

$$p = 2nkT. \tag{15.1.8}$$

We may now introduce the quantity N, the number of electrons per unit length of the cylinder, given by

$$N = \int_0^R \mathrm{d}r\, 2\pi r n. \tag{15.1.9}$$

We also introduce the total current I flowing in the z-direction,

$$I = \int \mathrm{d}r\, 2\pi r j_z. \tag{15.1.10}$$

On using (15.1.3), we find that

$$I = \tfrac{1}{2} R B_\phi(R). \tag{15.1.11}$$

Hence the relation (15.1.6) may be expressed as

$$2NkT = \tfrac{1}{2} I^2. \tag{15.1.12}$$

This may be expressed alternatively as

$$I = 2(NkT)^{1/2}. \tag{15.1.13}$$

This relation was first derived by Bennett (1934) and is known as Bennett's relation.

15.2 Stability analysis

We now consider a perturbation of a magnetostatic system such as a linear pinch. The approach is similar to that of studying MHD waves, except that we will not assume the unperturbed system to be uniform. We therefore begin with the ideal MHD equations (14.1.1) through (14.1.5), and again consider the perturbations (14.1.6), assuming that the unperturbed system is static. Then the perturbations $\delta\rho$, δp, etc., satisfy the equations (14.1.7) through (14.1.10).

We now introduce the vector $\boldsymbol{\xi}(\mathbf{x}, t)$ that represents the physical perturbation of the system. That is to say, we consider that the fluid element that, in the unperturbed system, is at position $\mathbf{x}$ at time t is, in the perturbed system, found at position $\mathbf{x} + \boldsymbol{\xi}(\mathbf{x}, t)$ at time t. To first order, the perturbed velocity may be written as

$$\delta\mathbf{v} = \frac{\partial}{\partial t}\,\delta\boldsymbol{\xi}. \qquad (15.2.1)$$

We also find, from the Lagrange expansion (10.1.11), that (in the linear approximation)

$$\delta\rho = -\nabla\cdot(\rho\boldsymbol{\xi}). \qquad (15.2.2)$$

We find that (15.2.1) and (15.2.2) are consistent with the continuity equation (14.1.1).

We may now find the pressure perturbation δp by means of (14.1.3), replacing $\partial/\partial t$ by δ and $\mathbf{v}$ by $\boldsymbol{\xi}$:

$$(\delta + \boldsymbol{\xi}\cdot\nabla)(p\rho^{-\gamma}) = 0. \qquad (15.2.3)$$

This yields

$$\delta p = -\boldsymbol{\xi}\cdot\nabla p + \frac{\gamma p}{\rho}(\delta p + \boldsymbol{\xi}\cdot\nabla\rho). \qquad (15.2.4)$$

On using (15.2.2), this becomes

$$\delta p = -\gamma p\nabla\cdot\boldsymbol{\xi} - (\boldsymbol{\xi}\cdot\nabla)p. \qquad (15.2.5)$$

On substituting (15.2.1) in (14.1.4) to obtain

$$\frac{\partial}{\partial t}\,\delta\mathbf{B} = \nabla\times\left(\frac{\partial\boldsymbol{\xi}}{\partial t}\times\mathbf{B}\right), \qquad (15.2.6)$$

we see that

$$\delta\mathbf{B} = \nabla\times(\boldsymbol{\xi}\times\mathbf{B}). \qquad (15.2.7)$$

Hence

$$\delta \mathbf{j} = \frac{1}{4\pi} \nabla \times [\nabla \times (\boldsymbol{\xi} \times \mathbf{B})]. \tag{15.2.8}$$

On combining the above expressions in the equation of motion (14.1.8), we finally obtain the equation

$$\rho \frac{\partial^2 \boldsymbol{\xi}}{\partial t^2} = \mathbf{F}(\boldsymbol{\xi}) \tag{15.2.9}$$

where

$$\mathbf{F} = \nabla[\gamma p \nabla \cdot \boldsymbol{\xi} + (\boldsymbol{\xi} \cdot \nabla)p] + \frac{1}{4\pi} \{[\nabla \times \nabla \times (\boldsymbol{\xi} \times \mathbf{B})] \times \mathbf{B}$$
$$+ (\nabla \times \mathbf{B}) \times [\nabla \times (\boldsymbol{\xi} \times \mathbf{B})]\}. \tag{15.2.10}$$

Since the unperturbed system is assumed to be static, we may Fourier analyse $\boldsymbol{\xi}(\mathbf{x}, t)$ in t. We must expect that there will be boundary conditions on $\boldsymbol{\xi}$ that will restrict possible values of ω to a set ω_n, $n = 1, 2, \dots$. If $\boldsymbol{\xi}^{(n)}$ is the eigenvector corresponding to the eigenvalue ω_n, then the general solution of (15.2.9) is expressible as

$$\boldsymbol{\xi}(\mathbf{x}, t) = \sum_n A_n \boldsymbol{\xi}^{(n)}(\mathbf{x}) e^{-i\omega_n t}. \tag{15.2.11}$$

Each mode satisfies the equation

$$-\omega_n^2 \rho \boldsymbol{\xi}^{(n)}(\mathbf{x}) = \mathbf{F}[\boldsymbol{\xi}^{(n)}(\mathbf{x})]. \tag{15.2.12}$$

It may be shown that $\mathbf{F}$ is Hermitian, from which it follows that ω_n^2 is always real. If (15.2.12) leads to any one mode for which $\omega_n^2 < 0$, then the system is unstable. Otherwise, it is stable.

15.3 Boundary conditions

As an example of the hydromagnetic stability analysis of Section 15.2, we consider the special case of the linear pinch shown in Fig. 15.1. We now suppose that the cylindrical plasma core contains a uniform magnetic field in the z-direction. We further suppose that the plasma core is surrounded by a vacuum region which has a magnetic field in the ϕ-direction only, and that the whole system is contained in a cylindrical conducting shell. We use the superscript i to refer to plasma parameters *internal* to the plasma, and the superscript e to refer to parameters *external* to the plasma.

The perturbation of the electric and magnetic fields in the vacuum region may be described in terms of δA^e, the perturbation of the vector potential, so that

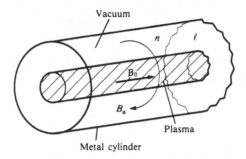

Fig. 15.1. A cylindrical pinch configuration, in which a plasma column, with an internal magnetic field, is confined within a cylindrical conductor.

$$\delta \mathbf{E}^e = -\frac{1}{c} \frac{\partial}{\partial t} \delta \mathbf{A}^e, \tag{15.3.1}$$

and

$$\delta \mathbf{B}^e = \nabla \times (\delta \mathbf{A}^e). \tag{15.3.2}$$

At the plasma surface, we require that the tangential component of the electric field be continuous. If we introduce $\mathbf{n}$ as the unit vector normal to the plasma pointing outward, this condition may be written as

$$\mathbf{n} \times \left[\delta \mathbf{E}^i + \frac{1}{c} (\delta \mathbf{v} \times \mathbf{B}^i) \right] = \mathbf{n} \times \left[\delta \mathbf{E}^e + \frac{1}{c} (\delta \mathbf{v} \times \mathbf{B}^e) \right]. \tag{15.3.3}$$

However, since we are assuming that the plasma has infinite conductivity

$$\delta \mathbf{E}^i + \frac{1}{c} (\delta \mathbf{v} \times \mathbf{B}^i) = 0. \tag{15.3.4}$$

On using (15.2.1) and (15.3.1), equation (15.3.3) takes the form

$$\mathbf{n} \times \left[\frac{1}{c} \frac{\partial}{\partial t} \delta \mathbf{A}^e + \frac{1}{c} \frac{\partial \boldsymbol{\xi}}{\partial t} \times \mathbf{B}^e \right] = 0. \tag{15.3.5}$$

This may be re-expressed as

$$\mathbf{n} \times \frac{\partial}{\partial t} \delta \mathbf{A}^e = - \left(\mathbf{n} \cdot \frac{\partial \boldsymbol{\xi}}{\partial t} \right) \mathbf{B}^e - (\mathbf{n} \cdot \mathbf{B}^e) \frac{\partial \boldsymbol{\xi}}{\partial t}. \tag{15.3.6}$$

However, in our model the external magnetic field is in the ϕ direction only, so that the second term on the right-hand side of (15.3.6) is zero. Hence the equation reduces to

$$\mathbf{n} \times \delta \mathbf{A}^e = -(\mathbf{n} \cdot \boldsymbol{\xi}) \mathbf{B}^e \tag{15.3.7}$$

at the boundary of the plasma, that we take to be $r = R$.

A further condition is imposed on $\delta \mathbf{A}^e$ at the conductor. Since the parallel component of electric field is zero,

$$\mathbf{n} \times \delta \mathbf{A}^e = 0. \qquad (15.3.8)$$

Since the normal component of magnetic field must be zero at the conductor, we also have the condition

$$\mathbf{n} \cdot (\nabla \times \delta \mathbf{A}^e) = 0, \qquad (15.3.9)$$

at the conductor. Between the plasma surface and the conductor, $\delta \mathbf{A}$ satisfies the vacuum wave equation

$$\frac{1}{c^2} \frac{\partial^2}{\partial t^2} \delta \mathbf{A}^e + \nabla \times (\nabla \times \delta \mathbf{A}^e) = 0. \qquad (15.3.10)$$

At the plasma boundary, the equation of motion reduces to the requirement that the total pressure be continuous across the boundary. This may be expressed as

$$p_t^i(\mathbf{x} + \boldsymbol{\xi}) = p_t^e(\mathbf{x} + \boldsymbol{\xi}), \qquad (15.3.11)$$

where p_t is the total pressure:

$$p_t = p + \frac{1}{8\pi} B^2. \qquad (15.3.12)$$

On substituting this expression for p_t in (15.3.11) and expanding by Taylor's theorem, we obtain

$$(1 + \boldsymbol{\xi} \cdot \nabla)\left[p + \frac{1}{8\pi} (\mathbf{B}^i)^2 + \delta p + \frac{1}{4\pi} \mathbf{B}^i \cdot \delta \mathbf{B}^i \right]$$

$$= (1 + \boldsymbol{\xi} \cdot \nabla)\left[\frac{1}{8\pi} (\mathbf{B}^e)^2 + \frac{1}{4\pi} \mathbf{B}^e \cdot \delta \mathbf{B}^e \right]. \qquad (15.3.13)$$

On using (15.2.5) and retaining only terms of first order, we obtain

$$-\gamma p \nabla \cdot \boldsymbol{\xi} + \frac{1}{4\pi} \mathbf{B}^i \cdot [\delta \mathbf{B}^i + (\boldsymbol{\xi} \cdot \nabla)\mathbf{B}^i] = \frac{1}{4\pi} \mathbf{B}^e \cdot [\delta \mathbf{B}^e + (\boldsymbol{\xi} \cdot \nabla)\mathbf{B}^e].$$

$$(15.3.14)$$

15.4 Internally homogeneous linear pinch

We now apply the above equations to a simple model of a linear pinch that is homogeneous, involving a uniform axial magnetic field of strength B_0. The only current carried by the plasma is a surface current that produces an external field B_ϕ. Clearly the force-balance condition requires that

$$P + \frac{1}{8\pi} B_0{}^2 = \frac{1}{8\pi} B_e{}^2, \tag{15.4.1}$$

where we denote by B_e the ϕ-component (the only non-zero component) of the external field $\mathbf{B}_e$ at the boundary of the plasma.

In this model, ∇p and $\nabla \times \mathbf{B}$ are zero within the plasma, so that (15.2.12) reduces to

$$-\omega^2 \rho \boldsymbol{\xi} = \gamma p \nabla(\nabla \cdot \boldsymbol{\xi}) + \frac{1}{4\pi} [\nabla \times \nabla \times (\boldsymbol{\xi} \times \mathbf{B}^i)] \times \mathbf{B}^i. \tag{15.4.2}$$

We may seek solutions of this differential equation and of the associated equation (15.3.10) and the boundary conditions by considering modes that are periodic in z and in ϕ:

$$\boldsymbol{\xi}(\mathbf{x}) \to \boldsymbol{\xi}(r) e^{i(kz + m\phi)}. \tag{15.4.3}$$

Study of the cylindrically symmetric mode $m = 0$ leads to the condition for 'sausage instability,' and discussion of the mode $m = 1$ leads to the condition for 'kink instability.' (See Fig. 15.2.) For simplicity, we here consider only the mode $m = 0$; that is, we consider only axially symmetric perturbations.

For this case, (15.4.2) yields

$$-\omega^2 \rho \xi_z = ik\gamma p \nabla \cdot \boldsymbol{\xi}, \tag{15.4.4}$$

and

$$-\omega^2 \rho \xi_r = \gamma p \frac{d}{dr}(\nabla \cdot \boldsymbol{\xi}) + \frac{1}{4\pi} B_z \left\{ \frac{d}{dr} \left[\frac{B_z}{r} \frac{d}{dr}(r\xi_r) \right] - k^2 B_z \xi_r \right\}. \tag{15.4.5}$$

(Since we are assuming cylindrical symmetry, the component ξ_ϕ may be ignored.) These equations may be written alternatively as

$$(v_s{}^2 k^2 - \omega^2)\xi_z = v_s{}^2 \frac{ik}{r} \frac{d}{dr}(r\xi_r) \tag{15.4.6}$$

and

$$(v_A{}^2 k^2 - \omega^2)\xi_r = (v_s{}^2 + v_A{}^2) \frac{d}{dr} \left[\frac{1}{r} \frac{d}{dr}(r\xi_r) \right] + ikv_s{}^2 \frac{d\xi_z}{dr}. \tag{15.4.7}$$

On eliminating the term ξ_r, we obtain the differential equation

$$\frac{d^2\xi_z}{dr^2} + \frac{1}{r} \frac{d\xi_z}{dr} - K^2 \xi_z = 0, \tag{15.4.8}$$

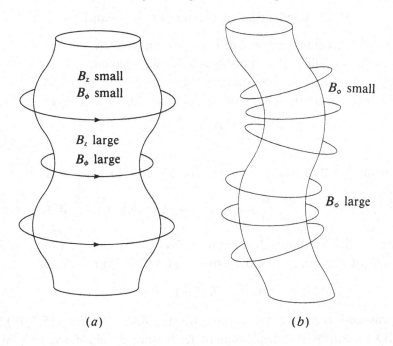

Fig. 15.2. (*a*) The $m = 0$ or 'sausage' mode. Where the radius is reduced, the external (B_ϕ) magnetic field is increased (a destabilizing effect), but the internal (B_z) magnetic field also is increased (a stabilizing effect). (*b*) The $m = 1$ or 'kink' mode. Once again, the changes in the external field are destabilizing, but the changes in the internal field are stabilizing.

where

$$K^2 = \frac{(v_s^2 k^2 - \omega^2)(v_A^2 k^2 - \omega^2)}{v_s^2 v_A^2 k^2 - (v_s^2 + v_A^2)\omega^2}, \qquad (15.4.9)$$

or, equivalently,

$$K^2 = k^2 \left[1 + \frac{(\omega/k)^4}{v_s^2 v_A^2 - (v_s^2 + v_A^2)(\omega/k)^2} \right]. \qquad (15.4.10)$$

Equation (15.4.8) has the form of Bessel's equation, and the appropriate solution, that is well behaved at $r=0$, is

$$\xi_z = I_0(Kr), \qquad (15.4.11)$$

where I_0 is the modified Bessel function of the first kind of order zero. We find from (15.4.6) and (15.4.7) that ξ_r is expressible as

$$\xi_r = -\frac{iK}{k} \left[\frac{v_s^2(v_A^2 k^2 - \omega^2) - v_A^2 \omega^2}{v_s^2(v_A^2 k^2 - \omega^2)} \right] I_0'(Kr). \qquad (15.4.12)$$

15.5 Application of the boundary conditions

We now apply the boundary conditions, set out in Section 5.3, to our study of the $m=0$ instability of a homogeneous linear pinch.

We see from (15.2.7) that the perturbation of the z-component of the magnetic field at the boundary of the plasma (at $r=R$) is given by

$$\delta B_z = \frac{B_0}{r} \frac{d}{dr}(r\xi_r). \tag{15.5.1}$$

Hence, using (15.2.7), the boundary condition (15.3.14) yields

$$-\gamma p \nabla \cdot \xi - \frac{B_0^2}{4\pi r} \frac{d}{dr}(r\xi_r) = \frac{1}{4\pi} \mathbf{B}_e \cdot (\nabla \times \delta \mathbf{A}^e + \xi_r \frac{d}{dr} \mathbf{B}^e). \tag{15.5.2}$$

Equation (15.3.7), that is the condition for the continuity of the tangential component of the electric field, shows that $\delta \mathbf{A}^e$ is expressible as

$$\delta \mathbf{A}^e = (\xi_r B_\phi^2, \delta A_r^e, 0). \tag{15.5.3}$$

In the vacuum region outside the plasma, $\delta \mathbf{A}^e$ satisfies (15.3.10), but for MHD situations the displacement term may be neglected so that $\delta \mathbf{A}^e$ satisfies simply

$$\nabla \times (\nabla \times \delta \mathbf{A}^e) = 0. \tag{15.5.4}$$

The z- and r-components of this equation are

$$\frac{1}{r} \frac{\partial}{\partial r}\left[r \frac{\partial \delta A_r^e}{\partial z} - \frac{\partial}{\partial r}(B_\phi^2 \xi_r)\right] = 0 \tag{15.5.5}$$

and

$$-\frac{\partial}{\partial z}\left[\frac{\partial \delta A_r^e}{\partial z} - \frac{\partial}{\partial r}(B_\phi^2 \xi_r)\right] = 0, \tag{15.5.6}$$

so that (since B_ϕ and ξ_r are functions only of r),

$$\frac{\partial \delta A_r^e}{\partial z} = f(r) \tag{15.5.7}$$

and

$$\frac{\partial \delta A_r^e}{\partial z} - \frac{\partial}{\partial r}(B_\phi^2 \xi_r) = g(z). \tag{15.5.8}$$

Since both terms on the left-hand side of (15.5.8) are functions of r only, g must be a constant. We may take this constant to be zero, since a nonzero value would merely represent a change in the value of the external field.

Hence, for the case being considered, there is no perturbation of the external magnetic field:

$$\delta \mathbf{B}^e \equiv \nabla \times \delta \mathbf{A}^e = 0. \tag{15.5.9}$$

We may now use (15.5.1), (15.5.9), and the fact that

$$B_\phi^2 \alpha r^{-1}, \tag{15.5.10}$$

to express the boundary condition (15.3.14) in the form

$$-\gamma p \nabla \cdot \boldsymbol{\xi} - \frac{B_0^2}{4\pi r} \frac{d}{dr}(r\xi_r) = -\frac{1}{4\pi r} B_e^2 \xi_r, \quad \text{at } r = R. \tag{15.5.11}$$

On using (15.4.5), this becomes

$$\left[\frac{\gamma \omega^2 p}{ikv_s^2} - \frac{B_0^2(v_s^2 k^2 - \omega^2)}{4\pi i \, kv_s^2}\right] \xi_z = -\frac{B_e^2}{4\pi r} \xi_r, \quad \text{at } r = R. \tag{15.5.12}$$

We may now use (15.4.10), (15.4.11) and (15.4.12) to obtain the following form of the dispersion relation relating the frequency ω and the wave-number k:

$$\omega^2 = v_A^2 k^2 - \frac{B_e^2}{B_0^2} v_A^2 \frac{K}{R} \frac{I_0'(KR)}{I_0(KR)}, \tag{15.5.13}$$

that may be written more explicitly as

$$\omega^2 = \frac{B_0^2 k^2}{4\pi\rho} - \frac{B_e^2 k^2}{4\pi\rho R^2}\left[\frac{KRI_0'(KR)}{I_0(KR)}\right]. \tag{15.5.14}$$

In order to solve this equation, K must be expressed in terms of ω and k by means of (15.4.9). Hence we must solve the equation numerically to obtain the relationship between ω and k. However, it is found that ω^2 takes only real values, so that the condition that the linear pinch is unstable against the $m = 0$ mode is

$$B_0^2 \leqslant \frac{B_e^2}{(KR)^2} \frac{KRI_0'(KR)}{I_0(KR)}. \tag{15.5.15}$$

This implies that there is a certain minimum value B_0 of the internal field that is required to stabilize the pinch against the cylindrically symmetric ($m = 0$) mode.

One may understand the physical mechanism of this instability (sometimes called the 'sausage' instability) by referring to Fig. 15.2. If the radius of the pinch were to be increased in some location, the internal gas pressure would not be affected, and the external magnetic pressure would drop, due to the relation (15.5.10). In the absence of an internal magnetic field, therefore, the perturbation would be unstable. However, if there is also an internal field, the internal magnetic pressure will decrease as the radius increases, since the total magnetic flux is fixed. If the internal field is strong enough, this decrease in the internal magnetic pressure can more than offset the decrease in external magnetic pressure, and the perturbation is then stable.

Problems

Problem 15.1. Show that, for an ideal MHD system in a static state, both **B** and **j** lie on surfaces of constant pressure.

Problem 15.2. Consider a linear pinch composed of fully ionized hydrogen. Denote the electron temperature and proton temperature by T_e and T_p. Assume that the density of electrons and of protons on the axis of the cylinder is n_0. Assume that the axial electron drift velocity, v_z, is independent of r, and neglect ion motion. Obtain the following expression for the electron density as a function of radius:

$$n(r) = \frac{n_0}{(1 + K^2 r^2)^2} \tag{1}$$

where

$$K^2 = \frac{\pi n_0 e^2 v_z^2}{2 c^2 k (T_e + T_p)}. \tag{2}$$

Problem 15.3. Consider a uniform perfectly conducting plasma of cylindrical symmetry. Assume that the temperature is T, the total current carried on the cylindrical surface of the plasma is I, and that the plasma has N electrons and N protons per cm of axial length. Show that 'Bennett's relation'

$$I^2 = 4NkT$$

is satisfied.

Problem 15.4. Consider a cylindrical pinch configuration of radius R, uniform electron density n, equal proton density, and uniform temperature T. Note Bennett's relation given in Problem 15.3.

(a) Calculate the rate of Joule heating per unit length, H (erg cm^{-1}s^{-1}), using the following approximate expression for the resistivity:

$$\eta = 10^{3.7} \, T^{-3/2}. \tag{1}$$

(b) Using the following approximate expression for bremsstrahlung radiation S (erg cm^{-3}s^{-1}),

$$S = 10^{-26.8} n^2 T^{1/2}, \tag{2}$$

calculate the energy loss rate per unit length L (erg cm^{-1} s^{-1}).

(c) If the current is increased progressively, the plasma arrives at a state at which Joule heating is balanced by bremsstrahlung radiation. Show that this condition sets a limit I_c to the total current, and obtain an expression for that value.

(d) Hence find the highest temperature that can be attained by the pinch for the value $N = 10^{18}$ cm^{-1}.

16

Variation principle for MHD systems

It was shown by Bernstein *et al.* (1958) that it is possible to study the stability of magnetostatic systems by means of an energy principle. They derived this energy principle from the equation of motion expressed in terms of the displacement vector $\xi_r(\mathbf{x}, t)$.

In this chapter, we study a variation principle that describes the behavior of MHD systems. This variation principle leads to an equation of motion equivalent to that which we obtained in Chapter 15. It also makes it possible to obtain expressions for energy, energy flux, momentum and momentum flux associated with small disturbances of an MHD system.

Some general theorems are presented in Section 16.1; we derive in Section 16.2 an expression for the change in magnetic field due to an arbitrary displacement of the plasma; a variation principle for MHD systems is presented in Section 16.3; and a variation principle for small-amplitude disturbances is presented in Section 16.4.

16.1 Variation principle for a spatially distributed system

The behavior of many spatially distributed systems, such as electromagnetic fields or fluids, may be described by variation principles. We write the variational equation as

$$\delta \int \int dt\, d^3x\, L = 0 \qquad (16.1.1)$$

where the Lagrangian function L is expressible as

$$L = L(q_\alpha, \dot{q}_\alpha, q_{\alpha,r}, \mathbf{x}, t) \qquad (16.1.2)$$

where $q_\alpha(\mathbf{x}, t)$ are the dynamical variables, and we use the notation

$$\dot{q}_\alpha = \frac{\partial q_\alpha}{\partial t}, \quad q_{\alpha,r} = \frac{\partial q_\alpha}{\partial x_r}, \text{etc.} \qquad (16.1.3)$$

We introduce the following conjugate variables, analogous to canonical momenta in simple dynamical systems:

$$p_{\alpha\tau} = \frac{\partial L}{\partial \dot{q}_\alpha}, \quad p_{\alpha r} = \frac{\partial L}{\partial q_{\alpha,r}}. \qquad (16.1.4)$$

Equation (16.1.1) may now be expressed as

$$\iint \mathrm{d}t\, \mathrm{d}^3x \left[\frac{\partial L}{\partial q_\alpha} \delta q_\alpha + p_{\alpha\tau} \delta \dot{q}_\alpha + p_{\alpha r} \delta q_{\alpha,r} \right] = 0, \qquad (16.1.5)$$

and the implication of (16.1.1) is that this expression should vanish for arbitrary perturbations δq_α subject to the restriction that these perturbations vanish on a bounding surface S, and are zero everywhere at selected initial and final times. We may integrate the expression (16.1.5) by parts to obtain

$$\int_{t_1}^{t_2} \int_V \mathrm{d}t\, \mathrm{d}^3x \left(\frac{\partial L}{\partial q_\alpha} - \frac{\partial p_{\alpha\tau}}{\partial t} - \frac{\partial p_{\alpha r}}{\partial x_r} \right) \delta q_\alpha$$

$$+ \int_V \mathrm{d}^3x [p_{\alpha\tau} \delta q_\alpha]_{t_1}^{t_2} + \int_{t_1}^{t_2} \mathrm{d}t \int \mathrm{d}S\, n_r p_{\alpha r} \delta q_\alpha = 0. \qquad (16.1.6)$$

On using the constraints on the perturbations δq_α, we obtain from this equation the Euler–Lagrange equations:

$$\frac{\partial p_{\alpha\tau}}{\partial t} + \frac{\partial p_{\alpha r}}{\partial x_r} - \frac{\partial L}{\partial q_\alpha} = 0. \qquad (16.1.7)$$

In terms of these variables, we may define an energy density E by

$$E = \dot{q}_\alpha p_{\alpha\tau} - L, \qquad (16.1.8)$$

where the summation convention is used, and an energy flux vector S_r by

$$S_r = \dot{q}_\alpha p_{\alpha r}. \qquad (16.1.9)$$

If we note that the 'total' time derivative of L is given by

$$\left(\frac{\partial L}{\partial t} \right)_{\text{tot}} = \frac{\partial L}{\partial q_\alpha} \dot{q}_\alpha + \frac{\partial L}{\partial \dot{q}_\alpha} \ddot{q}_\alpha + \frac{\partial L}{\partial q_{\alpha,r}} \dot{q}_{\alpha,r} + \frac{\partial L}{\partial t}, \qquad (16.1.10)$$

we find that

$$\frac{\partial E}{\partial t} + \frac{\partial S_r}{\partial x_r} = -\frac{\partial L}{\partial t}. \qquad (16.1.11)$$

Hence if the Lagrangian function L does not depend explicitly on time, the right-hand side of this equation is zero, and (16.1.11) may then be interpreted as an energy conservation equation.

In the same way, we may define a momentum density P_r by

$$P_r = q_{\alpha,r} p_{\alpha\tau} \tag{16.1.12}$$

and a momentum flux tensor by

$$T_{rs} = q_{\alpha,r} p_{\alpha s} - L\delta_{rs}. \tag{16.1.13}$$

By analogy with the derivation of (16.1.11), we obtain

$$\frac{\partial P_r}{\partial t} + \frac{\partial T_{rs}}{\partial x_s} = -\frac{\partial L}{\partial x_r}, \tag{16.1.14}$$

where the term on the right-hand side refers only to the explicit dependence of L on the spatial coordinates. If the Lagrangian function is independent of the spatial coordinates, then (16.1.14) is an equation of momentum conservation.

16.2 Convection of magnetic field

We plan to consider small departures of the magneto-plasma configuration from some known ('base') configuration, and then develop a variation principle for those small departures. We again describe these departures by means of a displacement vector $\xi(\mathbf{x}, t)$, but allow for possible time-variation of the base configuration as follows.

Suppose that there is a perturbation that has the effect of displacing the element of plasma that was at position $\mathbf{x}$ at time t (in the unperturbed 'realization') to be at the position $\tilde{\mathbf{x}}$ at time t (in the perturbed 'realization'). We write

$$\tilde{\mathbf{x}} = \mathbf{x} + \xi(\mathbf{x}, t). \tag{16.2.1}$$

We start with unperturbed density $\rho(\mathbf{x})$, etc., and we will need to calculate the density, etc., in the perturbed system at the position $\mathbf{x}$ defined by (16.2.1). Most of these calculations are simple, except for that of the magnetic field that we therefore discuss in this section.

If we write

$$u_r = B_r/\rho, \tag{16.2.2}$$

we see from (12.2.14) that, for an *infinitesimal* displacement,

$$\mathbf{u}_r(\mathbf{x}) = u_r(\mathbf{x}) + u_s(\mathbf{x})\frac{\partial \xi_r(\mathbf{x})}{\partial x_s}. \tag{16.2.3}$$

Since the variation principle for small-amplitude displacements will be quadratic in the amplitude of the displacement, we will need expressions for the perturbed magnetic field, etc., that are accurate to *second* order, not merely to first order. Hence we expect the necessity of calculating correction terms to (16.2.3). However, we now show that this expression is in fact exact, and therefore needs no correction terms.

Suppose we begin with a vector field u_r^α defined in a coordinate system x_r^α. Suppose that each point x_r^α of the plasma is moved to position x_r^β where the relationship between x_r^β and x_r^α is given by

$$x_r^\beta = x_r^\alpha + \xi_r^\alpha(\mathbf{x}^\alpha). \tag{16.2.4}$$

Now suppose that this displacement is followed by a second displacement such that the point at x_r^β is moved to position x_r^γ defined by

$$x_r^\gamma = x_r^\beta + \xi_r^\beta(\mathbf{x}^\beta). \tag{16.2.5}$$

Clearly the combined effect of both displacements is the same as a single displacement that we may write as

$$x_r^\gamma = x_r^\alpha + \xi_r^\gamma(\mathbf{x}^\alpha), \tag{16.2.6}$$

where

$$\xi_r^\gamma(\mathbf{x}^\alpha) = \xi_r^\alpha(\mathbf{x}^\alpha) + \xi_r^\beta(\mathbf{x}^\alpha + \xi^\alpha(\mathbf{x}^\alpha)). \tag{16.2.7}$$

Let us now *suppose* that (16.2.3) is exact. If that is the case, we should be able to calculate the vector u_r^γ either by applying the equation to step α–β specified by (16.2.5) and then to step β–γ specified by (16.2.6), or by applying the equation to the one step a–γ specified by (16.2.7). We shall show that both procedures yield the same result.

We see from (16.2.3) (assuming it to be exact) that step α–β, specified by (16.2.4), leads to the following equation for u_r^β:

$$u_r^\rho(\mathbf{x}^\rho) = u_r^\alpha(\mathbf{x}^\alpha) + u_s^\alpha(\mathbf{x}^\alpha) \frac{\partial \xi_r^\alpha(\mathbf{x}^\alpha)}{\partial x_s^\alpha}. \tag{16.2.8}$$

Similarly, step β–γ leads to the following equation for u_r^γ:

$$u_r^\gamma(\mathbf{x}^\gamma) = u_r^\beta(\mathbf{x}^\beta) + u_s^\beta(\mathbf{x}^\beta) \frac{\partial \xi_r^\beta(\mathbf{x}^\beta)}{\partial x_s^\beta}. \tag{16.2.9}$$

On combining (16.2.8) and (16.2.9), we see that

$$u_r^\gamma(\mathbf{x}^\gamma) = u_r^\alpha(\mathbf{x}^\alpha) + u_s^\alpha(\mathbf{x}^\alpha) \frac{\partial \xi_r^\alpha(\mathbf{x}^\alpha)}{\partial x_s^\alpha}$$
$$+ \left[u_s^\alpha(\mathbf{x}^\alpha) + u_t^\alpha(\mathbf{x}^\alpha) \frac{\partial \xi_s^\alpha(\mathbf{x}^\alpha)}{\partial x_t^\alpha} \right] \frac{\partial \xi_r^\beta(\mathbf{x}^\beta)}{\partial x_s^\beta}. \tag{16.2.10}$$

On the other hand, if (16.2.3) is exact, we expect that

$$u_r^\gamma(\mathbf{x}^\gamma) = u_r^\alpha(\mathbf{x}^\alpha) + u_s^\alpha(\mathbf{x}^\alpha) \frac{\partial \xi_r^\gamma(\mathbf{x}^\alpha)}{\partial x_s^\alpha}. \qquad (16.2.11)$$

By substituting (16.2.7) for u_r^γ in this equation, we find that (16.2.10) and (16.2.11) are identical. Hence we obtain the same result by going through two consecutive steps, or through one equivalent step, just as we would expect if (16.2.3) is exact.

We can now argue that (16.2.3) is indeed exact as follows. We know that the evolution of the vector field u_r is governed by (12.2.14). Hence (16.2.3) is exact if, for a displacement governed by the time-varying function $\xi_r(x, t)$, the vector determined by (16.2.3) satisfies (12.2.14) for all values of t. But we can now consider the three states corresponding to times $t = t_0$, $t = t_1$, and $t = t_1 + dt_1$. By inverting our earlier argument, we know that if (16.2.3) is applied to the step t_0 to t_1, and to the step t_0 to $t_1 + dt_1$, then the step from t_1 to $t_1 + dt_1$ is also governed by (16.2.3). But, for an infinitesimal time interval dt_1, (16.2.3) is equivalent to (12.2.14). Hence we have shown that the time evolution specified by (16.2.3), for a time-dependent displacement function, satisfies (12.2.14) for all values of t. This shows that (16.2.3) is indeed exact.

16.3 Variation principle for MHD motion

We now consider a plasma within the context of ideal MHD theory, taking into account the effect of a possible gravitational field. The Lagrangian function L may be expressed as

$$L = K - V_H - V_M - V_G \qquad (16.3.1)$$

where K denotes the kinetic energy density, V_H the hydrodynamic energy density, V_M the magnetic energy density, and V_G the gravitational energy density. We plan to describe the behavior of the system in terms of a time-dependent displacement vector $\boldsymbol{\xi}(\mathbf{x}, t)$. We suppose that we start with an unperturbed state of density $\rho(\mathbf{x})$, etc., and that the perturbation has the effect of displacing the element of plasma that was at position $\mathbf{x}$ at time t (in the unperturbed or 'base' 'realization') to the position $\tilde{\mathbf{x}}$ at time t (in the perturbed 'realization'). We write

$$\tilde{\mathbf{x}} = \mathbf{x} + \boldsymbol{\xi}(\mathbf{x}, t). \qquad (16.3.2)$$

By noting that an element of gas does work per unit mass

$$\int p \, \mathrm{d}(1/\rho) \qquad (16.3.3)$$

as it expands, and by adopting the adiabatic equation of state

$$p \propto \rho^{\gamma}, \qquad (16.3.4)$$

we find that the total 'hydrodynamic potential energy' of a perturbed system is given by

$$\int \mathrm{d}^3 x \, V_H = \frac{1}{\gamma - 1} \int \mathrm{d}^3 x \, \tilde{p}(\mathbf{x}). \qquad (16.3.5)$$

This may clearly be expressed alternatively as

$$\int \mathrm{d}^3 x \, V_H = \frac{1}{\gamma - 1} \int \mathrm{d}^3 \tilde{x} \, \tilde{p}(\tilde{\mathbf{x}}). \qquad (16.3.6)$$

On noting that

$$\tilde{\rho}(\tilde{\mathbf{x}}) \, \mathrm{d}^3 \tilde{x} = \rho(\mathbf{x}) \mathrm{d}^3 x \qquad (16.3.7)$$

and using (16.3.4), we see that (16.3.6) may be re-expressed as

$$\int \mathrm{d}^3 x \, V_H = \frac{1}{\gamma - 1} \int \mathrm{d}^3 x \, p(\mathbf{x}) \left(\frac{\tilde{\rho}(\tilde{\mathbf{x}})}{\rho(\mathbf{x})} \right)^{\gamma - 1}. \qquad (16.3.8)$$

Hence we obtain

$$V_H = \frac{1}{\gamma - 1} p J^{-(\gamma - 1)} \qquad (16.3.9)$$

where J is the Jacobian of the transformation defined by

$$J = \frac{\partial(\tilde{x}_1, \tilde{x}_2, \tilde{x}_3)}{\partial(x_1, x_2, x_3)}, \qquad (16.3.10)$$

since

$$\mathrm{d}^3 \tilde{x} = J \mathrm{d}^3 x \qquad (16.3.11)$$

and

$$\tilde{\rho}(\tilde{\mathbf{x}}) = J^{-1} \rho(\mathbf{x}). \qquad (16.3.12)$$

We find from (16.3.3) that

$$J = 1 + \xi_{r,r} + \tfrac{1}{2}(\xi_{r,r})^2 - \tfrac{1}{2}\xi_{r,s}\xi_{s,r}. \qquad (16.3.13)$$

We now consider the magnetic potential energy that, by analogy with (16.3.6), may clearly be expressed as

$$\int d^3x \, V_M = \frac{1}{8\pi} \int d^3\tilde{x} (\tilde{B}(\tilde{x}))^2. \qquad (16.3.14)$$

We now use (16.2.3), that may be written explicitly as

$$\frac{\tilde{B}_r(\tilde{x})}{\tilde{\rho}(\tilde{x})} = \frac{B_r(x)}{\rho(x)} + \frac{1}{\rho(x)} B_s(x) \xi_{r,s}(x). \qquad (16.3.15)$$

Hence, using (16.3.11) and (16.3.12), we find that V_M is expressible as

$$V_M = \frac{1}{8\pi} J^{-1} (B_r + B_s \xi_{r,s})^2. \qquad (16.3.16)$$

The total gravitational potential energy may be expressed as

$$\int d^3x \, V_G = \int d^3\tilde{x} \, \tilde{\rho}(\tilde{x}) \Phi(\tilde{x}) \qquad (16.3.17)$$

so that, using (16.3.7), we see that

$$V_G = \rho(x)\Phi(x + \xi). \qquad (16.3.18)$$

The total kinetic energy is expressible as

$$\int d^3x \, K = \tfrac{1}{2} \int d^3\tilde{x} \, \tilde{\rho}(\tilde{x})(\tilde{v}(\tilde{x}))^2. \qquad (16.3.19)$$

Since

$$\tilde{v}_r(\tilde{x}) = v_r(x) + \dot{\xi}_r(x) + v_s(x)\xi_{r,s}(x), \qquad (16.3.20)$$

we find that

$$K = \tfrac{1}{2}\rho(v_r + \dot{\xi}_r + v_s\xi_{r,s})^2. \qquad (16.3.21)$$

Note that we are here assuming that the gravitational field is not influenced by the perturbation. In helioseismology (the study of the interior of the Sun by the detection and analysis of normal modes of oscillation (Deubner and Gough, 1984), estimates of the properties (including the frequencies) of normal modes are usually made by calculations that neglect the dependence of the gravitational field on the perturbation, a procedure known as the 'Cowling approximation.' Such calculations could be made on the basis of the Lagrangian function given in (16.3.1).

By including an additional term in (16.3.1), one can incorporate into the calculation the influence of the perturbation on the gravitational field (see Problem 16.1). This extended form of the Lagrangian function would make it possible to carry out helioseismology calculations without adopting the restriction of the Cowling approximation.

16.4 Small-amplitude disturbances

We may now substitute the appropriate formulas on the right-hand side of (16.2.1). If we now regard ξ as a small quantity, we may expand the Lagrangian as

$$L = L^{(0)} + L^{(1)} + L^{(2)} + \ldots \qquad (16.4.1)$$

where $L^{(0)}$ is independent of ξ, etc. We then find the expressions

$$L^{(0)} = \frac{1}{2}\rho v^2 - \frac{1}{\gamma - 1}p - \frac{1}{8\pi}B^2 - \rho\Phi, \qquad (16.4.2)$$

$$L^{(1)} = \rho v_r \dot{\xi}_r + \rho v_r v_s \xi_{r,s} + p\xi_{r,r} + \frac{1}{8\pi}B^2\xi_{r,r} - \frac{1}{4\pi}B_r B_s \xi_{r,s} - \rho\Phi_{,r}\xi_r$$

$$(16.4.3)$$

and

$$L^{(2)} = \frac{1}{2}\rho(\dot{\xi}_r + v_s\xi_{r,s})^2 - \frac{1}{2}(\gamma - 1)p(\xi_{r,r})^2 - \frac{1}{2}p\xi_{r,s}\xi_{s,r} - \frac{1}{16\pi}B^2(\xi_{r,r})^2$$

$$- \frac{1}{16\pi}B^2\xi_{r,s}\xi_{s,r} + \frac{1}{4\pi}B_r B_s\xi_{r,s}\xi_{l,l} - \frac{1}{8\pi}B_s B_l\xi_{r,s}\xi_{r,l} - \frac{1}{2}\rho\Phi_{,rs}\xi_r\xi_s.$$

$$(16.4.4)$$

From now on, we will consider the simpler systems that are static in their unperturbed state, so that $\mathbf{v} = 0$ in the above expressions. Then the Euler–Lagrange equation corresponding to $L^{(1)}$ is found to be

$$-p_{,r} + \frac{1}{4\pi}B_s(B_{r,s} - B_{s,r}) - \rho\Phi_{,r} = 0, \qquad (16.4.5)$$

which is simply the force-balance equation for the unperturbed state. Similarly, we find from (16.4.4) that the equation of motion, in the linear approximation, takes the form

$$\rho\ddot{\xi}_r = \frac{\partial}{\partial x_r}\left[(\gamma - 1)p\xi_{s,s} + \frac{1}{8\pi}B^2\xi_{s,s} - \frac{1}{4\pi}B_s B_l\xi_{s,l}\right]$$

$$+ \frac{\partial}{\partial x_s}\left[p\xi_{s,r} + \frac{1}{8\pi}B^2\xi_{s,r} - \frac{1}{4\pi}B_r B_s\xi_{l,l} + \frac{1}{4\pi}B_s B_l\xi_{r,l}\right] - \rho\Phi_{,rs}\xi_s.$$

$$(16.4.6)$$

One may verify that (16.4.6) is equivalent to (15.2.9) and (15.2.10) for the special case that there is no gravitational field.

By using (16.1.8), one may verify, for the special case that the unperturbed state is static, that

$$E = K + V \tag{16.4.7}$$

where

$$V = V_H + V_M + V_G. \tag{16.4.8}$$

We find from (16.4.4), for the special case that the unperturbed system is static, that

$$K^{(2)} = \tfrac{1}{2}\rho\dot{\xi}_r^2 \tag{16.4.9}$$

and

$$V^{(2)} = \frac{1}{2}(\gamma - 1)p(\xi_{r,r})^2 + \frac{1}{2}p\xi_{r,s}\xi_{s,r} + \frac{1}{16\pi}B^2(\xi_{r,r})^2 + \frac{1}{16\pi}B^2\xi_{r,s}\xi_{s,r}$$

$$- \frac{1}{4\pi}B_r B_s \xi_{r,s}\xi_{l,l} + \frac{1}{8\pi}B_s B_l \xi_{r,s}\xi_{r,l} + \frac{1}{2}\rho\Phi_{,rs}\xi_r\xi_s. \tag{16.4.10}$$

Since $K^{(2)}$ is positive definite and $E^{(2)}$ is a constant of the motion, it follows that a mode is stable if $V^{(2)}$ is positive-definite for that mode. Hence investigation of the energy term $V^{(2)}$ is useful in studying the stability of MHD systems.

The above expressions may also be used to obtain estimates of the frequency of a mode. If we carry out the Fourier transformation in time by

$$\xi_r(\mathbf{x}, t) \rightarrow \int d\omega\, e^{-i\omega t}\xi_r(\mathbf{x}, \omega) \tag{16.4.11}$$

we find that

$$K^{(2)} = 2\pi \int d\omega\, \omega^2 G^{(2)}(\xi(\omega), \xi(-\omega)) \tag{16.4.12}$$

where

$$G^{(2)}(\xi(\omega), \xi(-\omega)) = \tfrac{1}{2}\rho\xi_r(\omega)\xi_r(-\omega) \tag{16.4.13}$$

and similarly for $V^{(2)}$. By considering a change of amplitude of any mode, we find from (16.1.1) that

$$L^{(2)} \equiv \omega^2 G^{(2)}(\xi(\omega), \xi(-\omega)) - V^{(2)}(\xi(\omega), \xi(-\omega)) = 0. \tag{16.4.14}$$

Hence

$$\omega^2 = \frac{V^{(2)}(\xi(\omega), \xi(-\omega))}{G^{(2)}(\xi(\omega), \xi(-\omega))}. \tag{16.4.15}$$

Now consider a perturbation in ξ, and suppose that a perturbation $\delta\xi$ leads to a perturbation $\delta(\omega^2)$ in the estimate of ω^2 given by (16.4.15). We see from (16.4.14) that

$$\delta(\omega^2)G^{(2)} + \omega^2 \delta G^{(2)} - \delta V^{(2)} = 0 \tag{16.4.16}$$

where $\delta G^{(2)}$ and $\delta V^{(2)}$ are the perturbations in $G^{(2)}$ and $V^{(2)}$ due to the perturbation $\delta\xi$. However, because of (16.1.1),

$$\delta K^{(2)} - \delta V^{(2)} = \omega^2 \delta G^{(2)} - \delta V^{(2)} = 0. \tag{16.4.17}$$

Hence we see from (16.4.16) that

$$\delta(\omega^2) = 0. \tag{16.4.18}$$

Hence, if we estimate the frequency of a mode by using (16.4.15), but the assumed mode is in error, the resulting error in the frequency is of second order.

Finally, we note that it is possible to compute by means of (16.1.8), (16.1.9), (16.1.12) and (16.1.13) the terms of the energy–momentum tensor (energy density, etc.) that may be derived from $L^{(2)}$, the second-order contribution to the Lagrangian function given by (16.4.4). These quantities will be of second order in the perturbation. However, it is important to note that the quantities so obtained are not necessarily identical to the second-order contributions to the exact energy–momentum tensor. The quantities derived from (16.4.4) involve solutions of the equations of motion calculated only in the linear approximation, whereas the second-order contribution to the quantities calculated from the exact Lagrangian function involve (in general) second-order nonlinear contributions to the solutions of the equations of motion. In order to avoid confusion, it is preferable to refer to the 'energy,' etc., calculated from (16.4.4) as 'pseudo-energy,' etc. (Sturrock, 1962). For instance, the pseudo-energy of a wave may be negative (Sturrock, 1960b), and certain instabilities of plasmas may conveniently be understood from this point of view.

Problems

Problem 16.1. Consider the Lagrangian function given by

$$L = K - V_\mathrm{H} - V_\mathrm{M} - V_\mathrm{G} - V_\mathrm{GF}, \tag{1}$$

where K, V_H, V_M, and V_G have the same meanings as earlier in the chapter, and the remaining term is given by

$$V_{GF} = -\frac{1}{8\pi} G^{-1} \left(\frac{\partial \Phi}{\partial x_r} \right)^2. \tag{2}$$

In this expression, that represents the potential energy of the gravitational field, G is the usual gravitational constant.

Verify that the Euler–Lagrange equation corresponding to the quantity Φ, now regarded as a dynamical variable, is the field equation (in the Newtonian approximation) for the gravitational field:

$$\frac{\partial^2 \Phi}{\partial x_r^2} = -4\pi G \rho. \tag{3}$$

Problem 16.2. Consider the magnetic-field configuration

$$B_3 = B_0, \quad B_1 = b B_0 \sin(k x_3). \tag{1}$$

(a) Verify that

$$x_1 = x_1(x_3) = \frac{b}{k} (1 - \cos(k x_3)) \tag{2}$$

traces a field line.

(b) Verify that the displacement

$$\xi_3(x_3) = Z, \quad \xi_1(x_3) = \frac{b}{k} [\cos(k x_3) - \cos(k(x_3 + Z))], \tag{3}$$

represents a displacement of a point on a given field line to another point on the same field line. Hence we expect that this displacement will leave the field configuration unchanged. Furthermore, it is clear that the displacement of equation (3) leaves the plasma density unchanged.

(c) Now use (16.3.14) to calculate the magnetic-field vector at the displaced position

$$\tilde{x}_3 = x_3 + \xi_3(x_3), \quad \tilde{x}_1 = x_1 + \xi_1(x_3). \tag{4}$$

Verify that, as expected, the magnetic-field vector so calculated is the same as the magnetic-field vector in the original magnetic-field configuration.

Problem 16.3. The expression (16.4.10) is not the same as the expression derived by Bernstein *et al.* (1958). Using our notation, and ignoring the gravitational term, the Bernstein expression is

$$V^{(2)} = \frac{1}{8\pi} [\nabla \times (\xi \times \mathbf{B})]^2 + \frac{1}{8\pi} \{\xi \times [\nabla \times (\xi \times \mathbf{B})]\} \cdot (\nabla \times \mathbf{B})$$

$$+ \frac{1}{2} \gamma p (\nabla \cdot \xi)^2 + \frac{1}{2} (\nabla \cdot \xi)(\xi \cdot \nabla p). \tag{1}$$

Show that the above expression is equivalent to the expression (16.4.10) (ignoring the gravitational term) either by using the equilibrium condition (16.4.5) and using the fact that two Lagrangian functions (say L_a and L_b) are equivalent (they describe the same physical system and yield the same equations of motion) if they differ only by the divergence of a function of the dynamical variables so that

$$L_b - L_a = \frac{\partial}{\partial x_r} (F_r(\xi_s, x_u, t)), \qquad (2)$$

or in any other way.

Problem 16.4. Consider the propagation of Alfvén waves in a uniform compressible plasma immersed in a uniform magnetic field **B** where $\mathbf{B} = (0, 0, B)$. Assume that the wave has frequency ω and wave vector **k** where $\mathbf{k} = (k_1, 0, k_3)$, and assume that the wave motion may be described by a displacement vector $\boldsymbol{\xi}$ where $\boldsymbol{\xi} = (0, \xi_2, 0)$. Suppose that

$$\xi_2 = A \sin(k_1 x_1 + k_3 x_3 - \omega t). \qquad (1)$$

Study the properties of this wave by means of the second-order Lagrangian function (16.4.4), setting $\mathbf{v} = 0$ and $\Phi = 0$.

(a) Verify that the dispersion relation is the usual dispersion relation for Alfvén waves.

(b) Calculate the energy density E by means of (16.1.8). Express E in the form

$$E = \bar{E} + \hat{E} \qquad (2)$$

where $\bar{E}$ is the (time-independent) average value of E and $\hat{E}$ is the remaining contribution that varies with frequency 2ω.

(c) Verify that $\bar{E}$ is the sum of two equal contributions, one due to kinetic energy and the other to magnetic energy.

(d) By using (16.1.9), calculate the energy flow vector S_r and express S_r in the form

$$S_r = \bar{S}_r + \hat{S}_r. \qquad (3)$$

(e) Verify that $\bar{S}_r$ and $\bar{E}$ are related by the group velocity u_r of the wave.

(f) Verify that $\hat{S}_r$ and $\hat{E}$ satisfy the relation derived in Appendix B:

$$k_r \hat{S}_r = \omega \hat{E}. \qquad (4)$$

(g) Obtain expressions for the nonzero components of the momentum and momentum flux, using (16.1.12) and (16.1.13).

(h) Separate each expression into average and periodic contributions, and show that the following relations are satisfied:

$$\bar{T}_{rs} = \bar{P}_r u_s \qquad (5)$$

and

$$k_s \hat{T}_{rs} = \omega \hat{P}_r. \qquad (6)$$

17

Resistive instabilities

17.1 Introductory remarks

In previous chapters, we have considered microscopic instabilities, such as the two-stream instability, and MHD instabilities. The third major category of instability is that of 'resistive instabilities.' They differ from two-stream instabilities in that they may be treated by fluid equations, and they differ from MHD instabilities in that they are due essentially to the fact that the resistivity is nonzero. A nonzero resistivity allows the magnetic-field lines to move independently of the plasma, so that the 'frozen flux theorem' is not applicable in this context. A crucial consequence of this effect is that 'magnetic reconnection,' shown schematically in Fig. 17.1, is allowed. In addition, field lines can 'vanish.' Consider, for instance, the azimuthal magnetic field set up in a linear pinch. If the resistivity is nonzero, the current will dissipate. As a result, the circular field lines will shrink in diameter and the innermost loops will shrink to a point and disappear (see Fig. 17.2). If there is no external energy supply, the decrease in magnetic energy is converted into joule heating.

17.2 Current sheet configuration

The resistive instability that is of special importance in astrophysics is the 'tearing-mode' instability that can develop in a current sheet configuration. We shall find that this instability leads to development, in a field configuration of the type shown in Fig. 17.3(a), of 'magnetic islands,' as shown in Fig. 17.3(b). The seminal treatment of this topic is the article by Furth, Killeen and Rosenbluth (1963). Our treatment follows more closely the more recent article by White (1983).

In the problem to be examined, the unperturbed system is not homo-

Fig. 17.1. A schematic representation of the concept of field-line 'reconnection.' (Figures reproduced with kind permission from Boyd and Sanderson 1969.)

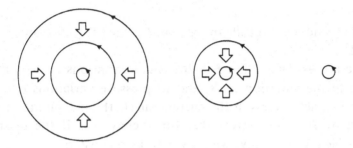

Fig. 17.2. Schematic representation of the vanishing of field lines due the effects of finite resistivity.

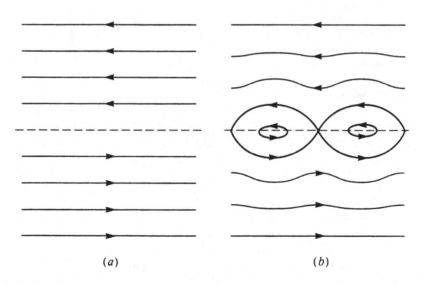

(a)   (b)

Fig. 17.3. Representation of the development of 'magnetic islands' in panel (b) from the initial field configuration shown in panel (a).

geneous. We adopt rectangular Cartesian coordinates x, y, z, and assume that the unperturbed magnetic-field configuration is given by

$$\mathbf{B} = (0, B_1 F(x), B_0) \qquad (17.2.1)$$

where

$$B_0 \gg B_1. \qquad (17.2.2)$$

Then $B_y(x)$ has the form shown in Fig. 17.4. We assume that the important variation of $F(x)$ is confined to a region of length L, so that

$$\left| \frac{dB_y}{dx} \right| \approx \frac{B_1}{L} \qquad (17.2.3)$$

in this region, and that (outside the region) $F(x) \approx 1$ for $x > L$ and $F(x) \approx -1$ for $x < -L$.

Since we are assuming that $B_0 \gg B_1$, small variations in B_0 can balance forces due to the variation in B_y and to pressure variations that develop because of magnetic stress in the current sheet, the magnitude of which is determined by B_y. The system therefore behaves as if the plasma were incompressible. Accordingly, we regard B_0 as a constant.

We now introduce the notation

$$\tilde{\mathbf{B}} = \mathbf{B} + \mathbf{b}, \quad \tilde{\mathbf{J}} = \mathbf{J} + \mathbf{j}, \qquad (17.2.4)$$

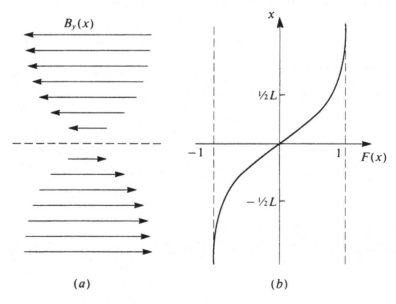

Fig. 17.4. Schematic representation (panel (*a*)) of the magnetic-field variation in the current sheet defined by the function $F(x)$ shown in panel (*b*).

where $\tilde{\mathbf{B}}$ indicates the total magnetic field vector, comprised of the unperturbed component $\mathbf{B}$ and the perturbation $\mathbf{b}$. Since we may ignore the displacement current, (2.1.4) reduces to

$$\nabla \times \mathbf{B} = 4\pi \mathbf{J}, \quad \nabla \times \mathbf{b} = 4\pi \mathbf{j}. \tag{17.2.5}$$

We see that the unperturbed form of the current density is given by

$$\mathbf{J} = \left(0, 0, \frac{1}{4\pi} B_1 \frac{dF}{dx} \right). \tag{17.2.6}$$

Since the resistivity is nonzero, there must be an electric field to maintain the current given by (17.2.6). We assume that the resistivity is isotropic; then

$$\mathbf{E} = \eta \mathbf{J}, \tag{17.2.7}$$

so that

$$\mathbf{E} = \left(0, 0, \frac{\eta}{4\pi} B_1 \frac{dF}{dx} \right). \tag{17.2.8}$$

In laboratory systems such as fusion devices, it is possible to apply an electric field and so maintain the plasma in essentially an unperturbed state. In astrophysical situations, much the same situation could arise if the ends of a flux tube were being rotated at exactly the same rate that the tube unwinds due to resistivity. However, a more likely situation is that a twisted magnetic-field configuration has been developed and the currents decay very slowly. Nevertheless, the very slow decay of the current leads to a slow rate of change of the magnetic field that leads to an electric field that is small but sufficient to maintain the system in an almost steady state. The fluctuations that we will be considering occur on a much more rapid time scale so that, even in astrophysical situations, it is appropriate to consider the unperturbed state as being static.

17.3 Evolution of the magnetic field

Assuming that the resistivity is isotropic and homogeneous, the evolution of the magnetic field is determined by (12.1.4) i.e. by

$$\frac{\partial \tilde{\mathbf{B}}}{\partial t} = \nabla \times (\mathbf{v} \times \tilde{\mathbf{B}}) + \frac{c\eta}{4\pi} \nabla^2 \tilde{\mathbf{B}}. \tag{17.3.1}$$

In this equation, $\mathbf{v}$ represents the perturbation in the velocity field; this is also the total velocity field, since the unperturbed state is assumed to be static.

We assume that all quantities are independent of z, even in the perturbed state. Hence we may write the total magnetic-field vector as

$$\tilde{\mathbf{B}} = \left(\frac{\partial \tilde{\Psi}}{\partial y}, -\frac{\partial \tilde{\Psi}}{\partial x}, B_0 \right). \tag{17.3.2}$$

Since all quantities are independent of z, it follows that

$$\tilde{\mathbf{B}} \cdot \nabla \tilde{\Psi} = 0, \tag{17.3.3}$$

so that $\tilde{\Psi}$ is constant along any field line. On introducing the same notation for $\tilde{\Psi}$ that we introduced in (17.2.4), we see that

$$\mathbf{B} = \left(\frac{\partial \Psi}{\partial y}, -\frac{\partial \Psi}{\partial x}, B_0 \right) \tag{17.3.4}$$

and

$$\mathbf{b} = \left(\frac{\partial \psi}{\partial y}, -\frac{\partial \psi}{\partial x}, 0 \right). \tag{17.3.5}$$

We see from equations (17.2.5) and (17.3.5) that the perturbation in the current density is given by

$$j_x = 0, \quad j_y = 0, \quad j_z = -\frac{1}{4\pi} \nabla^2 \psi \tag{17.3.6}$$

where

$$\nabla^2 = \frac{\partial^2}{\partial x^2} + \frac{\partial^2}{\partial y^2}. \tag{17.3.7}$$

The perturbation in the electric field may be obtained from

$$\frac{1}{c} \frac{\partial \mathbf{b}}{\partial t} = -\nabla \times \mathbf{e}. \tag{17.3.8}$$

Hence we find

$$\frac{1}{c} \frac{\partial^2 \psi}{\partial t \partial y} = -\frac{\partial e_z}{\partial y}, \tag{17.3.9}$$

$$-\frac{1}{c} \frac{\partial^2 \psi}{\partial t \partial x} = \frac{\partial e_z}{\partial x}, \tag{17.3.10}$$

$$0 = -\frac{\partial e_y}{\partial x} + \frac{\partial e_x}{\partial y}. \tag{17.3.11}$$

We see that (17.3.9) and (17.3.10) are both satisfied by the relation (17.3.12):

$$e_z = -\frac{1}{c} \frac{\partial \psi}{\partial t}. \tag{17.3.12}$$

In order to obtain expressions for e_x and e_y, we note that

$$\tilde{\mathbf{E}} + \frac{1}{c} \mathbf{v} \times \tilde{\mathbf{B}} = \eta \tilde{\mathbf{J}}, \tag{17.3.13}$$

that leads to the following equation for the perturbed quantities:

$$\mathbf{e} + \frac{1}{c} \mathbf{v} \times \mathbf{B} = \eta \mathbf{j}. \tag{17.3.14}$$

Assuming that the perturbation is such that $v_z \equiv 0$, we find that the x and y components of (17.3.14) yield

$$e_x = -\frac{1}{c} B_0 v_y, \quad e_y = \frac{1}{c} B_0 v_x, \tag{17.3.15}$$

since $j_x = 0$ and $j_y = 0$. On recalling our assumption that the medium is incompressible, we find that (17.3.11) is satisfied:

$$\frac{\partial e_y}{\partial x} - \frac{\partial e_x}{\partial y} = \frac{B_0}{c} \left(\frac{\partial v_x}{\partial x} + \frac{\partial v_y}{\partial y} \right) = 0. \tag{17.3.16}$$

The z component of (17.3.14) yields

$$e_z + \frac{1}{c} B_y v_x = \eta j_z. \tag{17.3.17}$$

Since the system behaves as if the plasma were incompressible, it is convenient to express the velocity field in terms of the stream function ϕ:

$$\mathbf{v} = \left(\frac{\partial \phi}{\partial y}, -\frac{\partial \phi}{\partial x}, 0 \right). \tag{17.3.18}$$

Hence, using (17.3.6) and (17.3.12), we find that (17.3.17) is expressible as

$$\frac{\partial \psi}{\partial t} = B_y \frac{\partial \phi}{\partial y} + \frac{c\eta}{4\pi} \nabla^2 \psi. \tag{17.3.19}$$

If the velocity field were known, (17.3.19) would enable us to calculate the perturbation in the magnetic field. Clearly we need another equation to determine the evolution of the velocity field.

17.4 Equation of motion

The equation of motion may be written in the form

$$\rho \frac{\partial \mathbf{v}}{\partial t} = \tilde{\mathbf{J}} \times \tilde{\mathbf{B}} - \nabla \tilde{P}, \tag{17.4.1}$$

where $\tilde{P}$ is the total pressure and we recall that the density is effectively constant. Hence the equation of motion for the small fluctuating quantities takes the form

$$\rho \frac{\partial \mathbf{v}}{\partial t} = \mathbf{J} \times \mathbf{b} + \mathbf{j} \times \mathbf{B} - \nabla p. \qquad (17.4.2)$$

The pressure is of no interest, and we may eliminate it from our calculations by taking the curl of the above equation:

$$\rho \nabla \times \frac{\partial \mathbf{v}}{\partial t} = \nabla \times (\mathbf{J} \times \mathbf{b} + \mathbf{j} \times \mathbf{B}). \qquad (17.4.3)$$

On expressing the velocity components in terms of ϕ by means of (17.3.18), we find that the x and y components of (17.4.3) are satisfied, and that the z component has the form

$$-\rho \frac{\partial}{\partial t} (\nabla^2 \phi) = \frac{\partial (J_z b_x)}{\partial x} + \frac{\partial (J_z b_y)}{\partial y} + \frac{\partial (B_y j_z)}{\partial y}. \qquad (17.4.4)$$

On using (17.2.5) and noting (17.3.5) and (17.3.6), we find that (17.4.4) can be expressed as

$$4\pi\rho \frac{\partial}{\partial t} (\nabla^2 \phi) = B_y \frac{\partial}{\partial y} (\nabla^2 \psi) - \frac{\mathrm{d}^2 B_y}{\mathrm{d}x^2} \frac{\partial \psi}{\partial y}. \qquad (17.4.5)$$

This equation is the required form of the equation of motion to accompany (17.3.19) to form a coupled set of partial differential equation.

17.5 The tearing mode

In order to study the topology of magnetic-field lines, it is convenient to note from (17.2.1) and (17.3.4) that $\Psi(x)$ is determined by

$$\frac{\mathrm{d}\Psi(x)}{\mathrm{d}x} = -B_1 F(x). \qquad (17.5.1)$$

Since the unperturbed state is uniform in y, we may implicitly Fourier transform in that variable. Since we are studying instabilities, we also assume that the perturbation grows exponentially in time. We therefore adopt the following form for $\psi(x, y, t)$,

$$\psi(x, y, t) = \psi_1(x) \cos(ky) e^{\gamma t}, \qquad (17.5.2)$$

and the following form for $\phi(x, y, t)$,

$$\phi(x, y, t) = \frac{\gamma}{kB_1} f_1(x) \sin(ky) e^{\gamma t}. \qquad (17.5.3)$$

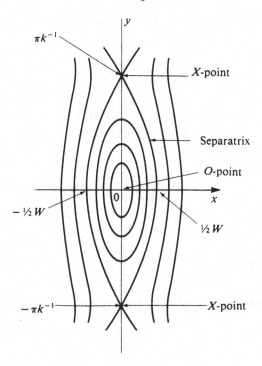

Fig. 17.5. Magnetic-field configuration in the tearing region. (Figure reproduced with kind permission from White 1983.)

Fig. 17.5 shows the magnetic-field configuration that results from the perturbation. The magnetic surfaces form 'islands' centered about the plane $x = 0$, repeated at intervals of $2\pi/k$. The division between the field lines forming islands and those that are open, extending to infinity, is known as the *separatrix*. The separatrices cross on the plane $x = 0$ at X-type neutral points. We see from Fig. 17.5 that there is also a series of O-type neutral points. The locations are given by

$$\left.\begin{array}{lll} X\text{-point}, & x = 0, & y = 2n\pi k^{-1}, \\[2mm] O\text{-point}, & x = 0, & y = (2n + 1)\pi k^{-1}. \end{array}\right\} \tag{17.5.4}$$

We may estimate the width W of the islands by noting that the value of $\bar{\Psi}$ at an X-type point must be the same as the value on the separatrix in the neighborhood of an O-type point, so that

$$\Psi(0) + \frac{1}{2}\frac{\mathrm{d}^2\Psi}{\mathrm{d}x^2}\left(\frac{W}{2}\right)^2 + \psi_1 = \Psi(0) - \psi_1. \tag{17.5.5}$$

(In this equation we have taken $t=0$ to drop the time-dependent term.) Hence

$$W = 4\left(-\psi_1 \Big/ \frac{d^2\Psi}{dx^2}\right)^{1/2}. \tag{17.5.6}$$

On substituting the expressions (17.5.2) and (17.5.3) in the differential equations (17.3.19) and (17.4.5), we obtained the coupled second-order ordinary differential equations

$$\psi_1 - Ff_1 = \frac{c\eta}{4\pi\gamma}\left(\frac{d^2\psi_1}{dx^2} - k^2\psi_1\right) \tag{17.5.7}$$

and

$$-\gamma^2 \frac{4\pi\rho}{B_1^2}\left(\frac{d^2 f_1}{dx^2} - k^2 f_1\right) = Fk^2\left(\frac{d^2\psi_1}{dx^2} - k^2\psi_1\right) - k^2\frac{d^2 F}{dx^2}\psi_1. \tag{17.5.8}$$

These equations can be simplified by making the following change of variables,

$$\xi = \frac{x}{L}, \tag{17.5.9}$$

and

$$\kappa = kL, \tag{17.5.10}$$

and introducing the following terms:

$$v_A^2 = \frac{B_1^2}{4\pi\rho}, \tag{17.5.11}$$

$$\tau_A = \frac{L}{v_A}, \tag{17.5.12}$$

and

$$\tau_R = \frac{4\pi}{c\eta}L^2. \tag{17.5.13}$$

Here ξ is the x-coordinate normalized so that the half-width of the sheet is unity, κ is the wave number normalized in units of L, v_A is the Alfvén speed, τ_A is the characteristic time for propagation of an Alfvén wave a distance equal to the width of the current sheet, and τ_R is the characteristic time for magnetic diffusion across the current sheet. With these definitions, the differential equations become

$$\psi_1 - F f_1 = \frac{1}{\gamma \tau_R} (\psi_1'' - \kappa^2 \psi_1) \qquad (17.5.14)$$

and

$$-\frac{\gamma^2 \tau_A^2}{\kappa^2} (f_1'' - \kappa^2 f_1) = F(\psi_1'' - \kappa^2 \psi_1) - F'' \psi_1 \qquad (17.5.15)$$

where a prime now refers to a derivative with respect to ξ. These equations must now be solved self-consistently both in the current sheet and far from it to obtain a complete solution over all space.

17.6 Solution of the differential equations

The coupled ordinary differential equations (17.5.14) and (17.5.15) may be solved numerically for any physical configuration, but there is no general analytic solution. In this section, we derive an approximate analytic solution for a particular case that is physically significant. We limit our analysis to systems for which the growth rates are intermediate between the ideal MHD time scale and the diffusive time scale. That is, we assume that

$$\tau_R^{-1} \ll \gamma \ll \tau_A^{-1}. \qquad (17.6.1)$$

We shall find, *a posteriori*, that this condition is satisfied.

The method of solution that we adopt is similar to that of boundary-layer theory in fluid dynamics, the process that is described in detail in Bender and Orszag (1978). It is assumed that the solution is slowly varying except in some boundary region near $x=0$ where the derivatives may be large. We separate the problem into constructing solutions in two separate regions, an 'internal' region that includes the boundary layer, and an 'external' region that excludes the boundary layer. Two different approximations are made in these regions. In the external region, the dependent variables are assumed to be slowly varying so that derivatives are of the same order of magnitude as the variables themselves. In the internal region, on the other hand, the derivatives are large but the region is sufficiently small that the coefficients of the differential equation can be approximated by constants. The result of these approximations is to replace the differential equations (17.5.14) and (17.5.15) by two simpler equations for which analytic solutions exist. The variables and their derivatives in the internal region are then matched in their asymptotic limit to the values of the variables and their derivatives in the external region, at what is a small value of the independent variable ξ. In this way, we construct an approximate solution over all space. We assume that the effect of resistivity is significant only in a narrow layer of width ξ_T, $\xi_T \ll 1$, where reconnection or tearing occurs.

In the external region, we neglect the effect of resistivity and assume that the length-scale for the solution is of order L. On using the second inequality in (17.6.1), we obtain from (17.5.14) the relation

$$\psi_1 = F f_1, \tag{17.6.2}$$

and from (17.5.15) the equation

$$F(\psi_1'' - k^2 \psi_1) = F'' \psi_1. \tag{17.6.3}$$

A convenient form for the function $F(\xi)$, that leads to an analytic solution of (17.6.3), is the form

$$F(\xi) = \tanh \xi. \tag{17.6.4}$$

The appropriate solutions of (17.6.3) are then

$$\psi_{1+}(\xi) = e^{-\kappa\xi}[1 + \kappa^{-1}\tanh\xi], \quad \xi > 0, \tag{17.6.5}$$

$$\psi_{1-}(\xi) = e^{\kappa\xi}[1 - \kappa^{-1}\tanh\xi], \quad \xi < 0.$$

We see that

$$\psi_{1+}(\xi) \to 1, \quad \psi_{1-}(\xi) \to 1, \quad \text{as } \xi \to 0, \tag{17.6.6}$$

so that the value of the variable $\psi(\xi)$ is continuous at $\xi = 0$. However, the derivative is discontinuous and we therefore introduce the notation

$$(\Delta\psi_1')_0 = \lim_{\varepsilon \to 0} [\psi_{1+}'(\varepsilon) - \psi_{1-}'(-\varepsilon)]. \tag{17.6.7}$$

Hence, introducing the notation

$$\Delta = \frac{(\Delta\psi_1')_0}{\psi_1(0)}, \tag{17.6.8}$$

we see from (17.6.5) that

$$\Delta = 2(\kappa^{-1} - \kappa). \tag{17.6.9}$$

The functions $\psi_1(\xi)$ and $f_1(\xi)$ are shown in Fig. 17.6 for both positive and negative values of Δ, for the model based on (17.6.4).

We now consider the solution of (17.5.14) and (17.5.15) in the internal region, where it is essential to take account of resistivity. We now make use of the fact that $\xi_T \ll 1$ and therefore use the approximation

$$F(\xi) \approx \xi, \quad \text{for } |\xi| \leqslant \xi_T \tag{17.6.10}$$

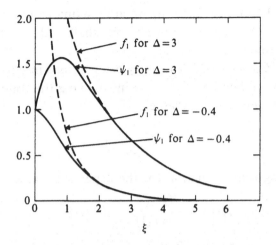

Fig. 17.6. Examples of the functional form of the magnetic-flux function ψ_1 and the velocity potential f_1, shown for a positive value of Δ and a negative value of Δ. (Figure reproduced with kind permission from White 1983.)

in this region. Based on our analysis of the behavior of $\psi_1(\xi)$ in the external region, we now assume that $\psi_1(\xi)$ is approximately constant within the tearing region, although it is of course necessary to take account of changes in $\psi_1'(\xi)$. Since the change of the derivative is determined by the width of the tearing region ξ_T, and since $\xi_T \ll 1$, we see that

$$|\psi_1''| \gg \kappa^2 |\psi_1|, \quad |f_1''| \gg \kappa^2 |f_1|. \tag{17.6.11}$$

Hence (17.5.14) now reduces to

$$\psi_1(0) - \xi f_1 = \frac{\psi_1''}{\gamma \tau_R}, \tag{17.6.12}$$

and (17.5.15) reduces to

$$\frac{\gamma^2 \tau_A^2 f_1''}{\kappa^2} = -\xi \psi_1''. \tag{17.6.13}$$

Since we are assuming that $\psi_1(\xi)$ is approximately constant in the tearing region, there is no problem in matching the values of $\psi_1(\xi)$ on the internal side and the external side of the boundary $\xi = \xi_T$.

The condition that the derivatives of $\psi_1(\xi)$ should match at the boundary $\xi = \xi_T$ may be expressed, as we see from (17.6.13), in the form

$$-\frac{\gamma^2 \tau_A^2}{\kappa^2 \psi_1(0)} \int_{-\infty}^{\infty} \frac{f_1''(\xi)}{\xi} \, d\xi = \Delta. \tag{17.6.14}$$

In writing the relation in this form, we are making assumptions about the asymptotic behavior of $f_1(\xi)$ as $\xi \to \pm\infty$. We shall find *a posteriori* that this assumption is validated.

Equations (17.6.12) and (17.6.13) may be combined into a single differential equation by eliminating $\psi_1''(\xi)$. On introducing the following transformation for the independent variable,

$$\xi = \left(\frac{\gamma T_A^2}{\kappa^2 T_R}\right)^{1/4} \zeta \tag{17.6.15}$$

and the following transformation for the dependent variable,

$$f_1(\xi) = -\left(\frac{\kappa^2 T_R}{\gamma T_A^2}\right)^{1/4} \psi_1(0)\chi(\zeta), \tag{17.6.16}$$

we find that $\chi(\zeta)$ satisfies the differential equation

$$\chi'' - \zeta^2\chi = \zeta, \tag{17.6.17}$$

in which a prime now denotes $d/d\zeta$.

One may verify that the solution of this equation may be expressed either as

$$\chi(\zeta) = -\tfrac{1}{2}\zeta \int_0^{\pi/2} d\theta \, \sin^{1/2}\theta \, \exp\left[-\tfrac{1}{2}\zeta^2 \cos\theta\right] \tag{17.6.18}$$

or, equivalently, as

$$\chi(\zeta) = -\tfrac{1}{2}\zeta \int_0^1 d\mu \, (1-\mu^2)^{-1/4} \exp\left[-\tfrac{1}{2}\zeta^2\mu\right]. \tag{17.6.19}$$

Clearly $\chi(\zeta)$ is an odd function of ζ, and we may verify that

$$\chi(\zeta) \to -\frac{1}{\zeta} \quad \text{as } |\zeta| \to \infty. \tag{17.6.20}$$

The matching condition (17.6.14) may now be expressed, in terms of the function $\chi(\zeta)$, as

$$\kappa^{-1/2}\gamma^{5/4}T_R^{3/4}T_A^{1/2}I = \Delta, \tag{17.6.21}$$

where

$$I = \int_{-\infty}^{\infty} \frac{d\zeta}{\zeta}\chi'' = \int_{-\infty}^{\infty} d\zeta(1 + \zeta\chi(\zeta)). \tag{17.6.22}$$

On substituting the expression (17.6.19) in the integral (17.6.22), we find that it is possible to carry out the integration over ζ to obtain the expression

$$I = \left(\frac{\pi}{2}\right)^{1/2} \int_0^1 d\mu \, \mu^{1/2}(1-\mu^2)^{-1/4}. \tag{17.6.23}$$

This integral is found to have the value

$$I = \pi \frac{\Gamma(\frac{3}{4})}{\Gamma(\frac{1}{4})} . \tag{17.6.24}$$

The form of the function $\chi(\zeta)$ that satisfies (17.6.17), that has been obtained by numerical integration, is shown in Fig. 17.7. Using this solution $\chi(\zeta)$, one may evaluate the integrand of the integral (17.6.22), and this quantity is shown in Fig. 17.8. We see that most of the contribution comes from $\zeta < 2$, and we therefore adopt the limit $\zeta_T = 2$ as the half-width of the tearing region. On using (17.6.15), we now obtain the following expression for ξ_T:

$$\xi_T = 2 \left(\frac{\gamma \tau_A^2}{\kappa^2 \tau_R} \right)^{1/4} . \tag{17.6.25}$$

On combining (17.6.9), (17.6.21) and (17.6.24), we finally obtain the following expression for the growth rate of the tearing mode:

$$\gamma = \left(\frac{2\Gamma(\frac{1}{4})}{\pi \Gamma(\frac{3}{4})} \right)^{4/5} S^{2/5} \tau_R^{-1} \kappa^{-2/5} (1 - \kappa^2)^{4/5}, \tag{17.6.26}$$

where

$$S = \frac{\tau_R}{\tau_A} . \tag{17.6.27}$$

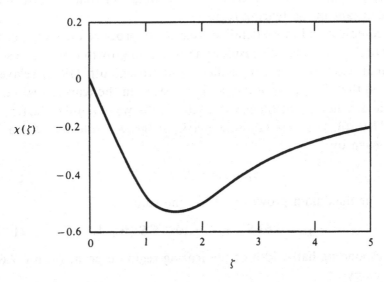

Fig. 17.7. The interior solution for the normalized velocity-potential function $\chi(\zeta)$. (Figure reproduced with kind permission from White 1983.)

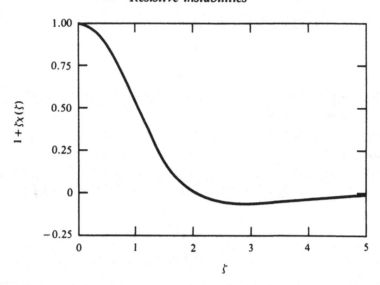

Fig. 17.8. The integral that appears in (17.6.22). Most of the contribution comes from the range $0 < \zeta < 2$. (Figure reproduced with kind permission from White 1983.)

We note from (17.6.26) that a severe drawback of the present treatment is that it leads to a growth rate γ that increases indefinitely as the assumed wave-number κ decreases to zero. This is clearly unphysical. This defect in our treatment may be attributed to the assumption that $\psi_1(\xi)$ is approximately constant in the tearing region.

More complete and more detailed treatments produce different formulas that lead to the more realistic result of a maximum growth rate. For instance, the original analysis by Furth, Killeen and Rosenbluth (1963) relaxes the assumption that $\psi_1(\xi)$ is approximately constant in the tearing region. As a result, the dependence of the growth rate on the wave number differs from that of (17.6.26). It is found that the maximum growth rate occurs at a wave-number given by

$$k_{\mathrm{m}} \approx 1.4\, S^{-1/4} L^{-1}, \tag{17.6.28}$$

and that the maximum growth rate has the value

$$\gamma_{\mathrm{M}} \approx 0.63\, S^{1/2} \tau_{\mathrm{R}}^{-1} \approx 10^{-0.20} S^{1/2} \tau_{\mathrm{R}}^{-1}. \tag{17.6.29}$$

The corresponding half-width of the tearing region is seen, from (17.6.25), to be given by

$$x_{\mathrm{T,M}} \approx 1.51\, S^{-1/4} L \approx 10^{0.18} S^{-1/4} L. \tag{17.6.30}$$

A related instability arises from a combination of resistive effects upon the gravitational interchange mode (Furth, Killeen and Rosenbluth 1963). Furthermore, similar instabilities occur if the infinite-conductivity approximation breaks down due to other effects, such as finite electron inertia (Cross and Van Hoven, 1976). For instance, a 'collision free' version of the tearing-mode instability has been developed by Laval, Pellat and Vuillemin (1966).

Problem

Problem 17.1. Consider a fully ionized hydrogen plasma of temperature T, for which the electron and proton densities are each n. Suppose that the plasma contains a current sheet in which the field, of strength B, changes sign in a region of half-width L.

(a) Find expressions for τ_R and τ_A, and hence S, expressed in terms of B, T, n and L. Use expression (11.6.10) for the resistivity of a fully ionized hydrogen plasma.

(b) Hence obtain expressions for γ_M and $x_{T,M}$ in terms of B, T, n and L.

(c) Justify the following expressions for the maximum current density j_M and the corresponding maximum electric field e_M strength that would develop in the tearing region:

$$j_M \approx \frac{1}{4\pi} B x_{T,M}^{-1}, \quad e_M \approx \frac{1}{c} \gamma_M B x_{T,M}. \tag{1}$$

(d) Verify that these estimates are related by the appropriate form of Ohm's law.

(e) Hence obtain expressions for j_M and e_M in terms of B, T, n and L.

(f) Hence estimate, in terms of the same parameters, the maximum Joule heating rate Q_M, using $Q_M = c e_M j_M$.

(g) On making the reasonable assumption that the energy that goes into Joule heating is being extracted from the magnetic field, obtain an estimate of the 'energy release rate' γ_{ERR}, defined by

$$Q = \frac{1}{8\pi} B^2 \tau_{ERR}^{-1} \tag{2}$$

Find an expression for γ_{ERR}. Can you relate this to γ_M?

(h) Typical values of the temperature, density and field strength in a solar active region are $T = 10^{6.4}$ K, $n = 10^9$ cm^{-3} and $B = 10^2$ G. The thickness of the current sheets is not known. If two flux tubes, originating in different regions of the photosphere, come into contact, L could be quite small. Consider the value $L = 10^6$ cm, and calculate the resulting values of γ_M, $x_{T,M}$, j_M, e_M and Q_M.

18

Stochastic processes

In previous chapters, we have studied the behavior of charged particles, using various approximations, in given electric and magnetic field configurations. For instance, we have studied waves in a medium that is assumed to be homogeneous and static. In reality, an astrophysical plasma (such as the solar wind) is neither homogeneous nor static. Furthermore, we have only incomplete information about the structure and time-evolution of the medium. Another important class of problems involves systems that are mildly unstable, in which perturbations grow from some initial small level to a finite but significant level. An example of such a situation is provided by the study of the growth of plasma oscillations in the quasilinear approximation (see, for instance, Drummond and Pines, 1962, or Nicholson, 1983).

In problems such as these, we must develop procedures for representing and perhaps calculating the fluctuations that arise in a plasma. In many cases, we can conveniently represent these fluctuations as a superposition of waves (normal modes) in the plasma. However, the individual particles in such a plasma may behave in a manner that differs in important respects from the behavior of particles in the simpler situations that we have studied so far.

In discussing collision theory, we took account of the fact that a plasma is highly random at the *microscopic* level that is relevant for the discussion of binary collisions, etc. We now introduce the concept that the plasma may also be random on a *macroscopic* level, and we study the implications of such randomness. We shall find that the resulting motions of charged particles are stochastic in nature. That is, they resemble 'Brownian motion' or a 'random-walk' process. In discussing the motion of particles in ordinary space, we shall find that randomness in the electromagnetic fields can lead to spatial diffusion. Even more important, when we come to consider the effect of randomness in electromagnetic fields on motion in velocity space, we shall find that particles can diffuse in velocity space – that is, they can experience 'stochastic acceleration.'

18.1 Stochastic diffusion

Before discussing diffusion in velocity space, it is convenient to explore briefly the simpler process of diffusion in ordinary space. We consider the simple situation of a static magnetic field in the x_3 direction and suppose that there is a time-varying but spatially uniform electric field in the x_1 and the x_2 directions so that

$$\mathbf{B} = (0,\ 0,\ B),$$
$$E = (E_1(t),\ E_2(t),\ 0). \tag{18.1.1}$$

We assume that the time scale for variation of the electric field is long compared with the gyroperiod. Then, at any instant, charged particles will experience the transverse drift velocity

$$\mathbf{v}_D = \frac{c}{B^2} \mathbf{E} \times \mathbf{B}, \tag{18.1.2}$$

that is

$$\left. \begin{aligned} v_{D,1}(t) &= \frac{c}{B} E_2(t), \\ v_{D,2}(t) &= -\frac{c}{B} E_1(t). \end{aligned} \right\} \tag{18.1.3}$$

Over a time interval t large compared with the gyroperiod, the displacement of a particle in the x_1 direction is given by

$$\Delta x_1(t) \equiv x_1(t) - x_1(0) = \frac{c}{B} \int_0^t dt'\, E_2(t'), \tag{18.1.4}$$

and there is a similar expression for $\Delta x_2(t)$.

We now wish to represent the stochastic properties of the electric field. We assume that the average value of the electric field is zero, so that

$$\langle E_1(t) \rangle = 0, \quad \langle E_2(t) \rangle = 0. \tag{18.1.5}$$

However, we assume that quadratic averages may be expressed as

$$\langle E_r(t) E_s(t+\tau) \rangle = \langle E_1^2 \rangle R(\tau)\, \delta_{rs}. \tag{18.1.6}$$

There are different ways to interpret these expressions. One possibility is to assume that the quantities within brackets are measured for many values of t over a very long period of time. Another possibility is to consider a hypothetical situation in which a certain experiment is repeated many times over with different values of E_1 and E_2 as functions of t. The former interpretation is closer to one's physical ideas about diffusion and random-walk processes, but the latter interpretation may be more clear-cut from a mathematical point of view. In writing the quadratic expression (18.1.6), we have assumed for convenience that the electric field is statistically isotropic.

We see from (18.1.4) and (18.1.5) that

$$\langle \Delta x_1(t) \rangle = 0, \quad \langle \Delta x_2(t) \rangle = 0. \tag{18.1.7}$$

On the other hand, we find that

$$\langle \Delta x_r \Delta x_s \rangle = \frac{c^2}{B^2} \int_0^t dt' \int_0^t dt'' \langle E_r(t') E_s(t'') \rangle. \tag{18.1.8}$$

The correlation between $E_1(t')$ and $E_1(t'')$ is a function of $t'' - t'$, as illustrated in Fig. 18.1, where the 'correlation time' τ_c is a measure of the width of the function $R(\tau)$, as indicated in Fig. 18.2. By changing variables, (18.1.8) becomes

$$\langle \Delta x_r \Delta x_s \rangle = \frac{c^2}{B^2} \int_0^t dt' \int_{-t'}^{t-t'} d\tau \langle E_r(t') E_s(t' + \tau) \rangle. \tag{18.1.9}$$

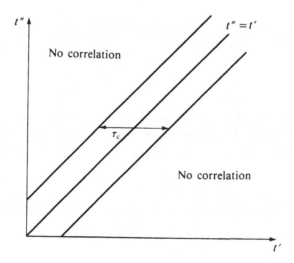

Fig. 18.1. It is assumed that the correlation between $E_1(t')$ and $E_1(t'')$ is significant only in a strip of width τ_c in the $t' - t''$ plane.

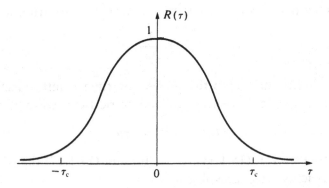

Fig. 18.2. It is assumed that the correlation function $R(\tau)$ is significant only for τ in the range $-\tau_c < \tau < \tau_c$.

We now suppose that $t \gg \tau_c$, and use expression (18.1.6). We then arrive at the following expression,

$$\langle \Delta x_r \, \Delta x_s \rangle \approx \frac{c^2}{B^2} t \delta_{rs} \langle E_1{}^2 \rangle \int_{-\infty}^{\infty} \mathrm{d}\tau \, R(\tau), \qquad (18.1.10)$$

which indicates that the mean-square displacement of a particle increases linearly with time, as is usual in a 'random-walk' process. On introducing the notation previously used in discussion of the Fokker–Planck equation (Section 10.2), we arrive at the result

$$\left\langle \frac{\Delta x_r \, \Delta x_s}{\Delta t} \right\rangle = \frac{c^2}{B^2} \delta_{rs} \langle E_1{}^2 \rangle \int_{-\infty}^{\infty} \mathrm{d}\tau \, R(\tau). \qquad (18.1.11)$$

The correlation function $R(\tau)$ is related to the power spectrum $S(\omega)$ by the relations

$$R(\tau) = \int \mathrm{d}\omega \, S(\omega) \mathrm{e}^{-i\omega\tau}, \qquad (18.1.12)$$

and

$$S(\omega) = \frac{1}{2\pi} \int \mathrm{d}\tau \, R(\tau) \, \mathrm{e}^{i\omega\tau}, \qquad (18.1.13)$$

where all integrals run from $-\infty$ to ∞ unless otherwise specified. Since

$$S(0) = \frac{1}{2\pi} \int \mathrm{d}\tau \, R(\tau) \qquad (18.1.14)$$

and since $R(0) = 1$, we see that

$$\int \mathrm{d}\omega \, S(\omega) = 1, \qquad (18.1.15)$$

so that we may define a characteristic width of the spectrum ω_0 by

$$S(0) = \frac{1}{2\omega_0}. \tag{18.1.16}$$

Clearly ω_0^{-1} will be similar to the correlation time τ_c introduced earlier.

If u is the mean energy density of the electric field, so that

$$\langle E_1^2 \rangle = \langle E_2^2 \rangle = 4\pi u, \tag{18.1.17}$$

we find from (18.1.11), (18.1.14), (18.1.16) and (18.1.17) that the Fokker–Planck coefficients are expressible as

$$\left\langle \frac{\Delta x_r \Delta x_s}{\Delta t} \right\rangle = 4\pi^2 \frac{c^2}{B^2} \frac{u}{\omega_0} \delta_{rs}. \tag{18.1.18}$$

On using the following form for the Fokker–Planck equation,

$$\frac{\partial n}{\partial t} + v_r \frac{\partial n}{\partial x_r} = -\frac{\partial}{\partial x_r} \left(\left\langle \frac{\Delta x_r}{\Delta t} \right\rangle n \right) + \frac{1}{2} \frac{\partial^2}{\partial x_r \partial x_s} \left(\left\langle \frac{\Delta x_r \Delta x_s}{\Delta t} \right\rangle n \right), \tag{18.1.19}$$

we find that, for the present problem, this equation may be expressed in the form of a diffusion equation,

$$\frac{\partial n}{\partial t} = D \nabla_\perp^2 n, \tag{18.1.20}$$

where

$$\nabla_\perp^2 = \frac{\partial^2}{\partial x_1^2} + \frac{\partial^2}{\partial x_2^2} \tag{18.1.21}$$

and the diffusion coefficient D is given by

$$D = 2\pi^2 \frac{c^2}{B^2} \frac{u}{\omega_0}. \tag{18.1.22}$$

If we make the definition of τ_c more precise by writing

$$\omega_0 \tau_c = 1, \tag{18.1.23}$$

(18.1.22) becomes

$$D = 2\pi^2 \frac{c^2}{B^2} u \tau_c. \tag{18.1.24}$$

Whereas a static electric field, transverse to a static magnetic field, leads to a steady drift motion, we now see that a stochastic electric field leads to spatial diffusion in the directions transverse to the magnetic field.

For the particular case that the particles all start at the origin, so that

$$n(x_1, x_2, 0) = N\delta(x_1)\delta(x_2), \tag{18.1.25}$$

we find that

$$n(x_1, x_2, t) = \frac{N}{4\pi Dt} \exp\left[-\frac{x_1^2 + x_2^2}{4Dt}\right]. \tag{18.1.26}$$

Hence, defining the mean-square displacement from the origin by

$$\langle r^2 \rangle = N^{-1} \int\int dx_1\,dx_2 (x_1^2 + x_2^2)\, n(x_1, x_2, t), \tag{18.1.27}$$

we find that

$$\langle r^2 \rangle = 4Dt. \tag{18.1.28}$$

The above discussion, due originally to Spitzer (1960), may be appropriate to a discussion of stochastic diffusion in a laboratory plasma, but we must be careful in applying it to an astrophysical situation. Suppose, for instance, that we are entitled to use the ideal MHD approximation. Then the transverse electric field is due to motion of the plasma transverse to the magnetic field:

$$\mathbf{E} = -\frac{1}{c}\mathbf{v} \times \mathbf{B}. \tag{18.1.29}$$

However, we see from (18.1.2) that

$$\mathbf{v}_D = \mathbf{v}. \tag{18.1.30}$$

That is to say, the random motion of the particles is simply the random motion of the plasma. Hence if there were random motion of the field lines, due perhaps to a spectrum of Alfvén waves, *but each field line remained in approximately the same position*, due for instance to 'line-tying' of foot-points in a fixed surface of high conductivity, then there would be no net spatial diffusion since particles originally on a field line remain tied to that field line.

It is fair to ask how our treatment breaks down in this instance. One possibility is that the mistake arises from confusing an ensemble average with a real time-averaging process, but a more likely possibility is that the representation of the stochastic properties of the field in terms simply of first-order and second-order correlations functions is a gross oversimplification that, in some instances, fails to represent adequately the physical properties of the electromagnetic fields.

18.2 One-dimensional stochastic acceleration

We next consider what is, perhaps, the simplest situation leading to stochastic acceleration. We consider a one-dimensional electric field but now allow for the possibility that it varies in space as well as in time (Sturrock, 1966). Considering only one dimension, we may write

$$\frac{dv}{dt} = \frac{d^2x}{dt^2} = \frac{q}{m} E(x, t). \tag{18.2.1}$$

We now represent the trajectory of a test particle by

$$x = X(t), \tag{18.2.2}$$

so that the equation of motion becomes

$$\frac{d^2X}{dt^2} = \frac{q}{m} E(X(t), t). \tag{18.2.3}$$

We now suppose that the electric field is – in a certain sense – weak, and expand the trajectory of the particle in powers of the amplitude of the electric field,

$$x = x_0 + v_0 t + X^{\mathrm{I}}(t) + X^{\mathrm{II}}(t) + \cdots, \tag{18.2.4}$$

where x_0 and v_0 are the initial values of x and v. Then we assign the initial conditions:

$$\left.\begin{aligned} X^{\mathrm{I}}(0) &= X^{\mathrm{II}}(0) = \ldots = 0, \\ \dot{X}^{\mathrm{I}}(0) &= \dot{X}^{\mathrm{II}}(0) = \ldots = 0. \end{aligned}\right\} \tag{18.2.5}$$

On substituting the expression (18.2.4) into the equation of motion (18.2.3) and then separating out terms of equal order, we obtain the following equations for X^{I} and X^{II}:

$$\ddot{X}^{\mathrm{I}} = \frac{q}{m} E(x_0 + v_0 t, t) \tag{18.2.6}$$

and

$$\ddot{X}^{\mathrm{II}} = \frac{q}{m} X^{\mathrm{I}} \frac{\partial E}{\partial x}(x_0 + v_0 t, t). \tag{18.2.7}$$

We now introduce notation that is an extension of (18.1.6) to two dimensions. We assume that the fluctuating electric field is statistically uniform in time and space, so that we may write (as a simple extension of (18.1.6))

$$\langle E(x, t) E(x + \xi, \tau + \tau) \rangle = \langle E^2 \rangle R(\xi, \tau). \tag{18.2.8}$$

The correlation function $R(\xi, \tau)$ is now related to the corresponding power spectrum $S(k, \omega)$ by

$$R(\xi, \tau) = \int \int dk \, d\omega \, S(k, \omega) \, e^{i(k\xi - \omega\tau)} \qquad (18.2.9)$$

and

$$S(k, \omega) = (2\pi)^{-2} \int \int d\xi \, d\tau \, R(\xi, \tau) \, e^{-i(k\xi - \omega\tau)}. \qquad (18.2.10)$$

The change in velocity of a particle over a time interval t is given, to first order, by

$$\Delta \dot{X}^{\text{I}} \equiv \dot{X}^{\text{I}}(t) - \dot{X}^{\text{I}}(0) = \frac{q}{m} \int_0^t dt' \, E(x_0 + v_0 t', t'). \qquad (18.2.11)$$

Hence

$$\langle (\Delta \dot{X}^{\text{I}})^2 \rangle = \frac{q^2}{m^2} \int_0^t dt' \int_0^t dt'' \langle E(x_0 + v_0 t', t') E(x_0 + v_0 t'', t'') \rangle. \qquad (18.2.12)$$

Proceeding as in Section 18.1 and assuming that the time t is long compared with the correlation time of the electric field, we obtain the following Fokker-Plank coefficient,

$$\left\langle \frac{(\Delta v)^2}{\Delta t} \right\rangle = \left(\frac{q}{m} \right)^2 \langle E^2 \rangle \int d\tau \, R(v\tau, \tau), \qquad (18.2.13)$$

where we have now dropped the subscript zero from v_0.

We may calculate the first Fokker-Planck coefficient $\langle \Delta v / \Delta t \rangle$ to second order by integrating (18.2.7) to obtain

$$\Delta \dot{X}^{\text{II}} \equiv \dot{X}^{\text{II}}(t) - \dot{X}^{\text{II}}(0) = \frac{q}{m} \int_0^t dt' \, X^{\text{I}}(t') \frac{\partial E}{\partial x} (x_0 + v_0 t', t'). \qquad (18.2.14)$$

On integrating the integral in (18.2.11) by parts and using the boundary conditions (18.2.5), we find that

$$\Delta X^{\text{I}} = \frac{q}{m} \int_0^t dt' \, (t - t') E(x_0 + v_0 t', t'). \qquad (18.2.15)$$

On substituting this expression into (18.2.14) and then taking the average values of quantities, we obtain

$$\langle \Delta v \rangle = \left(\frac{q}{m}\right)^2 \langle E^2 \rangle \int_0^{t'} dt' \int_0^{t'} dt''\, (t'-t'') \frac{\partial R}{\partial \xi}\, (v_0(t'-t''),\ t'-t'').$$

$$(18.2.16)$$

If we once again assume that the integration time is much greater than the coherence time, and note that $R(v\tau, \tau)$ is even in τ, (18.2.16) leads to the first Fokker–Planck coefficient:

$$\left\langle \frac{\Delta v}{\Delta t} \right\rangle = \frac{1}{2}\left(\frac{q}{m}\right)^2 \langle E^2 \rangle \frac{\partial}{\partial v}\left[\int d\tau\, R(v\tau, \tau)\right]. \qquad (18.2.17)$$

Finally, using the expressions (18.2.13) and (18.2.17), we find that the Fokker–Planck equation for $f(x, v, t)$,

$$\frac{\partial f}{\partial t} = -\frac{\partial}{\partial v}\left(\left\langle \frac{\Delta v}{\Delta t}\right\rangle f\right) + \frac{1}{2}\frac{\partial^2}{\partial v^2}\left(\left\langle \frac{(\Delta v)^2}{\Delta t}\right\rangle f\right), \qquad (18.2.18)$$

takes the form of a diffusion equation,

$$\frac{\partial f}{\partial t} = \frac{\partial}{\partial v}\left[D(v)\frac{\partial f}{\partial v}\right], \qquad (18.2.19)$$

where the diffusion coefficient is given by

$$D(v) = \frac{1}{2}\left(\frac{q}{m}\right)^2 \langle E^2 \rangle \int d\tau\, R(v\tau, \tau). \qquad (18.2.20)$$

We see from (18.2.9) that this may be expressed alternatively in terms of the energy spectrum as

$$D(v) = \pi\left(\frac{q}{m}\right)^2 \langle E^2 \rangle \int dk\, S(k, vk). \qquad (18.2.21)$$

On noting that the electric field energy density is now given by

$$u = \frac{1}{8\pi}\langle E^2 \rangle, \qquad (18.2.22)$$

(18.2.21) becomes, alternatively,

$$D(v) = 8\pi^2\left(\frac{q}{m}\right)^2 u \int dk\, S(k, vk). \qquad (18.2.23)$$

In Section 18.1, we found that the spatial diffusion could be represented by a diffusion equation (18.1.21), but this was recognizable as a diffusion equation only because the system had been assumed to be spatially homogeneous so that D was not a function of the independent variable x. In the present problem, by contrast, D is a function of the variable v but nevertheless we have found that the Fokker–Planck equation takes the form of a diffusion equation (equation (18.2.19)). The procedure that we have adopted is quite straightforward and could be applied to much more complex situations, that would also be found to lead to diffusion equations. However, our derivation does not show *why* in the present case or in more complicated cases the resulting equation should turn out to be a diffusion equation. There are more sophisticated approaches to the problem of stochastic acceleration that do lead naturally to a diffusion equation. One such approach is presented in the next chapter.

18.3 Stochastic diffusion, Landau damping and quasilinear theory

It is interesting to consider the special case that the fluctuating electric field that we are considering is a spectrum of plasma oscillations, and that the particles under discussion are the electrons of the plasma supporting these oscillations. If the total energy of the electron plasma is changing as a result of the stochastic process we are considering, this change must come at the expense of the energy of the plasma oscillations. In this way we may obtain once more, from a quite different perspective, the formula for Landau damping that was previously obtained in Section 9.6.

The total energy density of the system may be expressed as

$$w = \bar{w}_p + \hat{w}_p + \hat{w}_E, \qquad (18.3.1)$$

where $\bar{w}_p$ is the mean kinetic energy density of the electrons (ignoring the fluctuation due to the plasma oscillations), and $\hat{w}_p$, $\hat{w}_E$ are the electron kinetic energy density and electric-field energy density associated with the plasma oscillations. Since plasma oscillations represent an exchange of energy between the electrons and the electric field, we expect that

$$\hat{w}_p = \hat{w}_E \qquad (18.3.2)$$

and detailed calculations show this to be the case. Since the total energy is conserved, (18.3.1) and (18.3.2) lead to

$$\frac{d\bar{w}_p}{dt} = -2\frac{d\hat{w}_E}{dt}. \qquad (18.3.3)$$

Since

$$\frac{d\bar{w}_p}{dt} = \int dv \frac{1}{2} mv^2 \frac{\partial f}{\partial t}, \qquad (18.3.4)$$

where $f(v, t)$ represents the *mean* distribution function (averaged over a plasma-oscillation period), we may use (18.2.19) to obtain

$$\frac{d\bar{w}_p}{dt} = -\int dv \, mvD \frac{\partial f}{\partial v}, \qquad (18.3.5)$$

where we have carried out an integration by parts. If we spectrum-analyze the electric-field fluctuations in the form

$$\hat{w}_E = \int \int dk\, d\omega\, w(k, \omega), \qquad (18.3.6)$$

we see from (18.2.21) that

$$D = 8\pi^2 \left(\frac{q}{m}\right)^2 \int dk\, w(k, vk), \qquad (18.3.7)$$

so that (18.3.5) becomes

$$\frac{d\bar{w}_p}{dt} = -8\pi^2 \frac{q^2}{m} \int \int dk\, dv\, f'(v)\, v\, w(k, vk). \qquad (18.3.8)$$

With a change of variables, this may be rewritten as

$$\frac{d\bar{w}_p}{dt} = -8\pi^2 \frac{q^2}{m} \int \int dk\, d\omega\, \frac{\omega}{|k|k} f'(\omega/k)\, w(k, \omega). \qquad (18.3.9)$$

If we assume that the wave of wave-number k and frequency ω is growing at the rate $\gamma(k, \omega)$, so that

$$|E(k, \omega)|^2 \propto e^{2\gamma(k, \omega)t}, \qquad (18.3.10)$$

we see that

$$\frac{d\hat{w}_E}{dt} = \int \int dk\, d\omega\, 2\gamma(k, \omega)\, w(k, \omega). \qquad (18.3.11)$$

If we now use the relation (18.3.2) and note that the relationship we have obtained is true for any form of the function $w(k, \omega)$, we may equate coefficients of $w(k, \omega)$ in the two integrals (18.3.9) and (18.3.11) and so obtain the following expression for the growth rate:

$$\gamma(k, \omega) = 2\pi^2 \left(\frac{q^2}{m}\right) \frac{\omega}{|k|k} f'(\omega/k). \qquad (18.3.12)$$

This is equivalent to the expression for the rate of Landau damping of plasma oscillations that was previously derived ((equation (9.6.13)), if we make the assumption (Section 9.6) that the group velocity is much smaller than the phase velocity.

The derivation of the expression (18.3.7) for the diffusion coefficient was based on the assumption that the wave energy $w(k, \omega)$ covers a continuous range of frequency ω for any given value of the wave-number k. However, in considering the evolution of plasma oscillations in a plasma, waves that are not heavily damped conform to a dispersion relation that we may write as

$$\omega = \Omega(k). \qquad (18.3.13)$$

The wave energy density is then expressible as

$$w(k, \omega) = u(k)\delta(\omega - \Omega(k)),\qquad(18.3.14)$$

so that the expression (18.3.7) for the diffusion coefficient now becomes

$$D(v) = 8\pi^2 \left(\frac{q^2}{m}\right)\left[\frac{u(k)}{|v - d\Omega(k)/dk|}\right]_{k=k_v},\qquad(18.3.15)$$

where k_v satisfies the relation

$$vk_v = \Omega(k_v).\qquad(18.3.16)$$

The expression (18.3.15) for the diffusion coefficient is identical to that occurring in the quasilinear theory of plasma oscillations (see, for instance, Drummond and Pines, 1962), except that the group velocity occurs in (18.3.15), whereas in quasilinear theory it is usually assumed that the group velocity is small and may be neglected.

Problem

Problem 18.1 Particle acceleration may be effected also by electric fields directed transverse to an applied steady magnetic field. In this case, the most important contribution to acceleration is due to that part of the electric-field spectrum that is resonant with the gyromotion (or 'cyclotron' motion) of the electron, as in a cyclotron accelerator. We consider the motion of charged particles in a uniform static magnetic field $\mathbf{B} = (0, 0, B)$ and uniform but time-varying electric-field components E_x and E_y, for which $\langle E_x\rangle = \langle E_y\rangle = 0$.

Beginning with the equations of motion in the form

$$\begin{aligned}\frac{d^2x}{dt^2} - \omega_g \frac{dy}{dt} &= \frac{q}{m}E_x,\\ \frac{d^2y}{dt^2} + \omega_g \frac{dx}{dt} &= \frac{q}{m}E_y,\end{aligned}\qquad(1)$$

express the solution in the form

$$\begin{aligned}x &= X + \tfrac{1}{2}(Ae^{-i\omega_g t} + A^*e^{i\omega_g t}) = X + r\cos(\omega_g t + \alpha),\\ y &= Y - \tfrac{1}{2}i(Ae^{-i\omega_g t} - A^*e^{i\omega_g t}) = Y - r\sin(\omega_g t + \alpha),\end{aligned}\qquad(2)$$

where $A = re^{i\alpha}$.

(a) By using the additional conditions

$$\begin{aligned}\dot{X} + \tfrac{1}{2}(\dot{A}e^{-i\omega_g t} + \dot{A}^*e^{i\omega_g t}) &= 0,\\ \dot{Y} - \tfrac{1}{2}i(\dot{A}e^{-i\omega_g t} - \dot{A}^*e^{i\omega_g t}) &= 0,\end{aligned}\qquad(3)$$

obtain expressions for the velocity components v_x, v_y.

(b) Hence obtain the following equations for $\dot{X}$, $\dot{Y}$ and $\dot{A}$:

$$\left. \begin{aligned} \dot{X} &= \frac{c}{B} E_y, \quad \dot{Y} = -\frac{c}{B} E_x, \\ \dot{A} &= i\frac{c}{B} e^{i\omega_g t}(E_x + iE_y). \end{aligned} \right\} \tag{4}$$

(c) Assume that the time-varying electric field is statistically isotropic, so that $S_{xx} = S_{yy}$, and assume also that $S_{xy} = 0$. Consider the electric-field energy density

$$W^T = \frac{1}{8\pi} \langle E_x^2 + E_y^2 \rangle \tag{5}$$

and its spectrum

$$W^T = \int d\omega\, W(\omega). \tag{6}$$

Relate $W(\omega)$ to the Fourier transforms S_{xx}, etc., of the correlation functions $R_{xx}(t)$, etc., that are formed from the electric field components.

(d) Obtain an expression for $\langle |\Delta A|^2 / \Delta t \rangle$, in terms of $W(\omega)$.

(e) If w is the transverse kinetic energy of the particle, express w in terms of A.

(f) Consider the Fokker–Planck equation for $F(w, t)$, the distribution function for the particle energy, and find expressions for the Fokker–Planck coefficients $\langle \Delta w / \Delta t \rangle$ and $\langle (\Delta w)^2 / \Delta t \rangle$.

(g) Show that the Fokker–Planck equation takes the form of a diffusion equation.

(h) Show that the solution of the diffusion equation is a Maxwellian distribution for which the temperature increases linearly with time.

19

Interaction of particles and waves

We saw in the previous chapter that the equation governing the evolution of the particle distribution function has the form of a diffusion equation (18.2.19). When starting from the Fokker–Planck equation, it is not obvious why this should be the case.

In this chapter, we consider a homogeneous plasma and assume that the fluctuating electric and magnetic fields can be expressed as a distribution of normal modes of the plasma. We shall find that this approach leads directly to a diffusion equation. In addition, we shall find that the diffusion coefficient is related to the emission and absorption processes for the waves that comprise the normal modes.

Our analysis follows the *semi-classical* or *quasi-quantum-mechanical* approach due originally to Ginzburg (1939) and subsequently applied to plasma physics by Smerd and Westfold (1949), Pines and Schrieffer (1962), Melrose (1968) and Harris (1969). Our analysis will be restricted to the case that the unperturbed system contains no magnetic field. For a recent and more complete exposition of this procedure (that includes a possible magnetic field) see Melrose (1980).

19.1 Quantum-mechanical description

The relationship between emissivity, absorption and particle diffusion rests upon certain quantum-mechanical relationships. We therefore begin by describing the system in quantum-mechanical terms, taking note of the required relationships, and then making the transition to a classical description.

It is sufficient to consider a simple quantum-mechanical system with discrete energy levels (assumed non-degenerate) of energy E_1, E_2, ..., with particle populations n_1, n_2, ... (See Fig. 19.1.) A particle can make a

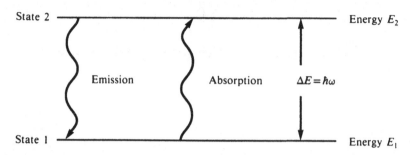

Fig. 19.1. Model of a two-level quantum-mechanical system, in which transitions can involve emission or absorption of a quantum of frequency ω.

transition from level 1 to level 2 by absorbing a photon of frequency ω such that

$$\hbar\omega = E_2 - E_1. \qquad (19.1.1)$$

Similarly, it can make a transition from level 2 to level 1 by emitting a photon of frequency ω.

If we now consider that the system, comprising the particles and the radiation, is in thermodynamic equilibrium at temperature T, photons will be distributed in energy according to the Planck distribution,

$$N(\omega) = [e^{\hbar\omega/kT} - 1]^{-1}, \qquad (19.1.2)$$

and the particle occupation numbers of the different levels will be related by the Boltzmann relation

$$\frac{n_2}{n_1} = e^{-(E_2 - E_1)/kT} = e^{-\hbar\omega/kT}. \qquad (19.1.3)$$

There are three distinct processes governing the interaction between the photons and the particles.

(1) Spontaneous emission

This occurs when a particle makes a downward transition accompanied by the emission of a photon, independently of the existing radiation field. We denote by $W_{2\to1}^s$ the transition probability per unit time per particle for spontaneous emission.

(2) Induced (stimulated) emission

This type of emission occurs in the presence of an existing radiation field and is proportional to the intensity of that field. The stimulated transition probability per unit time is therefore written as $NW^i_{2\to1}$, where N is the number of photons of the same wave-vector as that of the photon being emitted.

(3) Absorption

This occurs in the presence of a radiation field. The particle makes an upward transition by absorbing a photon. The transition probability per unit time per particle is proportional to the density of photons and is therefore written as $NW^a_{1\to2}$.

These three processes involve the *Einstein coefficients* $W^s_{2\to1}$, $W^i_{2\to1}$, $W^a_{1\to2}$. Ignoring possible degeneracy, and assuming that (19.1.1) holds for pairs (1, 2) and (2, 3) of levels, the rate equation for level 2 is

$$\frac{dn_2}{dt} = -n_2 NW^a_{2\to3} + n_1 NW^a_{1\to2} - n_2 W^s_{2\to1} + n_3 W^s_{3\to2} - n_2 NW^i_{2\to1} + n_3 NW^i_{3\to2}.$$

(19.1.4)

It is understood that the variables n, N, etc., are functions of time, as well as of any other arguments that may be shown explicitly. In thermodynamic equilibrium, the rate of upward transitions must equal the rate of downward transitions between two states. This is the *principle of detailed balancing* or the *principle of microscopic reversibility*. Hence we obtain the relation

$$\frac{n_1}{n_2} NW^a_{1\to2} = W^s_{2\to1} + NW^i_{2\to1}.$$

(19.1.5)

By using (19.1.2) and (19.1.3), we see that (19.1.5) yields

$$e^{\hbar\omega/kT} W^a_{1\to2} = [e^{\hbar\omega/kT} - 1] W^s_{2\to1} + W^i_{2\to1}.$$

(19.1.6)

However, this relationship holds for all temperatures. We may therefore separate terms independent of T and those dependent on T to obtain the following *Einstein relations*:

$$W^a_{1\to2} = W^s_{2\to1}, \quad W^i_{2\to1} = W^s_{2\to1}.$$

(19.1.7)

Hence we may introduce the single symbol $W_{1,2}$:

$$W_{1,2} = W^s_{2\to1} = W^i_{2\to1} = W^a_{1\to2}.$$

(19.1.8)

Equation (19.1.4) may now be rewritten in the simpler form

$$\frac{dn_1}{dt} = W_{2,3}\{-N(\mathbf{k})n_2 + [1+N(\mathbf{k})]n_3\} - W_{1,2}\{-N(\mathbf{k})n_1 + [1+N(\mathbf{k})]n_2\}.$$

(19.1.9)

19.2 Transition to the classical limit

We now wish to explore the process of passing from the quantum-mechanical description to the classical description. In order to do this, we need to relate the number density of particles and the number density of photons, etc., in the two descriptions.

We first consider the particle distribution. We suppose that the system is contained in a cube of size L. If the volume is V, then $V = L^3$. In the quantum mechanical description, we label the different states of a particle by $\{q\}$, and denote the number of particles in state $\{q\}$ by $F(q)$.

In the classical picture, we work with a particle distribution $f(\mathbf{p}, t)$ such that $f(\mathbf{p}, t)\,\mathrm{d}^3p\,\mathrm{d}^3x$ is the number of particles in the spatial volume d^3x and the volume d^3p of momentum space. Hence we see that

$$L^3 \int \mathrm{d}^3p\, f(\mathbf{p}, t) = \sum_q F(q), \tag{19.2.1}$$

where the integral and the summation may cover any region – small or large – that defines the same group of particles.

We now consider the photon distribution. Since we are considering waves in a cube of side L, the components of the wave vector k_r may assume only the discrete values

$$k_r = m_r(2\pi/L), \quad m_r = 0, \pm 1, \pm 2, \ldots \tag{19.2.2}$$

The momentum of each wave is $\hbar k$ and the energy is $\hbar \omega$. Clearly the number of photons with wave-vector in the range d^3k is given by

$$\mathrm{d}^3N(\mathbf{k}) = (L/2\pi)^3\, N(\mathbf{k})\mathrm{d}^3k. \tag{19.2.3}$$

Hence if $n_\phi(\mathbf{k})\,\mathrm{d}^3k\,\mathrm{d}^3x$ is the number of photons in the spatial volume d^3x that have wave-vectors in the range d^3k, we see that

$$n_\phi(\mathbf{k})L^3\,\mathrm{d}^3k = (L/2\pi)^3\, N(\mathbf{k})\mathrm{d}^3k, \tag{19.2.4}$$

so that

$$n_\phi(\mathbf{k}) = (2\pi)^{-3}\, N(\mathbf{k}). \tag{19.2.5}$$

In the classical picture, we normally represent the intensity of the radiation field by its energy density. If we denote by $u(\mathbf{k})\,\mathrm{d}^3k\,\mathrm{d}^3x$ the energy in volume d^3x of waves with wave-vectors in the range d^3k, then

$$u(\mathbf{k}) = n_\phi(\mathbf{k})\hbar\omega. \tag{19.2.6}$$

Hence we may relate the quantum-mechanical description $N(\mathbf{k})$ and the classical description $u(\mathbf{k})$ by

$$u(\mathbf{k}) = (2\pi)^{-3}N(\mathbf{k})\hbar\omega. \tag{19.2.7}$$

19.3 The three-state model: emission and absorption

We now consider three discrete particle states $\{q'\}$, $\{q\}$ and $\{q''\}$, for which the momentum will be denoted by $\mathbf{p}'$, $\mathbf{p}$ and $\mathbf{p}''$, respectively. (See Fig. 19.2.) These states are chosen so that they are separated by momentum $\hbar\mathbf{k}$ as follows:

$$\mathbf{p} - \mathbf{p}' = \hbar\mathbf{k}, \quad \mathbf{p}'' - \mathbf{p} = \hbar\mathbf{k}. \tag{19.3.1}$$

If the energies of the states are denoted by E', E and E'', then we have the additional relations

$$E - E' = \hbar\omega, \quad E'' - E = \hbar\omega, \tag{19.3.2}$$

where ω is related to $\mathbf{k}$ by the dispersion relation

$$\omega = \omega(\mathbf{k}). \tag{19.3.3}$$

In order to facilitate the transition from the quantum-mechanical description to the classical description, we introduce the operator D defined by

$$\mathrm{D} = \hbar k_r \frac{\partial}{\partial p_r}. \tag{19.3.4}$$

By using this operator, we may shift the argument of a function of $\mathbf{p}$ by an amount $\hbar\mathbf{k}$ as follows:

$$G(\mathbf{p} + \hbar\mathbf{k}) = e^{\mathrm{D}}G(\mathbf{p}). \tag{19.3.5}$$

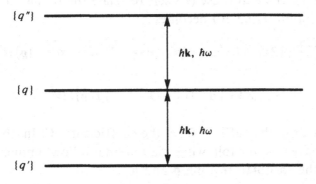

Fig. 19.2. Three levels of the multi-level system analyzed in Section 19.3.

In terms of this operator, the three momenta are related by

$$\left. \begin{array}{l} \mathbf{p}' = (1-D)\mathbf{p}, \\ \mathbf{p}'' = (1+D)\mathbf{p}. \end{array} \right\} \tag{19.3.6}$$

Retaining terms up to second order in the operator D, the energies are related by

$$\left. \begin{array}{l} E' = (1-D+\tfrac{1}{2}D^2)E, \\ E'' = (1+D+\tfrac{1}{2}D^2)E. \end{array} \right\} \tag{19.3.7}$$

On introducing the notation

$$W_{q,q'} = W^{\mathrm{s}}_{q \to q'} = W^{\mathrm{i}}_{q \to q'} = W^{\mathrm{a}}_{q' \to q}, \tag{19.3.8}$$

for the transition probabilities, we see that, to second order,

$$W_{q'',q} = (1+D+\tfrac{1}{2}D^2)\,W_{q,q'}. \tag{19.3.9}$$

On taking account of spontaneous and induced emission and of absorption, we find that the photon distribution evolves according to the equation

$$\frac{\partial N(\mathbf{k})}{\partial t} = \sum_q W_{q,q'}[1+N(\mathbf{k})]F(q) - \sum_q W_{q,q'}N(\mathbf{k})F(q'), \tag{19.3.10}$$

the terms in the first summation representing spontaneous and induced emission, and the term in the second summation representing absorption. For each value of q, q' is determined by the requirement that the first of equations (19.3.1) should be satisfied. [Remember that, for convenience, we denote variables N (etc.) simply as $N(\mathbf{k})$ rather than (as would be more correct) as $N(\mathbf{k}, t)$.]

We may convert the summation in the above equation into an integration by using (19.2.1). If we also use (19.2.7) to relate the radiation field to the classical description, (19.3.10) becomes

$$\frac{\partial u(\mathbf{k})}{\partial t} = (2\pi)^{-3}L^3\hbar\omega\int \mathrm{d}^3p\, f(\mathbf{p})\,W + L^3 u(\mathbf{k})\int \mathrm{d}^3p\, f(\mathbf{p})\,W$$

$$- L^3 u(\mathbf{k})\int \mathrm{d}^3p\,[(1-D+\tfrac{1}{2}D^2)f(\mathbf{p})]\,W, \tag{19.3.11}$$

where we now drop the suffixes from the coefficients W. In obtaining this equation, we have used the following relation (that follows immediately from (19.3.1)) for the Jacobian relating $\mathbf{p}$ and $\mathbf{p}'$:

$$\frac{\partial^3(p_1', p_2', p_3')}{\partial^3(p_1, p_2, p_3)} = 1. \tag{19.3.12}$$

On using the definition of D and keeping terms only to first order in h, we obtain

$$\frac{\partial u(\mathbf{k})}{\partial t} = \int d^3p \left[\frac{\hbar \omega L^3 W}{(2\pi)^3} \right] f(\mathbf{p}) + u(\mathbf{k}) \int d^3p \left[\frac{\hbar \omega L^3 W}{(2\pi)^3} \right] (2\pi)^3 \frac{k_r}{\omega} \frac{\partial f}{\partial p_r}.$$

$$(19.3.13)$$

From the first integral, it is clear that the term in brackets represents the emissivity per particle, for which we introduce the notation $\eta(\mathbf{p}, \mathbf{k})$, so that

$$\eta(\mathbf{p}, \mathbf{k}) = \frac{\hbar \omega L^3 W}{(2\pi)^3}. \qquad (19.3.14)$$

If we rewrite (19.3.13) in the form

$$\frac{\partial u(\mathbf{k})}{\partial t} = \varepsilon(\mathbf{k}) - \gamma(\mathbf{k}) u(\mathbf{k}), \qquad (19.3.15)$$

it is clear that ε represents the volume emissivity in erg cm^{-3} s^{-1}, and γ (in s^{-1}) represents the absorption rate. We now see that these two terms are related to the emissivity per particle by the two equations:

$$\varepsilon(\mathbf{k}) = \int d^3p\, f(\mathbf{p})\, \eta(\mathbf{p}, \mathbf{k}) \qquad (19.3.16)$$

and

$$\gamma(\mathbf{k}) = -(2\pi)^3 \frac{k_r}{\omega(\mathbf{k})} \int d^3p \frac{\partial f(\mathbf{p})}{\partial p_r} \eta(\mathbf{p}, \mathbf{k}). \qquad (19.3.17)$$

Hence, by going through the quantum-mechanical analysis, we have obtained a relationship between the classical expression for absorption and the classical expression for emission.

It is interesting to note from (19.3.17) that, if we consider a one-dimensional system, the absorption coefficient γ will be negative, representing instability, if $\partial f/\partial p > 0$. Hence our present analysis offers another interpretation of the condition for instability of a plasma (considering the spontaneous growth of Langmuir waves) that was derived in Chapter 9.

19.4 Diffusion equation for the particle distribution function

Equation (19.1.4) is an equation for the rate of change of the particle occupation number of a given state. When expressed in terms of the notation of the previous section, it takes the form

$$\frac{\partial F(q)}{\partial t} = -\sum_{\mathbf{k}} W_{q,q'}[1+N(\mathbf{k})]F(q) + \sum_{\mathbf{k}} W_{q,q'}N(\mathbf{k})F(q')$$

$$+ \sum_{\mathbf{k}} W_{q'',q}[1+N(\mathbf{k})]F(q'') - \sum_{\mathbf{k}} W_{q'',q}N(\mathbf{k})F(q). \tag{19.4.1}$$

These terms represent spontaneous and induced emission from $\{q\}$ to $\{q'\}$ and from $\{q''\}$ to $\{q\}$, and absorption from $\{q'\}$ to $\{q\}$ and from $\{q\}$ to $\{q''\}$.

We may make the transition from the quantum-mechanical formalism to the classical formalism by the same procedures that we used in Section 19.3. That is to say, we use (19.2.1) to replace $F(q)$ by $f(\mathbf{p})$, (9.2.7) to replace $N(\mathbf{k})$ by $u(\mathbf{k})$, and we use the operator D to relate $f(\mathbf{p}')$ and $f(\mathbf{p}'')$ to $f(\mathbf{p})$, and to relate $W_{q'',q}$ to $W_{q,q'}$. In this way, we find that (19.4.1) may be expressed as

$$\frac{\partial f(\mathbf{p})}{\partial t} = -\left(\frac{L}{2\pi}\right)^3 \int d^3k\, W\left[1+\frac{(2\pi)^3 u(\mathbf{k})}{\hbar\omega}\right]f(\mathbf{p})$$

$$+ \left(\frac{L}{2\pi}\right)^3 \int d^3k\, W\left[\frac{(2\pi)^3 u(\mathbf{k})}{\hbar\omega}\right]\left(1-D+\tfrac{1}{2}D^2\right)f(\mathbf{p})$$

$$+ \left(\frac{L}{2\pi}\right)^3 \int d^3k\left[1+\frac{(2\pi)^3 u(\mathbf{k})}{\hbar\omega}\right]\left(1+D+\tfrac{1}{2}D^2\right)(Wf(\mathbf{p}))$$

$$- \left(\frac{L}{2\pi}\right)^3 \int d^3k\left[\frac{(2\pi)^3 u(\mathbf{k})}{\hbar\omega}\right]\left[\left(1+D+\tfrac{1}{2}D^2\right)W\right]f(\mathbf{p}). \tag{19.4.2}$$

If we retain only terms of lowest order in $\hbar$, this reduces to

$$\frac{\partial f(\mathbf{p})}{\partial t} = \left(\frac{L}{2\pi}\right)^3 \int d^3k[-Wf(\mathbf{p})+(1+D)(Wf(\mathbf{p}))]$$

$$+ \left(\frac{L}{2\pi}\right)^3 \int d^3k\left[\frac{(2\pi)^3 u(\mathbf{k})}{\hbar\omega}\right]\left\{-Wf(\mathbf{p})+W\left(1-D+\tfrac{1}{2}D^2\right)f(\mathbf{p})\right\}$$

$$+ (1+D+\tfrac{1}{2}D^2)(Wf(\mathbf{p})) - [(1+D+\tfrac{1}{2}D^2)W]f(\mathbf{p}) \tag{19.4.3}$$

On examining the terms explicitly and using (19.3.14), we find that the above equation reduces to

$$\frac{\partial f}{\partial t} = \frac{\partial}{\partial p_r}(L_r f) + \frac{\partial}{\partial p_r}\left(D_{rs}\frac{\partial f}{\partial p_s}\right) \tag{19.4.4}$$

where

$$L_r(\mathbf{p}) = \int d^3k \frac{k_r}{\omega(\mathbf{k})}\eta(\mathbf{p},\mathbf{k}) \tag{19.4.5}$$

and

$$D_{rs}(\mathbf{p}) = (2\pi)^3 \int d^3k \frac{k_r k_s}{[\omega(\mathbf{k})]^2} \eta(\mathbf{p}, \mathbf{k}) u(\mathbf{k}). \qquad (19.4.6)$$

Clearly L_r represents the rate of loss of momentum to radiation, and D_{rs} represents a diffusion process resulting from the interaction of particles with the waves.

We saw earlier in this section that, by beginning with a quantum-mechanical description, it has been possible to relate the volume absorption coefficient to the emissivity per particle (by (19.3.17)). We now see that the same procedure has made it possible also to relate the momentum loss-rate of the plasma (19.4.5), and the momentum-space diffusion of the plasma (19.4.6) to the emissivity per particle.

If the collective (or coherent) processes in the plasma are much more important than the single-particle (or incoherent) processes, the first term on the right-hand side of (19.4.4) may be neglected with respect to the second term. This assumption was made implicitly in the last chapter when we discussed stochastic acceleration. We now see from (19.4.4) that, with this assumption, the quantum-mechanical approach fulfills the promise set out at the beginning of this chapter, in that it leads directly to a diffusion equation for the effect of a distribution of plasma waves on the particle distribution.

One of the main theories of particle acceleration in astrophysical contexts is that of stochastic acceleration by MHD turbulence or by electrostatic turbulence, that may be analyzed on the basis of (19.4.4). An important variant of this theory is that of acceleration at shock fronts (Bell 1978a,b). In this model, particles upstream of the shock are scattered by incoming waves, usually taken to be Alfvén waves. Particles downstream of the shock are scattered by Alfvén waves that are traveling backwards towards the shock front. An excellent review of these and other mechanisms of particle acceleration may be found in Arons, McKee and Max (1979). A more recent review of particle acceleration in solar flares has been presented by Heyvaerts (1981).

Problem

Problem 19.1. Consider (19.4.4) that determines the evolution of the distribution function. Assume that the term involving L_r can be neglected. Assume furthermore that the diffusion process is isotropic, so that the diffusion tensor may be written as

$$D_{rs} = D\,\delta_{rs}. \tag{1}$$

Assume that the particle distribution is initially isotropic.

(a) Show that the distribution function remains isotropic.

(b) The distribution may be represented alternatively as a distribution $g(U)$ in energy U. If the relation between U and p may be taken to be

$$U = p^2/2m, \tag{2}$$

find the relation between $g(U)$ and $f(p)$.

(c) Hence find the equation satisfied by $g(U)$.

(d) Now suppose that the system settles into a steady state. Assume also that D has a power-law form:

$$D \propto U^\alpha. \tag{3}$$

Find the resulting distribution $g(U)$.

(e) What physical process is going on that leads to this form of $g(U)$?

(f) Does this approach seem to offer a reasonable explanation for the development of power-law distributions by stochastic acceleration?

Appendix A

Units and constants

Units

The units used in this book are sometimes referred to as 'modified Gaussian units' (mgu). (See, for instance, Jackson (1962)). In this system, cgs units are used for non-electromagnetic quantities such as length, time, etc. Electrostatic units (esu) are used for clearly electrical quantities, such as electric potential, electric field strength and electric charge density. Electromagnetic units (emu) are used for clearly magnetic quantities such as magnetic field strength and magnetic flux. Current and current density are regarded as 'magnetic' quantities, since that is the role they play in Maxwell's equations, and are therefore measured in emu (whereas in Gaussian units current and current density are measured in esu).

The following table gives the relationship between quantities measured in mgu, gu, esu, emu, and mks units.

Time $\qquad$ $t(\text{mgu}) = t(\text{gu}) = t(\text{emu}) = t(\text{esu}) = t(\text{mks})$

Length $\qquad$ $x(\text{mgu}) = x(\text{gu}) = x(\text{emu}) = x(\text{esu}) = 10^2 x(\text{mks})$

Mass $\qquad$ $m(\text{mgu}) = m(\text{gu}) = m(\text{emu}) = m(\text{esu}) = 10^3 m(\text{mks})$

Mass density $\qquad$ $\rho(\text{mgu}) = \rho(\text{gu}) = \rho(\text{emu}) = \rho(\text{esu}) = 10^{-3} \rho(\text{mks})$

Energy $\qquad$ $U(\text{mgu}) = U(\text{gu}) = U(\text{emu}) = U(\text{esu}) = 10^7 U(\text{mks})$

Pressure $\qquad$ $p(\text{mgu}) = p(\text{gu}) = p(\text{emu}) = p(\text{esu}) = 10 p(\text{mks})$

Electric potential $\qquad$ $\phi(\text{mgu}) = \phi(\text{gu}) = 10^{-10.48} \phi(\text{emu}) = \phi(\text{esu})$
$= 10^{-2.48} \phi(\text{mks})$

Electric field strength $\qquad$ $E(\text{mgu}) = E(\text{gu}) = 10^{-10.48} E(\text{emu}) = E(\text{esu})$
$= 10^{-4.48} E(\text{mks})$

Electric charge $\qquad$ $Q(\text{mgu}) = Q(\text{gu}) = 10^{-10.48} Q(\text{emu}) = Q(\text{esu})$
$= 10^{9.48} Q(\text{mks})$

Charge density $\qquad$ $\zeta(\text{mgu}) = \zeta(\text{gu}) = 10^{-10.48} \zeta(\text{emu}) = \zeta(\text{esu})$
$= 10^{3.48} \zeta(\text{mks})$

Magnetic potential $\qquad$ $A(\text{mgu}) = A(\text{gu}) = A(\text{emu}) = 10^{-10.48} A(\text{esu})$
$= 10^6 A(\text{mks})$

Magnetic field strength $\qquad$ $B(\text{mgu}) = B(\text{gu}) = B(\text{emu}) = 10^{-10.48} B(\text{esu})$
$= 10^4 B(\text{mks})$

Magnetic flux $\Phi(\text{mgu}) = \Phi(\text{gu}) = \Phi(\text{emu}) = 10^{-10.48}\,\Phi(\text{esu})$
                        $= 10^{8}\,\Phi(\text{mks})$

Current $J(\text{mgu}) = 10^{-10.48}\,J(\text{gu}) = J(\text{emu}) = 10^{-10.48}\,J(\text{esu})$
                        $= 10^{-1}\,J(\text{mks})$

Current density $j(\text{mgu}) = 10^{-10.48}\,j(\text{gu}) = j(\text{emu}) = 10^{-10.48}\,j(\text{esu})$
                        $= 10^{-5}\,j(\text{mks})$

Conductivity $\sigma(\text{mgu}) = 10^{-10.48}\,\sigma(\text{gu}) = 10^{10.48}\,\sigma(\text{emu}) = 10^{-10.48}\,\sigma(\text{esu})$
                        $= 10^{-0.52}\,\sigma(\text{mks})$

Resistivity $\eta(\text{mgu}) = 10^{10.48}\,\eta(\text{gu}) = 10^{-10.48}\,\eta(\text{emu}) = 10^{10.48}\,\eta(\text{esu})$
                        $= 10^{0.52}\,\eta(\text{mks})$

Physical constants

The following is a short list of relevant physical constants, expressed in exponential decimal notation.

Speed of light $c = 10^{10.48}\ \text{cm s}^{-1}$
Gravitational constant $G = 10^{-7.18}\ \text{dyne cm}^2\,\text{g}^{-2}$
Boltzmann's constant $k = 10^{-15.86}\ \text{erg K}^{-1}$
Planck's constant $h = 10^{-26.18}\ \text{erg s}$
Modified Planck's constant $\hbar = 10^{-26.98}\ \text{erg s}$
Electron mass $m_e = 10^{-27.04}\ \text{g}$
Proton mass $m_p = 10^{-23.78}\ \text{g}$
Electron charge $e = 10^{-9.32}\ \text{mgu} = 10^{-9.32}\ \text{esu}$
Classical electron radius $r_e = 10^{-12.55}\ \text{cm}$
Thomson cross section $\sigma_T = 10^{-24.18}\ \text{cm}^2$

Conversion relations

The following are a few useful conversion relations:

1 ångstrom $= 10^{-8}\ \text{cm}$
1 micron $= 10^{-4}\ \text{cm}$
1 eV $= 10^{-11.80}\ \text{erg}$
1 eV $= 10^{4.06}\ \text{K}$
1 calorie $= 10^{7.62}\ \text{erg}$

Geophysical and astrophysical values

The following values are useful for geophysical and astrophysical calculations:

Mass of the Earth $M_\oplus = 10^{27.78}\ \text{g}$
Radius of the Earth $R_\oplus = 10^{8.80}\ \text{cm}$

Surface gravity of Earth $g_\oplus = 10^{2.99}$ cm s^{-2}

Mass of the Sun $M_\odot = 10^{33.30}$ g

Radius of the Sun $R_\odot = 10^{10.84}$ cm

Surface gravity of Sun $g_\odot = 10^{4.44}$ cm s^{-2}

Luminosity of Sun $L_\odot = 10^{33.58}$ erg s^{-1}

Astronomical unit $1\ AU = 10^{13.18}$ cm

Parsec $1\ pc = 10^{18.49}$ cm

Appendix B

Group velocity

One of the most important techniques of the analysis of physical systems is that of wave analysis, and one of the most important concepts in wave analysis is that of group velocity. Analysis of wave-propagating systems has shown that, where it is possible to define an 'energy propagation velocity,' this velocity is found to be identical with the group velocity. This identity has been referred to by Lighthill (1978) as a 'major conundrum.' Lighthill points out that it is peculiar that the energy propagation velocity, that is defined and calculated for waves of a fixed wave number $\mathbf{k}$ and fixed frequency ω, should be identical with the group velocity, that is defined in terms of the *changes* in frequency and wave vector in going from one wave to neighboring waves.

There is another aspect to this conundrum. Group velocity is a purely 'kinematic' concept, depending only on the dispersion relation relating the frequency to the wave vector of a wave. Many systems of quite different dynamical composition may have the same dispersion relation. On the other hand, quantities such as energy and energy flux depend upon the detailed dynamical structure of a system, and will be quite different for different physical systems even if they have the same dispersion relation. It is therefore a puzzle to understand why the relationship between two *dynamical* quantities should be expressible in terms of a *kinematic* quantity.

The identity of the energy-propagation velocity and the group velocity has been demonstrated for many specific systems. One important question is clearly to determine the most general conditions under which this identity is valid. It has been shown by Sturrock (1961a) and by Whitham (1974) that it is true for any system that can be derived from a variation principle. Is this the most general condition for the identity to be valid?

Furthermore, one may ask a more basic question. Rather than think in terms of the relationship between group velocity and 'energy propagation velocity,' we may regard the relationship as one between group velocity, energy density E, and energy flux S_r:

$$S_r = E u_r. \tag{B.1}$$

Are there other quantities, analogous to E and S_r, that are related by the group velocity? There are indeed; for instance, momentum flux and momentum density are so related (Sturrock, 1961a).

This leads us to pose the following question. What dynamical quantities are related by the group velocity, and under what conditions? We show in this appendix that *for any uniform wave-propagating system with a finite number of degrees of freedom, treated in the linear approximation, for which the frequency is real for a small range of wave vector, the time-averaged values of any two quantities that satisfy a conservation relation are related by the group velocity.* This wave-propagating system may or may not be defined by a variation principle.

We consider an infinite, uniform system with a finite number of degrees of freedom, so that the state of the system at any point x, t may be defined by a finite number of variables $\Psi_n(\mathbf{x}, t)$. We assume that, in the small-amplitude approximation, the system satisfies a set of linear differential equations expressible in the form

$$L_{mn}(\partial/x_r, \partial/t)\Psi_n(\mathbf{x}, t) = 0. \tag{B.2}$$

As usual, we use the Fourier-transform notation

$$\Psi_n(x, t) = \int\int d^3k\, dt\, e^{i(\mathbf{k}\cdot\mathbf{x}-\omega t)}\tilde{\Psi}_n(\mathbf{k}, \omega), \tag{B.3}$$

and we understand the usual inverse notation. Equation (B.2) then takes the form

$$L_{mn}(ik_r, -i\omega)\tilde{\Psi}_n(\mathbf{k}, \omega) = 0. \tag{B.4}$$

The condition that there should be a solution of nonzero amplitude is that the determinant of the matrix L_{mn} should be zero, that is,

$$D(\mathbf{k}, \omega) \equiv |L_{mn}(ik_r, -i\omega)| = 0. \tag{B.5}$$

This is the 'dispersion relation' of the system.

In general, there will be a finite number of values of ω for each value of $\mathbf{k}$ that we may express as

$$\omega = \Omega_\alpha(\mathbf{k}), \tag{B.6}$$

where α enumerates the 'modes' of the system. For present purposes, we may consider one particular mode and we may therefore drop the suffix α. It is necessary, for what follows, that ω be real in a small but finite neighborhood of a wave vector $\mathbf{k}_0$.

Focusing attention on this mode, we consider the wave function defined by

$$\Psi_n(\mathbf{x}, t) = \int d^3k\, e^{i(\mathbf{k}\cdot\mathbf{x}-i\Omega(\mathbf{k})t)}A(\mathbf{k})\tilde{\Psi}_n(\mathbf{k}) + \text{c.c.}, \tag{B.7}$$

where the 'amplitude' $A(\mathbf{k})$ is to be nonzero only in a small neighborhood of the value $\mathbf{k}_0$. This ensures that $\Omega(\mathbf{k})$ takes only real values in the range of values of $\mathbf{k}$ for which the integrand is nonzero. Assuming that the basic physical quantities $\Psi_n(\mathbf{x}, t)$ are chosen to be real, it is necessary to include in (B.7) the complex conjugate term, denoted by 'c.c.'

We now express $\mathbf{k}$ as

$$\mathbf{k} = \mathbf{k}_0 k + k_1\boldsymbol{\kappa}, \tag{B.8}$$

where k_1 denotes the extent of the small region in which $A(\mathbf{k})$ is nonzero. Then we may express $A(\mathbf{k})$ as

$$A(\mathbf{k}) \equiv A(\mathbf{k}_0 + k_1\boldsymbol{\kappa}) = k_1^{-3}\tilde{G}(\boldsymbol{\kappa}), \tag{B.9}$$

where $\tilde{G}(\mathbf{k}) \neq 0$ only for $|\kappa| \leqslant 1$. Equation (B.7) may now be rewritten as

$$\Psi_n(\mathbf{x}, t) = \int d^3\kappa \exp[i(\mathbf{k}_0 + k_1\kappa) \cdot \mathbf{x} - i\Omega(\mathbf{k}_0 + k_1\kappa)t]\,\tilde{G}(\kappa)\,\tilde{\Psi}_n(\mathbf{k}_0 + k_1\kappa) + \text{c.c.}$$

(B.10)

We now regard k_1 as a small quantity and expand in powers of k_1. The dominant terms are

$$\Psi_n(\mathbf{x}, t) = e^{i(\mathbf{k}_0 \cdot \mathbf{x} - i\omega_0 t)}\tilde{\Psi}_n(\mathbf{k}_0) \int d^3\kappa\, e^{ik_1\kappa_r(x_r - u_r(\mathbf{k}_0)t)}\,\tilde{G}(\kappa) + \text{c.c.},$$

(B.11)

where

$$\omega_0 = \Omega(\mathbf{k}_0)$$

(B.12)

and $u_r(\mathbf{k})$ is the *group velocity* defined by

$$u_r(\mathbf{k}) = \frac{\partial \Omega(\mathbf{k})}{\partial k_r}.$$

(B.13)

Equation (B.11) may be re-expressed as

$$\Psi_n(\mathbf{x}, t) = G(k_1(\mathbf{x} - \mathbf{u}t))\,\tilde{\Psi}_n(\mathbf{k}_0)e^{i(\mathbf{k}_0 \cdot \mathbf{x} - \omega_0 t)} + \text{c.c.},$$

(B.14)

where $\mathbf{u}$ denotes the value $\mathbf{u}(\mathbf{k}_0)$. Clearly, (B.14) represents a wave packet with spatial extent of order k_1^{-1}, that propagates with velocity $\mathbf{u}$.

We now consider any two quantities $E(\mathbf{x}, t)$ and $S_r(\mathbf{x}, t)$ that depend quadratically on the amplitudes of the wave functions and satisfy the conservation relation

$$\frac{\partial E}{\partial t} + \frac{\partial S_r}{\partial x_r} = 0.$$

(B.15)

E may be energy density and S_r the energy flux. On the other hand, E and S_r may be replaced by a component of the momentum density and the corresponding set of components of the momentum-flux tensor:

$$\frac{\partial P_r}{\partial t} + \frac{\partial T_{rs}}{\partial x_s} = 0.$$

(B.16)

In what follows, the identity of E and S_r as energy density and energy flux is irrelevant. We need only the fact that each quantity depends quadratically on the wave amplitude, and that the two quantities satisfy the conservation relation (B.15).

Since E and S_r depend quadratically on the functions given by (B.14), we see that they may be expressed as

$$E(\mathbf{x}, t) = \bar{E}(\mathbf{X}) + \hat{E}(\mathbf{X}, \phi)$$

(B.17)

and

$$S_r(\mathbf{x}, t) = \bar{S}_r(\mathbf{X}) + \hat{S}_r(\mathbf{X}, \phi)$$

(B.18)

where

$$\mathbf{X} = \mathbf{x} - \mathbf{u}t \qquad (B.19)$$

and

$$\phi = \mathbf{k} \cdot \mathbf{x} - \omega t. \qquad (B.20)$$

Clearly, the terms $\bar{E}(\mathbf{X})$ and $\bar{S}_r(\mathbf{X})$ are the slowly varying contributions that could be formed by averaging over a cycle of the oscillation, and $\hat{E}(\mathbf{X}, \phi)$ and $\hat{S}_r(\mathbf{X}, \phi)$ are the periodic contributions that will oscillate with frequency 2ω. The phase-averaged values of $\hat{E}(\mathbf{X}, \phi)$ and $\hat{S}_r(\mathbf{X}, \phi)$ are zero. At this point, we have chosen for simplicity to drop the suffix '0' on $\mathbf{k}_0$, etc.

Equations (B.19) and (B.20) enable us to make a transformation of independent variables from $\mathbf{x}, t$ to $\mathbf{X}, \phi$. On noting that

$$\begin{pmatrix} \mathrm{d}X_r \\ \mathrm{d}\phi \end{pmatrix} = \begin{pmatrix} \delta_{rs} & -u_r \\ k_s & -\omega \end{pmatrix} \begin{pmatrix} \mathrm{d}x_s \\ \mathrm{d}t \end{pmatrix}, \qquad (B.21)$$

we see that the derivatives are related by

$$\begin{pmatrix} \partial/\partial x_r \\ \partial/\partial t \end{pmatrix} = \begin{pmatrix} \delta_{rs} & k_r \\ -u_s & -\omega \end{pmatrix} \begin{pmatrix} \partial/\partial X_s \\ \partial/\partial \phi \end{pmatrix}. \qquad (B.22)$$

Hence equation (B.15) may be re-expressed in terms of $\mathbf{X}$ and ϕ in the form

$$\left(-u_r \frac{\partial}{\partial X_r} - \omega \frac{\partial}{\partial \phi} \right) [\bar{E}(\mathbf{X}) + \hat{E}(\mathbf{X}, \phi)] + \left(\frac{\partial}{\partial X_r} + k_r \frac{\partial}{\partial \phi} \right) [\bar{S}_r(\mathbf{X}) + \hat{S}_r(\mathbf{X}, \phi)] = 0. \qquad (B.23)$$

On averaging over ϕ, we find that the steady contribution to (B.23) is

$$\frac{\partial}{\partial X_r} [S_r(\mathbf{X}) - E(\mathbf{X}) u_r] = 0. \qquad (B.24)$$

On introducing the notation

$$F(\mathbf{X}) = |G(\mathbf{X})|^2, \qquad (B.25)$$

we see from (B.14) that $\bar{E}(\mathbf{X})$ and $\bar{S}_r(\mathbf{X})$ may be expressed as

$$E(\mathbf{X}) = F(\mathbf{X}) \bar{e} \quad \text{and} \quad S_r(\mathbf{X}) = F(\mathbf{X}) \bar{s}_r, \qquad (B.26)$$

where $\bar{e}$ and $\bar{s}_r$ are independent of the form of the wave packet. Equation (B.24) may therefore be rewritten as

$$\frac{\partial F(\mathbf{X})}{\partial X_r} (\bar{s}_r - \bar{e} u_r) = 0. \qquad (B.27)$$

Since this must be true for all forms of $F(\mathbf{X})$, we see that

$$\bar{s}_r = \bar{e} u_r, \qquad (B.28)$$

so that the slowly varying, or phase averaged, contributions to $E(\mathbf{x}, t)$ and $S_r(\mathbf{x}, t)$, are related by

$$\bar{S}_r = \bar{E}u_r. \tag{B.29}$$

It is interesting to note that we may obtain a second relation from (B.23). On subtracting out the terms that do not depend upon ϕ, and on noting that terms arising from the operation $\partial F/\partial X_r$ will be small since the packet is assumed to be much larger than a wavelength, we obtain the relation

$$\frac{\partial}{\partial \phi}[k_r \hat{S}_r(\mathbf{X}, \phi) - \omega \hat{E}(\mathbf{X}, \phi)] = 0. \tag{B.30}$$

On recalling that the phase-averaged values of $\hat{E}(\mathbf{X}, \phi)$ and $\hat{S}_r(\mathbf{X}, \phi)$ are zero, we see from (B.30.) that

$$k_r \hat{S}_r = \omega \hat{E}. \tag{B.31}$$

In the case of a one-dimensional system, or in the case of an isotropic system for which $\hat{\mathbf{S}}$ is parallel to $\mathbf{k}$, equation (B.33) implies that the rapidly varying contribution to the energy 'moves' at the phase velocity. However, in general this interpretation is not valid.

Finally, it is worth pointing out that, in a nonlinear dynamical system (that can be represented by a Lagrangian function or by a Hamiltonian function), the relations (B.29) and (B.31) will apply only to the components of the energy-momentum tensor that are derived from the second-order contribution to the Lagrangian function (that leads to the linearized equations for the system). That is, these relations apply to the 'pseudo-energy' and 'pseudo-momentum,' not to the energy and momentum as they are given by the full nonlinear Lagrangian function (Sturrock, 1962). This resolves the puzzle that the energy–momentum tensor derived from the second-order contribution to the Lagrangian function (the 'pseudo-energy–momentum tensor') may be asymmetric, whereas a real energy–momentum tensor must be symmetric.

Appendix C

Amplifying and evanescent waves, convective and absolute instability

This appendix also deals with the relationship between kinematic and dynamical concepts concerning the wave analysis of physical systems. We again consider an infinite uniform system with a finite number of degrees of freedom, so that there are a finite number of modes. We here consider a one-dimensional representation so that any variable $\psi_n(x, t)$ is expressible as

$$\psi_n(x, t) = \int \int dk \, d\omega \, e^{i(kx - \omega t)} \tilde{\psi}_n(k, \omega), \tag{C.1}$$

with the usual inverse relationship. The dispersion relation,

$$D(k, \omega) = 0, \tag{C.2}$$

may be solved either to give the frequency ω as a function of k,

$$\omega = \Omega_\alpha(k), \tag{C.3}$$

or to give the wave vector k as a function of frequency,

$$k = K_\alpha(\omega), \tag{C.4}$$

where α enumerates the modes.

Let us first consider a single wave with a definite wave number and a definite frequency, and suppose that the frequency ω is real and that the wave vector k is complex. It is known, from the analysis of physical systems, that such a wave may represent two quite different types of behavior. On the one hand, it may be an 'evanescent' wave, in which case the amplitude of a 'signal' or 'excitation' necessarily decays away from the source of the signal. On the other hand, the wave may represent an amplifying wave, in which case the disturbance can grow with distance from the source. One may suspect that the distinction between the two types of growing wave may be extracted from the dispersion relation, since two physical systems that have the same dispersion relation have the same wave-propagation properties. However, it is clearly not possible to distinguish between these two types of growing wave by examining simply a wave with a single value of the frequency.

Let us therefore consider a 'wave packet.' Suppose that a signal is excited at a certain location, say $x = 0$, but the excitation is localized in time. If we consider that the packet is composed of waves with values of frequency ω confined to a small but finite band of width $\Delta\omega$, then the excitation will extend over all time. Nevertheless, it can begin by being small, become finite only for a time interval of order $(\Delta\omega)^{-1}$, and then become small again. If

$$|\psi_n(x, t)| \to 0 \quad \text{as} \quad t \to \pm\infty, \tag{C.5}$$

we call it a 'time-like' packet that we denote by T. If (C.5) is not satisfied, we say that it is not 'time-like,' and denote it by $\bar{T}$.

We may describe the spatial properties of the packet in a similar manner. If

$$|\psi_n(x, t)| \to 0 \quad \text{as} \quad x \to \pm\infty, \tag{C.6}$$

we call the packet 'space-like,' and denote it by S. If the condition (C.6) is not satisfied, we say that it is not 'space-like,' and denote it by $\bar{S}$.

The distinction between evanescent and amplifying waves may now be understood as follows. If the dispersion relation is such that, in a small range of frequency, any time-like packet is also a space-like packet, then it is possible to 'launch' disturbances that propagate away from the source and grow in amplitude as they propagate. This property of the system may be identified with the property that it may support amplifying waves. Wave packets of this type, that we may denote by 'ST,' have the form shown in Fig. C.1.

On the other hand, the dispersion relation may be such that, for a small range of frequency, time-like packets are not space-like. A 'contour diagram' for such packets is shown in Fig. C.2. It is clear that a system that supports wave packets of this type cannot be used to 'launch' a disturbance, finite in time, that grows as it propagates. We may identify systems that support evanescent waves with those that have dispersion relations that support '$\bar{S}T$' types of wave packets.

There is a similar classification of instabilities. Suppose that a system is excited at time $t = 0$, the excitation being space-like. We consider a packet of waves with a small but finite range of wave vector k for which ω is complex, and consider the mode for which $\omega_i > 0$. Then the disturbance will clearly grow in time. Nevertheless, there are two possible types of wave packet. One is identical to that shown in Fig. C.1.

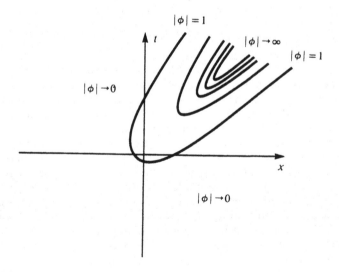

Fig. C.1. Schematic representation of a time-like packet that is also a space-like packet (ST).

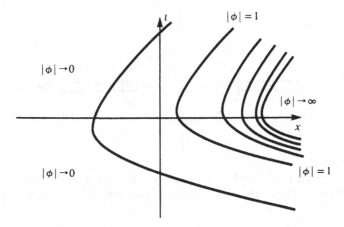

Fig. C.2. The case of a time-like packet that is not also a space-like packet ($\bar{S}T$).

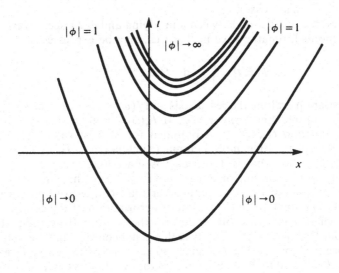

Fig. C.3. The case of a space-like packet that is not also a time-like packet ($S\bar{T}$).

In this case, the space-like packet is also a time-like packet. The other possibility is shown in Fig. C.3. In this case, the dispersion relation has the property that a space-like packet is not time-like ($S\bar{T}$). In the case shown in Fig. C.1 (ST), the disturbance at any particular point, say $x=0$, finally decays to zero. On the other hand, in the case shown in Fig. C.3 ($S\bar{T}$), the disturbance at any point, say $x=0$, grows as a function of time. Systems that exhibit ST wave packets are said to exhibit a 'convective' instability. Systems that exhibit $S\bar{T}$ wave packets are said to exhibit a 'non-convective' or 'absolute' instability.

The preceding paragraphs are intended to clarify the distinction between growing and amplifying waves (in the case of growth in space) and between convective and absolute instabilities (in the case of growth in time). It has also been argued that it should be possible to distinguish these cases simply by examination of the relevant

Appendix C

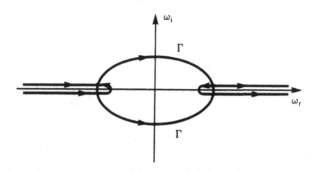

Fig. C.4. In this case the curve Γ, that maps out the (complex) frequencies $\omega = \Omega(k)$ as k takes all real values, bridges the band of frequencies for which k is complex.

dispersion relation. This is indeed possible, but in general it is not a simple matter. For an extensive discussion of this topic, the reader is referred to Bers (1983). In this appendix, we point out a simple approach (Sturrock, 1958) that is limited in application, but useful in many cases.

Consider first the distinction between amplifying and evanescent waves. We consider a quasi-monochromatic wave packet that may be expressed as

$$\psi(x, t) = \int_{-\infty}^{\infty} \mathrm{d}\omega\, f(\omega) \mathrm{e}^{\mathrm{i}(K(\omega)x - \omega t)}, \tag{C.7}$$

where the integration is along the real ω-axis and $f(\omega)$ is chosen to be sharply peaked at some particular frequency, say $\omega = \omega_0$. If $K_i(\omega)$, the imaginary part of $K(\omega)$, is bounded, the function $f(\omega) \mathrm{e}^{\mathrm{i}K(\omega)x}$ is bounded for all real values of ω and finite values of x. Hence, by Riemann's Lemma (Titchmarsh, 1937), the wave packet satisfies condition (C.6); that is, the packet is space-like.

We may deform the contour of integration in (C.7) so that the integration runs from $\omega = -\infty$ to $\omega = +\infty$ by some path other than real ω-axis, provided that the integrand in (C.7) is analytic in the region of deformation of the contour. We need to determine whether it is possible to deform the contour from the real ω-axis to another contour Γ along which $K(\omega)$ is real. Such a contour may be said to 'bridge the gap' between ω_1 and ω_2. Fig. C.4 gives an example of the curve Γ mapped out by $\omega = \Omega(k)$ as k takes all real values from $-\infty$ to $+\infty$. In this case, it is clearly possible to deform the contour integration of (C.7) from the real axis to Γ. In this case, time-like packets constructed by (C.7) are also space-like packets. This corresponds to the case shown in Figure C.1 (ST) so that such a system exhibits amplifying waves.

On the other hand, Fig. C.5 shows an example of a dispersion relation for which the curve Γ mapped out by $\omega = \Omega(k)$, for real values of k, does not bridge the gap between frequencies ω_1 and ω_2. In this case, a time-like packet is not also a space-like packet. This corresponds to the case shown in Fig. C.2 $(\bar{S}T)$. Hence such a dispersion relation supports evanescent waves.

We may suspect, from energy arguments, that if a stable system exhibits spatially growing waves, those waves must be evanescent. This intuition is in accord with our interpretation with Fig. C.5. If the contour Γ lies on the real ω axis, yet there is a range of values of ω for which k is complex, then it is clearly not possible to bridge the gap between ω_1 and ω_2 by integration along the contour Γ.

The distinction between convective and absolute instabilities may be interpreted in a similar manner. We now suppose that $\omega(k_0)$ is complex for some value k_0 of

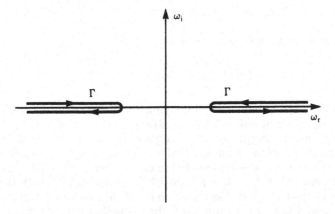

Fig. C.5. In this case the curve Γ, that maps out the frequencies $\omega = \Omega(k)$ as k takes all real values, does not bridge the band of frequencies for which k is complex.

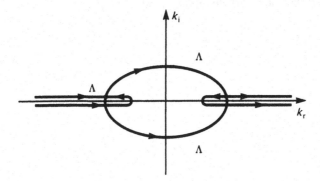

Fig. C.6. In this case the curve Λ, that maps out the (complex) wave numbers $k = K(\omega)$ as ω takes all real values, bridges the wave-number band for which ω is complex.

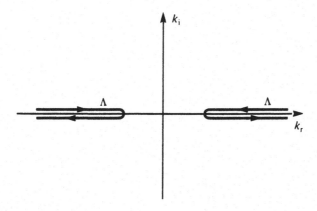

Fig. C.7. In this case the curve Λ, that maps out the wave numbers $k = K(\omega)$ as ω takes all real values, does not bridge the wave-number band for which ω is complex.

the wave vector k. We consider that $\omega_i > 0$, so that the amplitude of the wave grows in time.

We now consider the wave packet given by

$$\psi(x, t) = \int_{-\infty}^{\infty} dk\, g(k) e^{i(kx - \Omega(k)t)}, \tag{C.8}$$

where $g(k)$ is nonzero only in a small but finite neighborhood of k_0. Then, if $\Omega_i(k)$ is finite for this range of values of k, (C.8) represents a space-like packet.

The question now is whether or not it is possible to deform the contour of integration in the complex k-plane from the real axis to the contour $\Lambda(\omega)$ mapped out as ω takes all real values from $-\infty$ to ∞. In the case shown in Fig. C.6, this is possible. In the case shown in Fig. C.7, it is not possible. A system for which the dispersion relation gives rise to a contour Λ as shown in Fig. C.6 is such that a space-like packet may also be time-like (ST). In this case, the instability is convective.

In the case shown in Fig. C.7, the contour Λ is such that it is not possible to deform the contour of integration from one over all real values of the wave vector to one over all real values of the frequency. Hence a space-like packet will not be a time-like packet $(S\bar{T})$. In this case, the instability is absolute.

References

Alfvén, H. 1942. On the existence of electromagnetic-hydrodynamic waves. *Ark. f. Mat., Astr. o. Fysik*, **29B**, No. 2.

Aly, J.J. 1984. On some properties of force-free magnetic fields in infinite regions of space. *Ap. J.*, **283**, 349.

Arons, J., McKee, C., and Max, C. (eds.) 1979. Particle acceleration in astrophysics. *AIP Conference Proceedings*, No. 56. New York: American Institute of Physics.

Athay, R.G. 1976. *The Solar Chromosphere and Corona: Quiet Sun*. Dordrecht, Holland: Reidel.

Barkhausen, H. 1919. Zwei mit Hilfe der neuen Verstarker entdeckte Erscheinungen. *Phys. Z.*, **20**, 401.

Bateman, G. 1980. *MHD Instabilities*. Cambridge, Massachussetts: The MIT Press.

Bell, A.R. 1978a. The acceleration of cosmic rays in shock fronts. I. *M.N.R.A.S.*, **182**, 147.

Bell, A.R. 1978b. The acceleration of cosmic rays in shock fronts. II. *M.N.R.A.S.*, **182**, 443.

Bender, C.M., and Orszag, S.A. 1978. *Advanced Mathematical Methods for Scientists and Engineers*, p. 419. New York: McGraw-Hill.

Bennett, W.H. 1934. Magnetically self-focusing streams. *Phys. Rev.*, **45**, 890.

Bernstein, I.B., Frieman, E.A., Kruskal, M.D., and Kulsrud, R.M. 1958. An energy principle for hydromagnetic stability problems. *Proc. Roy. Soc., A*, **244**, 17.

Bers, A. 1983. *Handbook of Plasma Physics*, Vol. 1, p. 451. Space-time evolution of plasma instabilities – absolute and convective. Eds. A.A. Galeev and R.N. Sudan. New York: North-Holland.

Birn, J., Goldstein, H., and Schindler, K. 1978. A theory of the onset of solar eruptive processes. *Solar Phys.*, **57**, 81.

Boyd, T.J.M., and Sanderson, J.J. 1969. *Plasma Dynamics*. New York: Barnes and Noble.

Buneman, O. 1959. Dissipation of currents in ionized media. *Phys. Rev.*, **115**, 503.

Case, K.M. 1959. Plasma oscillations. *Ann. Phys. (N.Y.)*, **7**, 349.

Chodura, R., and Schluter, A. 1981. A 3D code for MHD equilibrium and stability. *J. Comput. Phys.*, **41**, 68.

Courant, R., and Hilbert, D. 1953. *Methods of Mathematical Physics*, Vol. 1, p. 175. New York: Interscience.

Cross, M.A., and Van Hoven, G. 1976. High-conductivity magnetic tearing instability. *Phys. Fluids*, **19**, 1591.

Deubner, F-L., and Gough, D. 1984. Helioseismology: Oscillations as a diagnostic of the solar interior. *Ann. Rev. Astron. Astrophys.*, **22**, 593.

Dreicer, H. 1959. Electron and ion runaway in a fully ionized gas. I. *Phys. Rev.*, **115**, 238.

Dreicer, H. 1960. Electron and ion runaway in a fully ionized gas. II. *Phys. Rev.*, **117**, 329.

Drummond, W.E., and Pines, D. 1962. Nonlinear stability of plasma oscillations. *Nucl. Fusion Suppl.*, Part **3**, 1049.

Dungey, J.W. 1953. Conditions for the occurrence of electrical discharges in astrophysical systems. *Phil. Mag.*, **44**, 725.

Eckersley, T.L. 1935. Musical atmospherics. *Nature*, **135**, 104.

Fermi, E. 1949. On the origin of cosmic radiation. *Phys. Rev.*, **75**, 1169.

Furth, H.P. 1968. *Advances in Plasma Physics*, Vol. 1, p. 67. Minimum-average-B stabilization for toruses. Eds. A. Simon and W.B. Thompson. New York: Interscience.

Furth, H.P., Killeen, J., and Rosenbluth, M.N. 1963. Finite-resistivity instabilities of a sheet pinch. *Phys. Fluids*, **6**, 459.

Ginzburg, V.L. 1939. The quantum theory of radiation of an electron uniformly moving in a medium. *J. Phys. (USSR)*, **2**, 441.

Gold, T., and Hoyle, F. 1958. On the origin of solar flares. *M.N.R.A.S.* **120**, 89.

Goldstein, H. 1950. *Classical mechanics*. Cambridge Mass.: Addison-Wesley.

Gurnett, D.A., Kurth, W.S., Cairns, I.H., and Granroph, L.J. 1990. Whistlers in Neptune's magnetosphere: evidence of atmospheric lightning. *J.G.R.*, **95**, 20,969.

Harris, E.G. 1969. *Advances in Plasma Physics*, Vol. 3, p. 157. Classical plasma phenomena from a quantum mechanical viewpoint. Eds. A. Simon and W.B. Thompson. New York: Interscience.

Helliwell, R.A. 1965. *Whistlers and Related Ionospheric Phenomena*. Stanford University Press.

Hess, W.N. 1968. *The Radiation Belt and Magnetosphere*. Waltham, Mass.: Blaisdell.

Heyvaerts, J. 1981. *Solar Flare Magnetohydrodynamics*, p. 429. Particle acceleration in solar flares. Ed. E.R. Priest. New York: Gordon and Breach.

Jackson, J.D. 1962. *Classical Electrodynamics*. New York: Wiley.

Klimchuk, J.A., and Sturrock, P.A. 1989. Force-free magnetic fields: is there a 'loss of equilibrium'? *Ap. J.*, **345**, 1034.

Kundu, M.R. 1965. *Solar Radio Astronomy*. New York: Interscience.

Landau, L.D. 1946. On the vibrations of the electronic plasma. *J. Phys. (USSR)*, **10**, 25.

Laval, G., Pellat, R., and Vuillemin, M. 1966. Instabilités électromagnetiques des plasma sans collisions. *Proceedings of the Conference on Plasma Physics and Controlled Nuclear Fusion Research (Int. A.E.A., Vienna)*, Vol. **2**, 259.

Lighthill, J. 1978. *Waves in Fluids*, p. 258. Cambridge University Press.

Low, B.C. 1977. Evolving force-free magnetic fields. I. The development of the preflare stage. *Ap. J.*, **212**, 234.

Manchester, R.N., and Taylor, J.H. 1977. *Pulsars*. San Francisco: W.H. Freeman.

Melrose, D.B. 1968. The emission and absorption of waves by charged particles in magnetized plasmas. *Astrophys. Space Sci.*, **2**, 171.

Melrose, D.B. 1980. *Plasma Astrophysics. Vol. 1. The Emission, Absorption and Transfer of Waves in Plasmas*. New York: Gordon and Breach.

Montgomery, D.C., and Tidman, D.A. 1964. *Plasma kinetic theory*. New York: McGraw-Hill.

Nicholson, D.R. 1983. *Introduction to Plasma Theory*. New York: Wiley.

Nyquist, H. 1932. Regeneration theory. *Bell System Tech. J.*, **11**, 126.

Parker, E.N. 1955. The formation of sunspots from the solar toroidal field. *Ap. J.*, **121**, 491.

Penrose, O. 1960. Electrostatic instabilities of a uniform non-Maxwellian plasma. *Phys. Fluids*, **3**, 258.

Pierce, J.R. 1950. *Traveling-Wave Tubes*. Princeton, New Jersey: Van Nostrand.

Pines, D., and Schrieffer, J.R. 1962. Approach to equilibrium of electrons, plasmons, and phonons in quantum and classical plasmas. *Phys. Rev.*, **125**, 804. .

Priest, E.R., and Milne, A.M. 1980. Force-free magnetic arcades relevant to two-ribbon solar flares. *Solar Phys.*, **65**, 315.

Sakurai, T. 1979. A new approach to the force-free field and its application to the magnetic field of solar active regions. *Pub. Astr. Soc. Japan*, **31**, 209.

Scarf, F.L., Gurnett, D.A., and Kurth, W.S. 1979. Jupiter plasma waves observations: An initial Voyager 1 overview. *Science*, **204**, 991.

Smerd, S.F., and Westfold, K.C. 1949. The characteristics of radio-frequency radiation in an ionized gas. *Phil. Mag.*, **40**, 831.

Spitzer, L. 1960. Particle diffusion across a magnetic field. *Phys. Fluids*, **3**, 659.

Spitzer, L. 1962. *Physics of Fully Ionized Gases*. (Second Edition). New York: Interscience.

Stix, T.H. 1962. *The Theory of Plasma Waves*. New York: McGraw-Hill.

Storey, L.R.O. 1953. An investigation of whistling atmospherics. *Phil. Trans. Roy. Soc. London, Ser. A*, **246**, 113.

Sturrock, P.A. 1955. *Static and Dynamic Electron Optics*, p. 159. Cambridge University Press.

Sturrock, P.A. 1958. Kinematics of growing waves. *Phys. Rev.*, **112**, 1488.

Sturrock, P.A. 1960a. Generalizations of the Lagrange expansion with application to physical problems. *J. Math. Phys.*, **1**, 405.

Sturrock, P.A. 1960b. In what sense do slow waves carry negative energy? *J. Appl. Phys.*, **31**, 2052.

Sturrock, P.A. 1961a. Energy–momentum tensor for plane waves. *Phys. Rev.*, **121**, 18.

Sturrock, P.A. 1961b. *Plasma Physics* (ed. J.E. Drummond; New York: McGraw-Hill), p. 124. Amplifying and evanescent waves, convective and nonconvective instabilities.

Sturrock, P.A. 1962. *Plasma Hydromagnetics*, p. 47, Energy and momentum in the theory of waves in plasmas. Ed. D. Bershader. Stanford University Press.

Sturrock, P.A. 1966. Stochastic acceleration. *Phys. Rev.*, **141**, 186.

Sturrock, P.A. 1991. Maximum energy of semi-infinite magnetic-field configurations. *Ap. J.*, **380**, 655.

Sturrock, P.A., and Woodbury, E. 1967. *Plasma Astrophysics*, p. 155. Force-free magnetic fields and solar filaments. Ed. P.A. Sturrock. London: Academic Press.

Tandberg-Hanssen, E., and Emslie, A.G. 1988. *The Physics of Solar Flares*. Cambridge University Press.

Taylor, J.B. 1974. Relaxation of toroidal plasma and generation of reverse magnetic fields. *Phys. Rev. Lett.*, **33**, 1139.

Thompson, W.B. 1962. *An introduction to Plasma Physics*. Oxford: Pergamon.

Titchmarsh, E.C. 1937. *Fourier Integrals*. Oxford: Clarendon Press.

Tonks, L., and Langmuir, I. 1929. Oscillations in ionized gases. *Phys. Rev.*, **33**, 195.

Van Kampen, N.G. 1955. On the theory of stationary waves in plasmas. *Physica*, **21**, 949.

Weinberg, S. 1962. Eikonal methods in magnetohydrodynamics. *Phys. Rev.*, **126**, 1899.

White, R.B. 1983. Resistive instabilities and field line reconnection. *Handbook of Plasma Physics*. Eds. M.N. Rosenbluth and R.Z. Sagdeev. *Vol. 1: Basic Plasma Physics I*, p. 611. Eds. A.A. Galeev and R.N. Sudan.

Whitham, G.B. 1974. *Linear and Nonlinear Waves*, Chapter 11. New York: Wiley.

Wild, J.P., Murray, J.D., and Rowe, W.C. 1954. Harmonics in the spectra of solar radio disturbances. *Australian J. Phys.*, **7**, 439.

Woltjer, L. 1958. A theorem on force-free magnetic fields. *Proc. Nat. Acad. Sci.*, **44**, 489.

Yang, W.H., Sturrock, P.A., and Antiochos, S. 1986. Force-free magnetic fields: the magneto-frictional method. *Ap. J.*, **309**, 383.

Author index

Subject index